The Philosophy of Nature and Philosophy of Physics in the Writings of Marian Smoluchowski

STUDIES IN PHILOSOPHY, CULTURE AND CONTEMPORARY STUDIES

Edited by Bogusław Paź

VOLUME 41

Jan Grzanka

The Philosophy of Nature and Philosophy of Physics in the Writings of Marian Smoluchowski

Translated by Ewan Jones

PETER LANG

Berlin · Bruxelles · Chennai · Lausanne · New York · Oxford

Bibliographic Information published by the Deutsche Nationalbibliothek
The Deutsche Nationalbibliothek lists this publication in the Deutsche Nationalbibliografie; detailed bibliographic data is available online at http://dnb.d-nb.de.

Library of Congress Cataloging-in-Publication Data
Names: Grzanka, Jan author | Jones, Ewan, translator
Title: The philosophy of nature and philosophy of physics in the writings of Marian Smoluchowski / Jan Grzanka ; translated by Ewan Jones.
Other titles: Między fizyką a filozofią. English
Description: New York : Peter Lang, [2026] | Series: Studies in philosophy, culture and contemporary studies, 2196-0151 ; 41 | Originally published in Polish as Między fizyką a filozofią : filozofia przyrody i filozofia fizyki w pismach Mariana Smoluchowskiego. | Includes bibliographical references.
Identifiers: LCCN 2025027759 (print) | LCCN 2025027760 (ebook) | ISBN 9783631941188 hardback | ISBN 9783631941195 pdf | ISBN 9783631941614 epub
Subjects: LCSH: Smoluchowski, Marian, 1872-1917 | Philosophy of nature | Physics--Philosophy | Physicists--Poland--Biography
Classification: LCC QC16.S73 G7913 2026 (print) | LCC QC16.S73 (ebook)
LC record available at https://lccn.loc.gov/2025027759
LC ebook record available at https://lccn.loc.gov/2025027760

A photograph of Marian Smoluchowski, from the Smoluchowski family collection.

ISSN 2196-0151
ISBN 978-3-631-94118-8 (Print)
ISBN 978-3-631-94119-5 (E-PDF)
ISBN 978-3-631-94161-4 (E-PUB)
DOI 10.3726/b23108

© 2026 Peter Lang Group AG, Lausanne (Switzerland)
Published by Peter Lang GmbH, Berlin (Germany)
info@peterlang.com

All rights reserved.

All parts of this publication are protected by copyright.
Any utilization outside the strict limits of the copyright law, without the permission of the publisher, is forbidden and liable to prosecution. This applies in particular to reproductions, translations, microfilming, and storage and processing in electronic retrieval systems.

This publication has been peer reviewed.

www.peterlang.com

For Ela and Pola

Table of Contents

CHAPTER 1
Marian Smoluchowski's place in Polish and world science
1.1. Smoluchowski's position in science as a physicist — 13
    1.1.1. Smoluchowski in citation lists — 13
    1.1.2. A Nobel for Smoluchowski? — 18
1.2. Smoluchowski in Polish science — 25
    1.2.1. Memory of Smoluchowski — 25
    1.2.2. Smoluchowski in publications — 28

CHAPTER 2
Preliminaries to the philosophy of Marian Smoluchowski
2.1. The beginnings of Smoluchowski's philosophy — 37
2.2. Smoluchowski's philosophy against the backdrop of the main trends in 19th-century philosophy — 41
2.3. Marian Smoluchowski's worldview — 54
    2.3.1. Marian Smoluchowski—materialist, atheist? — 54
    2.3.2. General characteristics of materialism — 55
    2.3.3. The genesis of attributing materialistic views to Smoluchowski — 56
    2.3.4. Alleged Marxism and materialism — 62
    2.3.5. Smoluchowski's philosophical predilections — 65
    2.3.6. Substantive analysis of Władysław Krajewski's arguments — 70
    2.3.7. Smoluchowski's silence about God — 82
2.4. Summary — 86

CHAPTER 3
Smoluchowski's approach to the methodology and epistemology of science

- 3.1. Fields and issues of physics ... 89
    - 3.1.1. The subject of physics ... 89
    - 3.1.2. Reflection on Aristotle's philosophy of nature ... 92
- 3.2. Methods of physics ... 100
    - 3.2.1. The experiment in the history of science ... 100
    - 3.2.2. The induction method ... 104
    - 3.2.3. Reflections on the mathematics of nature ... 108
- 3.3. The explanatory function of physical theories ... 115
- 3.4. The role of analogy in the methodology of physics ... 119
- 3.5. Convention and experience ... 123
    - 3.5.1. The concept of simplicity of scientific theories ... 123
    - 3.5.2. Poincaré's conventionalism ... 129
- 3.6. The monistic interpretation of nature ... 138
    - 3.6.1. Richard Avenarius, Ernst Mach and Pierre Duhem ... 138
    - 3.6.2. Smoluchowski and monism ... 147
- 3.7. Attributes of atomic theory ... 150
    - 3.7.1. Fluctuations ... 150
    - 3.7.2. Opalescence ... 154
    - 3.7.3. Perpetual motion ... 156

CHAPTER 4
Transformation of thermodynamics into the kinetic theory of matter at the turn of the 19th and 20th centuries

- 4.1. Reflections on physics at the end of the 19th century ... 163
- 4.2. Selective calendar of discoveries in 19th century physics ... 171
- 4.3. Conclusions after the transformation ... 199

CHAPTER 5
Smoluchowski's approach to the philosophical aspects of causality and chance

- 5.1. Causality ... 205
- 5.2. Probability calculus ... 217

5.3. Epistemic causality — 226
5.4. Ontic causality — 241

CHAPTER 6
Smoluchowski's *utility criterion*
6.1. The principle of economy of thought — 253
    6.1.1. Mach—economy of thought — 253
    6.1.2. Ockham's razor — 256
    6.1.3. Imagery in science — 261
6.2. The utility criterion — 270
    6.2.1. Pragmatism in methodology — 270
    6.2.2. The concept of truth in science — 275
6.3. The principles of utility in philosophy — 279
    6.3.1. Utility according to Charles Sanders Peirce and William James — 279
    6.3.2. Relation of the *utility criterion* to the concept of truth — 281
    6.3.3. Theory of abductive reasoning and other explanations — 285
    6.3.4. Thomas Kuhn—analysis of the structure and nature of knowledge — 288
6.4. Scientific realism and the *utility criterion* — 293
    6.4.1. Hilary Putnam's scientific realism — 293
    6.4.2. The *utility criterion* and social need — 300
6.5. Induction and the *utility criterion* — 304
    6.5.1. Hume's inductive scepticism — 304
    6.5.2. Peter Lipton's induction and the *utility criterion* — 308
    6.5.3. John D. Norton's inductive inference — 313
6.6. Summary of considerations on *utility theory* — 322

CHAPTER 7
Smoluchowski's contribution to the development of the kinetic theory of matter
7.1. Ludwig Boltzmann's influence on Smoluchowski's views — 325
7.2. Atomic-kinetic theory at the turn of the 19th and 20th centuries — 338
7.3. Controversy around the Brownian motion breakthrough — 346
7.4. Comparison of the works of Einstein and Smoluchowski — 364

TABLE OF CONTENTS

7.5. Perrin's experimental evidence 387
7.6. Summary 396

Conclusion 403

Index of people 449

CHAPTER 1

Marian Smoluchowski's place in Polish and world science

1.1. Smoluchowski's position in science as a physicist
1.1.1. Smoluchowski in citation lists

The year 2017 saw the hundredth anniversary of the death of outstanding Polish physicist Marian Smoluchowski. Over the more than a hundred years since his death, this great scholar has been widely forgotten. Few researchers, even among physicists, are able to cite any of Smoluchowski's scientific achievements on an *ad hoc* basis, and outside science circles his name is not associated with physics at all. Scientific, biographical or memorial publications about Smoluchowski have appeared so rarely over the last hundred years that the memory of him has been erased. The hundredth anniversary of his death passed practically unnoticed. Meanwhile, had it not been for his untimely death, Smoluchowski would probably have become a Polish Nobel Prize winner. His scientific achievements are impressive and undoubtedly entitled him to receive the prize. This monograph is primarily an attempt to demonstrate the contribution of the physicist Smoluchowski to philosophy, and especially to the philosophy of nature, but also to raise awareness of the position he occupies in the world of modern science, despite the passage of over a hundred years since his death.

Smoluchowski's presence in today's science is impressive. His equations are among the foundational equations of the theory of stochastic processes. In his book *Stochastic Processes in Physics and Chemistry*, Nico van Kampen writes that the Chapman-Kolmogorov equation, which relates to homogeneous Markov processes, is also often called the Smoluchowski equation but he will not use that name because it is used interchangeably with several other closely related but non-identical equations[1]. Elsewhere, he states that the Fokker-Planck equation, which describes the temporal evolution of the probability density function, is also sometimes called the 'Smoluchowski equation'[2]. The Smoluchowski continuity equation is one of the most frequently used in descriptions of sedimentation and coagulation processes. This is the equation into which Fick's law (the diffusion law) was inserted, and in which the operation of the external field was taken into account. Smoluchowski was the first to note that the phenomenological laws of macroscopic diffusion could be applied to probability. The diffusion process is the result of the superposition of the Brownian motion of individual particles of a substance[3]. The Smoluchowski continuity equation is used in a wide variety of contexts, ranging from theoretical work to industrial applications, for example for industrial water purification, calculations determining the formation of soot deposits in aircraft engines, milk coagulation, the formation of gel barriers, the growth of nanotubes, granulocyte aggregation, and leukocyte adhesion. This equation is often referred to as the 'Smoluchowski equation' (for example, when measuring zeta potential and deviations of the current-time relationship in electroosmotic flow). It also appears as the 'Smoluchowski limit' in applications of diffusion-limited chemical reaction kinetics and in the study of the class of systems of stochastic differential equations describing diffusion phenomena[4].

Smoluchowski's works are cited in publications on the issues of simulation of potassium channels in cell membranes, liquid mechanics (droplet collisions), protein electrophoresis, rotational diffusion, the impact of viscosity on electron transfer kinetics, analysis of the effects of cholera toxins, information thermodynamics (numerical calculations) and Brownian dynamics.

Marian Smoluchowski is the most cited Polish scientist in world scientific literature for the period from 1996 to 2001 – in scientific works globally he was

---

[1] See: N.G. van Kampen, *Procesy stochastyczne w fizyce i chemii* (*Stochastic processes in physics and chemistry*), ed. trans. Ł.A. Turski M. Dudyński, M. Ekiel-Jeżewska, D. Śledziewska-Błocka, Warsaw 1990, p. 79.
[2] Idem, p. 186.
[3] See: A. Fuliński, *Współczesne zastosowania równań Smoluchowskiego* (*Contemporary applications of the Smoluchowski equations*) in: *Marian Smoluchowski. From atomic theory to modern physics*, ed. A. Strzałkowski, Kraków 2003, p. 23.
[4] Idem, p. 25.

cited 553 times. Also, in the article *Contemporary applications of Smoluchowski's equations*, Andrzej Fuliński points out that in the years 1996–2002[5], Smoluchowski was cited 836 times, though this statistic does not include those publications which only referred to the term 'Smoluchowski's equation'. It follows that not only would the number of citations of Smoluchowski not decrease but would even increase.

Publications citing Smoluchowski's works or referring to his equation could recently be found in a variety of journals, from the most serious physical journals – both general (such as *Physical Review Letters* and *Physical Review*) and specialised (*Applied Physics Letters, Astronomy and Astrophysics, Colloid Journal, Journal of Aerosol Science, Journal of Chemical Physics*, 'Journal of Crystal Growth or Journal of Fluid Mechanics), through influential chemical journals (such as Dyes Pigments, *Industrial & Engineering Chemistry Research, The Journal of Physical Chemistry, The Chemical Society of Japan* or *Journal of Colloid and Interface Science*), biological ones (for example, Biochemistry, Biochemical Engineering Journal, The Journal of Biochemistry Molecular Biology and Biophysics* and *Journal of Biomedical Engineering*), to publications of the calibre of *Archive for Rational Mechanics and Analysis, Journal of Engineering for Gas Turbines and Power, Statistics – Theory and Methods, Computational Mechanics* or *Journal of Computational Physics*[6]. This list of scientific journals is impressive, especially considering that they relate to the achievements of a scholar who lived a hundred years ago and that physics has been one of the most dynamically developing fields of science over the last century.

In the article *ζ Potential and the Smoluchowski equation*, Marek Kosmulski writes that the name Smoluchowski is associated in modern science mainly with the ζ (zeta) electrokinetic potential and the Smoluchowski equation, which allows the calculation of this potential on the basis of measurable quantities, for example electrophoretic mobility. This is not the physicist's only, or even most important, scientific achievement, but due to the renaissance of interest in colloidal particles—currently called nanoparticles—it is in this context that Smoluchowski's name very often appears. In the last decade, the correctly spelled name 'Smoluchowski' has appeared on average in 1,500 publications per year, published by Elsevier, Springer, the

---

[5] Idem, p. 24.
[6] Idem, p. 25.

American Chemical Society or Wiley (excluding all other publishers). The search for scientific literature referring to Smoluchowski's achievements is very difficult, as his name is often misspelt. According to Kosmulski, articles indexed by the Web of Science include the following spellings of his surname: Schmoluchowski, Schmolukowski, Smolucbowski, Smoluchoski, Smoluchovski, Smoluchowiski, Smoluchowsi, Smoluchowsky, Smoluhovski, Smoluhowski, Smolukhovskii, Smolukhovskiy, Smolukhovsky, Smolukhowski, Smolukovski, Smolushovski—and this list, according to Kosmulski, is probably only the tip of the iceberg[7].

According to the Web of Science, the number of citations of Smoluchowski's works in the years 1894–2014 amounted to 7,235. For comparison, in a slightly longer period, i.e. in the years 1880–2015, the number of references to the works of Maria Skłodowska-Curie (1867–1934) is 1,582[8].

The numbers given by the Web of Science are constantly growing and will continue to grow. This is made possible by the situation in industry, especially in chemistry and pharmacy, where a large part of the current research concerns coagulation processes. The basis for this research is the formulae and rules developed by Smoluchowski. The most frequently cited article is from 1917, in which he presented a new theory of colloid solidification, published after the author's death in 1918 in *Zeitschrift für Physikalische Chemie* (Journal of Physical Chemistry) entitled *Study of a Mathematical Theory for the Coagulation Kinetics of Colloidal Solutions*. This article has been cited more than 5,500 times. *Three discourses on diffusion, Brownian movements and the coagulation of colloid particles*, given in 1916 at the Wolfskehl Congress in Göttingen, are still being cited. To this day, they are considered the best introduction to the issues of coagulation.

In 2005, Jorge E. Hirsch, an Argentinean physicist working at the University of California, San Diego, proposed an index reflecting the distribution of citations to a particular scientist's publications and the number of their best papers. The h-index is a way of measuring scientific achievements taking into

---

[7] See: ‚M. Kosmulski, *Potencjał ζ i równanie Smoluchowskiego (ζ Potential and the Smoluchowski equation)*, ‚PAUza Akademicka. The weekly publication of the Polish Academy of Arts and Sciences' 2017, No. 380–381, p. 7.

[8] Wikipedia, *Web of Science*, https://en.wikipedia.org/wiki/Web_of_Science (access: 10/03/2020).

account the number of publications and the number of citations. The *h-index* (*index h, Hirsch index, Hirsch number*) for a given author is the number of publications cited ≥ *h* times[9]. This indicator also shows that Marian Smoluchowski is the most cited Polish scientist. The Hirsch index for his publications was 29 at the time of writing, while Maria Skłodowska-Curie's was 16.

In 1990, two scholars from the Max Planck Institute in Stuttgart— philosopher Werner Marx (1910–1994) and physicist Manuel Cardona (1934– 2014)—began a research project called 'Blasts from the past'. The study was to answer the question of how the importance or usefulness of a scientific article can be measured.

The researchers decided to check how many times a given publication was on the reference lists of other articles. According to the researchers, the number of citations cannot be equated with the overall importance or usefulness of a publication, which applies especially to recent articles, the long-term significance of which may not yet be clear. This also applies to many older articles that are not cited because their results are now so well known that they appear in textbooks without the sources being given. It would be easy to theorise and speculate on this topic, but as the idea's initiators have stated, there is a satisfactory way to proceed, especially in the case of physics: collecting and analysing data. They analysed a huge proportion of the publications that have appeared in modern times, until 1930. Nearly ten thousand names of scientists—mainly in the exact sciences—were taken into consideration and the citation of their papers in contemporary scientific publications in the years 1990–2003 was examined.

The results of the research indicate the extremely high position of Marian Smoluchowski, whose works are referred to very often today. Out of ten thousand scientists, Smoluchowski was in sixth position, being the first person on the list without a Nobel Prize:

---

[9] See: M. Kapczyński, *Indeks Hirscha zastosowanie oraz metody obliczania* (*The Hirsch index, application and calculation methods*), Thomson Reuters (Scientific), July 2, 2012, https://puss.pila.pl/uploads/indeks_h_zastosowanie_metody_obliczania.pdf (accessed: March 10, 2020).

1. Albert Einstein (1921 Nobel in physics) – 3,025 citations (died 1955; lived 76 years);
2. Peter Debye (1936 Nobel in chemistry) – 1,592 citations (died 1966; lived 82 years);
3. Max Born (1954 Nobel in physics) – 1,575 citations (died 1970; lived 88 years);
4. Irving Langmuir (1932 Nobel in chemistry) – 1,564 citations (died 1957; lived 76 years);
5. William John Strutt (1904 Nobel in physics) – 1,503 citations (died 1919; lived 77 years);
6. Marian Smoluchowski – 1,356 citations (died 1917; lived 45 years)[10].

Although Smoluchowski never won a Nobel Prize and lived for 30–40 years less than the other scholars mentioned, the number of publications of the first five scientists listed after Einstein is similar. In this context, an obvious question arises—what would the scientific achievements of our physicist have been if he had lived to an old age?

1.1.2. A Nobel for Smoluchowski?

Another issue that arises after analysing Marx and Cardona's results is the question of whether Smoluchowski had a chance of being awarded the Nobel Prize. To answer this, it is helpful to realise that at least three Nobel Prize winners—Richard Zsigmondy (1865–1929, Nobel Prize in Chemistry in 1925), Jean Baptiste Perrin (1870–1942, Nobel Prize in Physics in 1926) and Theodor Svedberg (1884–1971; Nobel Prize in Chemistry in 1926)—received the Nobel Prize for the results of work in which they directly or indirectly used Smoluchowski's research. Bogdan Cichocki writes that he believes Smoluchowski did not receive the Nobel Prize because he died young and the prize is not awarded posthumously. In 1925, the Nobel Prize in Chemistry was awarded to Richard Zsigmondy, a professor at the University of Graz in Austria, for his research in the field of colloids. His work was experimental and closely related to Smoluchowski's theoretical achievements. The following year, as many as two Nobel Prizes were indirectly related to the great Polish academic. In the field of physics, the prize was awarded to Frenchman Jean Baptiste Perrin for experimental work

---

[10] See: W. Marx, M. Cardona, *Blasts from the Past*, 'Physics World' 2004, vol. 17, no. 2.

on Brownian motion, confirming the validity of the Einstein-Smoluchowski molecular theory, while in the field of chemistry the prize went to Theodor Svedberg from Uppsala, Sweden, whose experimental work on suspensions ran concurrently to Smoluchowski's theoretical considerations related to them[11].

Richard Zsigmondy received the Nobel Prize "for his demonstration of the heterogenous nature of colloid solutions and for the methods he used, which have since become fundamental in modern colloid chemistry"[12]. Smoluchowski's contribution to this research is spectacular. Zsigmondy, being an experimentalist, needed a theoretical foundation for his research and took it from Smoluchowski. During his Nobel lecture in Stockholm, he said:

> "This induced me to ask the theoretical physicist M. von Smoluchowski to derive an experimentally verifiable formula by which the presence of spheres of attraction could be deduced from the speed of coagulation; von Smoluchowski willingly agreed to my suggestion and gave, in addition, a complete theory of coagulation on a mathematical basis[13]."

It is therefore likely that it could have been a joint prize for both scientists.

Theodor Svedberg received the Nobel Prize "for his work on disperse systems"[14]. Dispersion systems are colloidal dispersed systems, composed of at least two phases, of which at least one is a strongly fragmented material, dispersed in a second phase of a continuous nature, called a dispersion medium. Svedberg and his colleagues published a series of papers devoted to the issue of fluctuation. In the introduction to the first article—*A new method of testing the validity of Boyle-Gay-Lussac's law for colloidal solutions*[15]—he

---

[11] See: B. Cichocki, *Nagroda Nobla* (*The Nobel Prize*), 'Delta' 1997, No. 12.

[12] The Nobel Prize, *Richard Zsigmondy. Facts*, https://www.nobelprize.org/prizes/chemistry/1925/zsigmondy/facts/ (access: 30/04/2020) – wherever not otherwise stated, texts are translated by the author.

[13] R.A. Zsigmondy, *Properties of Colloids*, Nobel Lecture, Chemistry 1922–1941, Singapore, New Jersey, London, Hong Kong 1999, p. 53.

[14] The Nobel Prize, *Theo Svedberg. Facts*, https://www.nobelprize.org/prizes/chemistry/1926/svedberg/facts/ (access: 30/04/2020).

[15] T. Svedberg, K. Inouye, *Eine neue Methode zur Prüfung der Gültigkeit des Boyle-Gay-Lussac'schen Gesetzes für kolloide Lösungen* (*A new method for testing the validity of the Boyle-Gay-Lussac law for colloidal solutions*), 'Zeitschrift für Physikalische Chemie-Stöchiometrie und Verwandtschaftslehre' (Journal of Physical Chemistry-Stoichiometry and Related Sciences) 1911, H. 77, pp. 145–190.

writes that to interpret the results of the measurements obtained, a theory was needed that would be able to link the observed quantities with the concepts of the kinetic theory of matter. Svedberg found this theory in two of Smoluchowski's works. These were two articles, the first of which, *Über Unregelmässigkeiten in der Verteilung von Gasmolekülen und deren Einfluss auf Entropie und Zustandsgleichung*[16] (*On irregularities in the distribution of gas molecules and their impact on entropy and on the equation of state*), was included in a commemorative publication of 20 February 1904 to celebrate the 60th birthday of Ludwig Boltzmann (1844–1906). In it, Smoluchowski formulated the theory of particle density fluctuations in gas and considered the impact of fluctuations on the equation of state. In the second text,[17] from 1907, entitled *A theory of kinetic opalescence of gases in a critical state and other related phenomena*, he dealt with the application of his theory to the phenomenon of gas opalescence in the vicinity of a critical state and to the phenomenon of a blue sky and other related effects in which the influence of molecular fluctuations is revealed on a macroscopic scale[18].

Bronisław Średniawa (1917–2014) writes in a paper devoted to both scholars:

> The collaboration between Smoluchowski and Svedberg lasted seven years, from 1907 to 1914. In those years, the two conducted scientific correspondence with one another. Beginning in 1908, Smoluchowski cited and discussed, sometimes critically, Svedberg's results in all his works on Brownian motion and fluctuations. In his works devoted to these issues, Svedberg also mentioned Smoluchowski's publications and letters[19].

---

[16] M. Smoluchowski, *Über Unregelmäßigkeiten in der Verteilung von Gasmolekülen und deren Einfluß auf Entropie und Zustandsgleichung* (*On irregularities in the distribution of gas molecules and their influence on entropy and the equation of state*), in: *Festschrift Ludwig Boltzmann gewidmet zum sechzigsten Geburtstage* (*Commemorative book dedicated to Ludwig Boltzmann on his sixtieth birthday*), Leipzig 1904, pp. 626–641.

[17] Idem, *Theory of kinetic opalescence in a critical state and other related phenomena*, Dissertations of the Faculty of Mathematics and Natural Sciences of the Academy of Arts and Sciences, series A, vol. XLVII, Kraków 1907, pp. 179–199.

[18] See: B. Średniawa, *Rola współpracy Mariana Smoluchowskiego i Teodora Svedberga w prowadzonych w pierwszych latach XX wieku badaniach ruchów Browna i fluktuacji* (*The role of collaboration between Marian Smoluchowski and Theodor Svedberg in studies on Brownian motion and fluctuations conducted in the early years of the twentieth century*), 'Postępy Fizyki' (Progress of Physics), 1991, vol. 42, book. 4, p. 441.

[19] Idem, p. 451.

In a speech at the Nobel Prize awards ceremony on May 19, 1927, Svedberg presented the results of research he had conducted in the years 1923–1926 (i.e. a few years after Smoluchowski's death). In discussing them he did not have the opportunity to mention collaborating with Smoluchowski. It was mentioned by Professor Henrik Gustaf Söderbaum, secretary of the Royal Academy of Sciences. Presenting Svedberg's merits, he said:

> As we have recently heard, Einstein evolved a theory for this so-called Brownian movement which was then developed to a high degree by the now late Smoluchowski… If we now consider a very small volume fraction, the result is that, as Smoluchowski has calculated in detail, the number of particles present simultaneously within this volume can change from one moment to another. Svedberg and his collaborators have been able to confirm this extremely interesting conclusion that a 'few-molecular' system having definite limits within a large volume of a material with a definite mean temperature may contain a varying number of particles, partly by counting the colloidal particles, partly in the case of solutions of radioactive substances by counting the number of so-called scintillations[20].

Jean Baptiste Perrin, the great French physicist, received the Nobel Prize in 1926 "for his work on the discontinuous structure of matter, and especially for his discovery of sedimentary equilibrium"[21]. The work concerned research that experimentally confirmed the Einstein-Smoluchowski theory. In his essay *On Thermodynamic Fluctuations and Brownian* Motion, Smoluchowski writes: "Among the experimenters who undertook this research, Svedberg and Perrin and their colleagues should be mentioned above all. "This last scholar in particular managed to verify with great accuracy the theoretical formulae related to Brownian motion." In 1913, Perrin published *Atoms*[22]. He described in it the scope of research, as well as the conclusions and thoughts resulting from the experiments carried out, intended to *de facto* falsify the Einstein-Smoluchowski theory. In *Atoms*, Perrin repeatedly cites Smoluchowski's research work in the field of the theories of Brownian motion, fluctuations and opalescence, which were a theoretical complement to Einstein's previously published works. Separately, Perrin devoted the entire fifth

---

[20] Idem, p. 452.
[21] The Nobel Prize, *Jean Baptiste Perrin. Facts*, https://www.nobelprize.org/prizes/physics/1926/perrin/facts/ (access: 30/04/2020).
[22] J. Perrin, *Les Atomes* (Atoms), Paris 1913.

chapter, *Fluctuations, Smoluchowski's Theory* to Smoluchowski's fluctuations[23]. In it, he describes in a matter-of-fact way the theoretical results obtained by Smoluchowski, which he used in his experiments:

> We have already indicated one of these phenomena in speaking of the definite though very feeble thermal inequalities which are produced spontaneously and continuously in spaces of the order of a micron, and which are, indeed, a second aspect of the Brownian movement itself. These thermal fluctuations, of the order of a *thousandth* of a degree of such volumes, seem in practice to be inaccessible to our measurements. The density of a fluid in equilibrium, like its temperature or molecular excitation, should vary from point to point. A cubic micron, for example, will contain sometimes a larger and sometimes a smaller number of molecules. Smoluchowski has drawn attention to these spontaneous inequalities, and has been able to calculate the fluctuation in density…[24].

Elsewhere in *Les Atomes (Atoms)* we read:

> No longer confining himself to the case of rarefied substances, Smoluchowski succeeded a little later (…) in calculating the mean density fluctuation for any fluid whatever, and proved that, even with condensed fluids, the fluctuations should become noticeable in spaces visible under the microscope when the fluid is near the critical state. He thus succeeded in explaining the enigmatic *opalescence* which is always shown by fluids in the neighbourhood of the critical state. This opalescence, which is absolutely stable, indicates a permanent condition of fine-grained heterogeneity in the fluid[25].

In *Les Atomes*, Perrin repeatedly recalls Smoluchowski's research, thus proving the importance of his theoretical deliberations and the scientist's contribution to proving the atomistic structure of matter. However, it should be noted that the works of Einstein were crucial for Perrin: *Über die von der molekularkinetischen Theorie der Wärme geforderte Bewegung von in ruhenden Flüssigkeiten suspendierten*[26] (*On the movement of small particles suspended in stationary liquids*

---

[23] Idem, *Atoms*, New York 1916, p. 134.
[24] Ibid.
[25] Idem, p. 135.
[26] A. Einstein, *Über die von der molekularkinetischen Theorie der Wärme geforderte Bewegung von in ruhenden Flüssigkeiten suspendierten Teilchen* (*On the Movement of Small Particles Suspended in Stationary Liquids Required by the Molecular-Kinetic* Theory of Heat), 'Annalen der Physik' ('Annals of Physics'), 1905, vol. 322, No. 8, pp. 549–560.

required by the molecular-kinetic theory of heat) of May 11, 1905, *and Zur Theorie der Brownschen Bewegung (On the Theory of Brownian Motion)* of May 19, 1906.[27] This approach of Perrin to the work of both scientists demonstrates that shortly after the publications, Smoluchowski was not treated as equal to Einstein in being responsible for the discovery of Brownian motion, despite the fact that—as it may seem—Smoluchowski's approach to the proof of Brownian motion should have been more persuasive to Perrin as it presented mathematical evidence of these movements from the microscopic perspective.

Smoluchowski's research used in the works of the aforementioned Nobel Prize winners are not the only examples of the scholar's scientific achievements that could have brought him the Nobel Prize. A discovery that brought about an almost paradigmatic change in physics was undoubtedly the development of the foundations in the field of stochastic processes and the introduction of probabilistics to the exact sciences. This was an achievement worthy not only of the Nobel Prize, thanks to which the Polish scientist's name should be permanently inscribed in the history of science.

Mark Kac notes:

> Smoluchowski may not have been aware of it but he begun writing a new chapter of Statistical Physics which in our times goes by the name of Stochastic Processes... (...) The novelty and originality of the Smoluchowski approach lie in his bold replacement of an impossibly difficult dynamical problem (...) with a relatively simple stochastic process[28].

Smoluchowski also provided the first correct equations linking the ζ potential to measurable quantities, i.e. electrophoretic mobility, flow potential and pressure difference in electro-osmosis. In world literature, these are known as the Smoluchowski equations[29], which also qualified him for the Nobel Prize.

Many years later, a great advocate of Smoluchowski, Indian astrophysicist, mathematician and Nobel Prize winner Subrahmanyan

---

[27] Idem, *Zur Theorie der Brownschen Bewegung (On the theory of Brownian motion)*, 'Annalen der Physik' ('Annals of Physics'), 1906, vol. 324, No. 2, pp. 371–381.
[28] M. Kac, *Marian Smoluchowski and the Evolution of Statistical Physics*, in: *Polish Men of Science. Marian Smoluchowski. His Life and Scientific Work*, ed. R.S. Ingarden, Warsaw 1986, p. 17.
[29] See: M. Kosmulski, *ζ Potential and the Smoluchowski equation*, op. cit., p. 7.

CHAPTER 1

Chandrasekhar (1910–1995), argued that only the Polish physicist's premature death prevented him from winning the Nobel Prize.

> In 1973 Chandrasekhar was awarded the Marian von Smoluchowski Medal of the Polish Physical Society in appreciation of his contributions to stochastic methods in physics and astrophysicsand, especially, the Review of Modern Physics 1943 article which covered Smoluchowski's contributions. In his speech at the award ceremony, Chandrasekhar noted that the Nobel prizes in chemistry awarded to R. Zsigmondy in 1925 and to T. Svedberg in 1926 were for experimental confirmation of Smoluchowski's theoretical predictions on colloidal and disperse systems and that if Smoluchowski had been still alive he would certainly have been a Nobel laureate himself[30].

Unlike another Polish physicist, Karol Olszewski (1846–1915), Smoluchowski was never nominated for the Nobel Prize. Olszewski had two nominations in physics and one in chemistry and is often cited in Polish literature as the only outstanding European physicist working in Poland at the turn of the 19th and 20th centuries. Considering how much his research work gained in importance in the first ten years after his death, Smoluchowski would have had a better chance of winning the prize.

An interesting fact that should be mentioned is the first Solvay Conference, which took place in Brussels in October 1911. It was a meeting of the most outstanding minds in physics and chemistry, devoted to the most important issues of science. This and subsequent Conferences are named after Ernest Solvay—a Belgian industrialist, passionate about science, who organised and financed the event. Among the twenty-one invited guests were such scientists as Albert Einstein (1879–1955), Maria Skłodowska-Curie, Max Planck (1858–1947) and Ernest Rutherford (1871–1937), as well as Jules Henri Poincaré (1854–1912). The event was chaired by Dutch Nobel laureate Hendrik Anton Lorentz (1853–1928). Smoluchowski did not participate in either this or the next Conference in 1913, which was rather surprising given that at that time he was one of the most outstanding physicists in Central and Eastern Europe. The argument presented in this context that physicists from the nations of Central Europe or from those belonging to the multinational empires of the time, i.e. Prussia and Austria-Hungary, is

---

[30] B. Duplantier, *Brownian motion, "diverse and undulating"*, Einstein, 1905–2005: Poincaré Seminar 2005, 'Progress in Mathematical Physics' 2006, vol. 47, pp. 251–252.

less than convincing as Emil Warburg (1846–1931) from Berlin participated in it. It is much more convincing that the main topic of the conference was the theory of radiation and quanta, i.e. issues that were not directly related to Smoluchowski's research, but this is an issue we shall return to in subsection 7.5, 'Perrin's Experimental Evidence.'

1.2. Smoluchowski in Polish science

1.2.1. Memory of Smoluchowski

Smoluchowski was not only an outstanding physicist conducting research at the universities of Vienna, Paris, Glasgow and Berlin, and a scholar working with Gabriel Lippmann (1845–1921), William Thomson, Lord Kelvin (1824–1907) and Warburg. He was a person moving in the circles of European scientific elites, working with the most outstanding European physicists and participating in scientific research that had an impact on the development of modern science. He was a forerunner of many methodological solutions in the field of conducting scientific research. We are indebted to him for our basic successes in promoting and improving the teaching of science in secondary and university education. His philosophical interests deserve special attention, in particular his deliberations in the fields of the philosophy of nature and the philosophy of science, which were often ahead of their time.

The recollections of many people from his social circle, as well as those who wrote about him based on the memories of others, are astonishing in their unanimity. An almost unreal figure emerges from them in which it is difficult to find blemishes or flaws. Undoubtedly, Smoluchowski's untimely and unexpected death, as is usually the case in such situations, prompted more benevolent, idealising recollections, but monographs and articles, such as the extract from an essay by Stanisław Loria (1883–1958) cited below, written on the 35th anniversary of the scholar's death, depict a unique figure. Loria wrote:

> the development of Smoluchowski's scientific work and the content and value of his most accurate dissertations draw a profile of this most outstanding Polish physicist as a researcher of pioneering initiative, equipped not only with extraordinary abilities, but also with willpower and the ability to overcome difficulties. Dedicated with all his soul to science, however, he was not of the erudite type, carefully collecting and storing information in case of need. Faced with a clearly formulated problem,

he did not waste time looking to see whether there existed ready-made solutions that may have slipped his mind, but pursued his own, short, sometimes bumpy and difficult, path to the goal. At the same time, he was characterised above all by the courage and apposite instinct of the seeker and the will to accomplish a task through his own creativity and assiduous work[31].

A thesis that well defines the essence of Marian Smoluchowski's thought was included in an article by Władysław Kapuściński (1898–1979), who noted that in the history of the rational knowledge of nature there are two trends that constantly complement one another and intertwine in the ongoing process of learning about nature. The first is the development of detailed subject research, consisting of discovering the laws of nature based on experiment. In the 17th century, this trend led to the creation of the so-called exact sciences. The second trend is philosophical research, which has developed as the philosophy of nature. For a significant period of the development of European thought, the philosophical trend played a dominant role, the situation only having changed in modern times. Research in the exact sciences, carried out with great success, accelerated the development of science, at the same time weakening the philosophical trend. This does not mean that philosophical reflection has disappeared from the view of supporters of rational knowledge of nature[32]. Both intertwining epistemological trends can easily be found in Smoluchowski's works. It is obvious that *strictly* scientific research is dominant, but theoretical considerations are supported by elements of the philosophy of science, methodology, epistemology and the philosophy of nature. Smoluchowski's philosophical reflections do not appear on the periphery of his work in the field of physics but form an important, integral part of his scientific activity, without which his achievements in the field of physics would not be fully understood or may even not exist at all. They can be seen, for example, in the 1923 article *On the Concept of Chance*

---

[31] S. Loria, *Marian Smoluchowski i jego dzieło (1872–1917)* (*Marian Smoluchowski and his work (1872–1917)*), 'Postępy Fizyki' ('Progres in Physics'), 1953, vol. IV, book. 1, p. 38.

[32] See: W. Kapuściński, *Poglądy filozoficzne Mariana Smoluchowskiego* (*Marian Smoluchowski's philosophical views*), 'Fizyka i Chemia' ('Physics and Chemistry'), 1953, No. 4 (28), p. 200.

and the Origin of the Laws of Physics Based on Probability[33], in which physics, mathematics and philosophy merge into one another.

In trying to characterise Smoluchowski's philosophical views, we encounter several problems, one of which is the dispersion of philosophical thoughts. Smoluchowski included philosophical reflections in numerous lectures, dissertations and essays, as well as in specialised works in the field of physics. They can be found both in scientific papers and in popular science works addressed to a wider audience. They are not always expressed directly, but an in-depth reading can reveal them between the lines. They are usually found in works strictly related to the problems of physics. Smoluchowski speaks most extensively and directly on philosophical topics in the introduction to the Physics section of his Self-Study Handbook.

Despite his many achievements, Marian Smoluchowski, as already mentioned, is not a well-known or appreciated figure in Polish science, not to mention in philosophy. In 1952, Stanisław Loria wrote about this state of affairs:

> Unfortunately, it must be said that Smoluchowski's fame does not match his merits and the importance of his achievements for the development of the methods and cognitive capabilities of modern physics. Apart from a relatively small handful of trained physicists, among the broad mass of Polish intelligentsia today there are few educated people who know as much as they should about Smoluchowski[34].

Unfortunately, little has changed in this respect over the years. Among philosophers, there is even less awareness of Smoluchowski's achievements. One of the reasons for this situation is the small number of publications that have appeared in the nearly one hundred years since his untimely death. Such sparse interest in the general achievements of the Polish scientist is incomprehensible, and particularly surprising is the lack of publications on the Polish market relating to Smoluchowski's research used in modern science in the field of the application of probabilistics in physics, stochastic processes,

---

[33] M. Smoluchowski, *O pojęciu przypadku i pochodzeniu praw fizyki opartych na prawdopodobieństwie (On the concept of chance and the origin of laws of physics based on probability)*, „Wiadomości Matematyczne' („Mathematical News') 1923, vol. 27, z. 2, pp. 1–26.

[34] S. Loria, *Marian Smoluchowski (1872–1917). Wspomnienie i próba charakterystyki (Marian Smoluchowski (1872–1917). Recollections and an attempt at characterization)*, 'Problemy' ('Problems'), 1952, No. 12, p. 794.

fluctuation processes, opalescence or the ever-increasing importance of his theories regarding coagulation processes.

1.2.2. Smoluchowski in publications

It should be emphasised at the outset that the bibliography devoted to Smoluchowski or referring to his scientific achievements is extremely modest. It is possible that the following discussion does not include all studies, however the purpose of this summary is not to create an exhaustive compendium of Smoluchowski's works, but rather to illustrate the scale of publications devoted to him. It should also be added that some of the works mentioned here contain numerous interpretative simplifications and even conscious manipulations.

Tadeusz Godlewski (1878–1921)—a physicist and professor at the Lviv Polytechnic—produced a brochure entitled *Maryan Smoluchowski* published by *Mathematical News*. His life and scientific work[35] had previously featured in *Muzeum* magazine in May 1918. The author's intention was to recall Smoluchowski and his achievements in the fields of physics and mathematics.

In 1921, one of the issues of *Taternik*, the magazine of the Tourist Section of the Polish Tatra Society,[36] was dedicated to the memory of Marian Smoluchowski—the mountaineer. However, these works form part of the *sui generis* epigraphic literature.

After Smoluchowski's death, his article *Several remarks on physical analogies, especially in theories of electric currents, thermal currents and the phenomenon of diffusion*, appeared in *Mathematical News*[37]. In 1923, with the consent of his widow, Zofia Smoluchowska, a translation from German was also published in *Mathematical News* of Smoluchowski's extremely important work *On the concept of chance and the origin of the laws of physics based on probability*.

---

[35] T. Godlewski, *Maryan Smoluchowski. Jego życie i działalność naukowa* (*Maryan Smoluchowski. His life and scientific activity*), 'Wiadomości Matematyczne' ('Mathematical News'), 1919, vol. 23, books. 1–3, pp. 1–36.

[36] 'Taternik (Mountaineer). A unit of the Tourist Section of the Polish Tatra Society, 1915–1921, Kraków 1921.

[37] M. Smoluchowski, *Kilka uwag o analogiach fizycznych, zwłaszcza w teoriach prądów elektrycznych, prądów cieplnych i zjawiska dyfuzji* (*Several remarks on physical analogies, especially in theories of electric currents, thermal currents and diffusion phenomena*), 'Wiadomości Matematyczne' ('Mathematical News'), 1918, vol. XXII, pp. 167–176.

In the 1920s, the Polish Academy of Arts and Sciences in Kraków decided to collect and publish all Marian Smoluchowski's works by, which had been developed by professors Władysław Natanson (1864–1937) and Jan Jakub Stock (1881–1925). Individual volumes were published in 1924, 1927 and 1928 under the title *The Writings of Marian Smoluchowski*[38]. This is the most serious collection to date—it includes 90 works.

The 1950s were exceptional for publications devoted to Smoluchowski. In 1951, a short article by Armin Teske, *Marian Smoluchowski*, appeared in the third issue of the bimonthly magazine for teachers *Physics and Chemistry*[39]. A year later, Władysław Krajewski (1919–2006) published an article entitled *The Great Physicist and Materialist Philosopher (On the 80th Birthday of Marian Smoluchowski)*. The article appeared in the *People's Tribune*[40], and was later reprinted in seven other newspapers. In the second half of 1952, the *Philosophical Thought* quarterly printed an essay, *Marian Smoluchowski as a philosopher-materialist*, by the same author[41]. While Armin Teske's article recalled Smoluchowski the person, Krajewski focused on his worldview, considering and analysing the physicist's beliefs in the light of Marxist ideology. Krajewski developed this concept in a book entitled *The Worldview of Marian Smoluchowski*, published four years later[42].

The culmination of celebrations of the 80th anniversary of the scholar's birth was the last (published in 1952), twelfth issue of the popular scientific monthly magazine *Problems*, which contained three articles devoted to

---

[38] *The Writings of Marian Smoluchowski. Commissioned by the Polish Academy of Arts and Sciences*, vol. 1, Kraków 1924, vol. 2, Kraków 1927, vol. 3, Kraków 1928.
[39] A. Teske, *Marian Smoluchowski*, 'Fizyka i Chemia' ('Physics and Chemistry'), 1951, year IV, May-June, No. 3 (17).
[40] W. Krajewski, *Wielki fizyk i filozof materialista. W 80-lecie urodzin Mariana Smoluchowskiego* (*The Great Physicist and Materialist Philosopher. On the 80th birth anniversary of Marian Smoluchowski*), 'Trybuna Ludu' ('People's Tribune'), 1952, No. 148. The article was then reprinted in the 'Słowo Ludu' ('Word of the People') daily in 1952, No. 129, and in the magazines: 'Świat i My' ('The World and Us') in 1952, No. 28, 'Widnokrąg' ('Horizon') in 1952, No. 24, 'Głos Szczeciński' ('The Szczecin Voice') in 1952, No. 23, 'Nowiny Rzeszowskie' ('Rzeszów News') in 1952, No. 23, 'Słowo Tygodnia' ('Word of the Week') in Rzeszów in 1952, No. 14, and 'Problemy' ('Problems') in 1952, No. 12.
[41] W. Krajewski, *Marian Smoluchowski jako filozof-materialista* (*Marian Smoluchowski as a materialist philosopher*), 'Myśl Filozoficzna' ('Philosophical Thought') 1952, No. 4, pp. 232–248.
[42] Ibid., *The Worldview of Marian Smoluchowski*, Warsaw, 1956.

Smoluchowski by Loria and Krajewski as well as extracts (in the form of an article) from the *Self-Study Handbook*.

In 1953, three more articles were published about Smoluchowski. In the bimonthly magazine *Physics and Chemistry*, Władysław Kapuściński published an article entitled *Marian Smoluchowski's philosophical views*[43]. In it, he recalled Smoluchowski the person and the philosophical aspects of the science he practised. Kapuściński confirms that a discussion of Smoluchowski's philosophical views is difficult as there is no separate publication in which he presented his views in a structured form.

In October 1952, at a ceremonial scientific gathering of the Poznań Branch of the Polish Physical Society to mark the 80th birth anniversary and 35th death anniversary of the scientist, Stanisław Loria gave a lecture entitled *Marian Smoluchowski*. The presentation was included, in a slightly altered form, in the *Yearbook of the Polish Physical Society* entitled *Marian Smoluchowski and his work*[44], as well as in the journal *Progress of Physics*, devoted to the dissemination of physical knowledge[45]. Loria's article was extended to include the issues of density fluctuations, thermodynamic fluctuations and Brownian motion. Loria highlights that he wanted to give a faithful picture of Smoluchowski's original mindset and emphasise the significance his work in the context of the scientific issues of the time the physicist lived in.

Kazimierz Gostkowski of the Institute of Physics at the Silesian University of Technology is the author *of Several Recollections of Marian Smoluchowski*[46]. The article is a short memoir essay of a few pages in length, remembering the great physicist.

The above articles appeared in the general and specialist press, and the pretext for writing them was supposed to be the 80th anniversary of Marian Smoluchowski's birth, though the publication was most likely political in nature. However, a direct contribution to their creation was the appearance

---

[43] W. Kapuściński, *Marian Smoluchowski's philosophical views*, op. cit.

[44] S. Loria, *Marian Smoluchowski (1872–1917). Wspomnienie i próba charakterystyki* (*Marian Smoluchowski (1872–1917). Recollections and an attempt at characterization*), 'Rocznik Polskiego Towarzystwa Fizycznego' ('Yearbook of the Polish Physical Society'), 1953, vol. 4, book. 1.

[45] Ibid., *Marian Smoluchowski and his work (1872–1917)*, op. cit.

[46] K. Gostkowski, *Kilka wspomnień o Marianie Smoluchowskim* (*Several recollections of Marian Smoluchowski*), 'Postępy Fizyki' ('Progress in Physics') 1953, vol. IV, book. 2.

of a small mention in an article by Russian physicist L.I. Storchak in *Woprosy Fiłosofii (Problems in Philosophy)*[47].

In 1955 and 1956, three books were published about Marian Smoluchowski. Armin Teske wrote a biography of the scholar—*Marian Smoluchowski Life and works*[48]—published in 1955. Władysław Krajewski published the book *The Worldview of Marian Smoluchowski* in 1956. The same year saw the publication of *A Selection of Philosophical Writings*[49], featuring several works by Marian Smoluchowski, the editor of which was also Władysław Krajewski.

In 1958, Zygmunt Klemensiewicz's article *Marian Smoluchowski, a memoir from forty years ago*, was printed in *Kosmos* magazine[50].

In 1965, the Polish Physical Society honoured the memory of Marian Smoluchowski by creating a medal in his name, which is awarded—no more than once a year—for outstanding scientific achievements and for contributing to the development of a selected field of physics.

In 1979, Izydora Dąmbska published an essay, *On the meta-scientific views of Władysław Natanson and Marian Smoluchowski* in the quarterly *Problems of Science*[51]. This was an attempt to analyse the scholar's epistemology.

In 1991, an important work by Bronisław Średniawa was published entitled *The role of collaboration between Marian Smoluchowski and Theodor Svedberg in studies of Brownian motion and fluctuations conducted in the first years of the*

---

[47] Ł. Storczak, Дискуссия о природе физического знания (*Diskussija o prirodie fiziczieskowo znanija — Discussion on the nature of physical knowledge*), "Вопросы философии" ("Woprosy Fiłosofii") 1948, No. 1, p. 206.

[48] A. Teske, *Marian Smoluchowski. Życie i twórczość* (*Marian Smoluchowski. Life and works*), Warsaw 1955

[49] M. Smoluchowski, *Wybór pism filozoficznych* (*Selection of Philosophical Writings*), Warsaw, 1956.

[50] Z. Klemensiewicz, *Marian Smoluchowski, wspomnienie sprzed lat czterdziestu* (*Marian Smoluchowski, a memoir from forty years ago*), 'Kosmos' ('Cosmos') 1958, series B, No. 4, pp. 95–107.

[51] I. Dąmbska, *O poglądach metanaukowych Władysława Natansona i Mariana Smoluchowskiego* (*On the meta-scientific views of Władysław Natanson and Marian Smoluchowski*), 'Zagadnienia Naukoznawstwa' ('Problems of Science') 1979, vol. XV, book. 1 (57), pp. 3–11.

twentieth century[52], in which the author describes the important period, from 1907 to 1914, of cooperation betweenthe two scientists—the theoretical physicist Smoluchowski and the experimentalist physicist Svedberg—on experiments related to the study of Brownian motion.

On the 130th anniversary of Marian Smoluchowski's birth and the 85th anniversary of his death, the Commission of the History of Science at the Polish Academy of Arts and Sciences organized a scientific session on 17 May 2002 entitled'Marian Smoluchowski—from atomic theory to contemporary physics'. The outcome of this one-day scientific session was the publication in 2003 by the Polish Academy of Sciences and the Commission for the History of Science of *Marian Smoluchowski—from atomistic theory to contemporary physics*, compiled by Adam Strzałkowski and containing six essays on Smoluchowski and an exhibition catalogue from 2002[53].

The Museum of the Jagiellonian University organised an exhibition presenting the life and works of Marian Smoluchowski in his various areas of interest. The exhibition 'Marian Smoluchowski (1872–1917). Physicist, mountaineer, and romantic of science' was held at the Jagiellonian University's Collegium Maius[54].

In 2000, a book appeared in English on the Polish publishing market with a preface by Roman Ingarden (1893–1970), entitled *Marian Smoluchowski, His Life and Scientific Work*[55], authored by Subrahmanyan Chandrasekhar, Mark Kac and Roman Smoluchowski.

Andrzej Fuliński, professor of physics at the Jagiellonian University, wrote the article *Contemporary applications of Smoluchowski's equations*, in which he

---

[52] B. Średniawa, *Rola współpracy Mariana Smoluchowskiego i Teodora Svedberga w prowadzonych w pierwszych latach XX wieku badaniach ruchów Browna i fluktuacji* (*The role of collaboration between Marian Smoluchowski and Theodor Svedberg in studies on Brownian motion and fluctuations conducted in the early years of the twentieth century*), op. cit., pp. 423–455.

[53] See: A. Fuliński, *Współczesne zastosowania równań Smoluchowskiego* (*Contemporary applications of the Smoluchowski equations*), op. cit., p. 23

[54] *Marian Smoluchowski 1872–1917. Fizyk, taternik, romantyk nauki*) *Marian Smoluchowski 1872–1917. Physicist, mountaineer, romantic of science*), catalogue of the temporary exhibition Collegium Maius, 17 May–14 July 2002.

[55] S. Chandrasekhar, M. Kac, R. Smoluchowski, *Marian Smoluchowski. His Life and Scientific Work*, ed. R.S. Ingarden, Warsaw 2000.

showed the use of Smoluchowski's equations in contemporary science and technology, raising readers' awareness to the extent of the great physicist's scientific heritage through specific examples[56].

In 2003, issue 2 (35) of online magazine *Zwoje* included several articles about Marian Smoluchowski. In his essay *The Importance of Marian Smoluchowski's Works for Subatomic Physics*[57], Andrzej Budzanowski highlights Smoluchowski's achievements in the field of applying probability to free Brownian particles, which due to their universality can be transferred to the world of nuclear and subnuclear processes. Budzanowski points out that after the discovery of other forms of matter, for example nucleic (nuclear) matter and further substructures, i.e. quarks and gluons, it can be expected that Smoluchowski's deliberations will be applied in nuclear collisions, during which highly excited states of nuclear matter are produced, including the critical state, and finally in the search for the original state of the matter of the Universe, i.e. the state of quark-gluon plasma.

In 2007, the OBI publishing house issued in its series *Źródła* (*Sources*). *Philosophy of Nature – Philosophy of science*, the book *Smoluchowski, Natanson and others*, as the third volume of its *Kraków Philosophy of Nature in the interwar period* series[58]. The publication contains four essays about Smoluchowski. The authors of the works—Krzysztof Starzec and Paweł Polak—discussed several extremely important issues of Smoluchowski's science, such as the concept of chance, the role of probability calculus in physics, his approach to the principle of causality, the issue of inductive and deductive research in physics, and even the problem of identification with materialistic views.

---

[56] A. Fuliński, *Współczesne zastosowania równań Smoluchowskiego* (*Contemporary applications of the Smoluchowski equations*), op. cit., p. 23

[57] A. Budzanowski, *Znaczenie prac Mariana Smoluchowskiego dla fizyki subatomowej* (*The Importance of Marian Smoluchowski's Works for Subatomic Physics*), in: *Marian Smoluchowski. Od teorii atomistycznej do fizyki współczesnej* (*Marian Smoluchowski. From atomic theory to modern physics*), op. cit.

[58] K. Starzec, *Marian Smoluchowski – teoria nauki a działalność naukowa* (*Marian Smoluchoski – scientific theory and scientific practice*), w: *Krakowska filozofia przyrody w okresie międzywojennym* (*Kraków Philosophy of Nature in the interwar period*), vol. 3, *Smoluchowski–Natanson–inni* (*Smoluchowski–Natanson–others*), ed. M. Heller, J. Mączka, Kraków–Tarnów 2007.

In 2008, Małgorzata Stawarz published the article *The starting point of Marian Smoluchowski's philosophical deliberations on chance and probability*[59].

For many years, a populariser of Marian Smoluchowski and his achievements has been Bogdan Cichocki, who publishes articles in *Delta*, a monthly popular science magazine published by the University of Warsaw covering mathematics, physics, information science and astronomy.

The quantity of these publications, although incomplete for a scholar of this calibre and the almost hundred years that have passed since his death, is astonishing in its modesty, especially the fact that to date only three books have been published on Smoluchowski. The listing of these publications clearly shows how few studies describing this scholar's achievements we are dealing with.

It should be emphasised that the Polish scientific community's most intense period of interest in Smoluchowski's work was the 1950s, which were extremely difficult for Polish science. Regardless of the political reasons for this interest, there were a number of scientific publications thanks to which the outstanding scholar was not completely forgotten. Following the relatively abundant period of the 1950s, for the next four decades (with a few exceptions) Smoluchowski was virtually forgotten. It is only in the new millennium that renewed interest in the achievements of the Polish physicist has been noticeable.

The hundredth anniversary of Smoluchowski's death inspired the scientific community to create numerous articles and essays on this outstanding physicist. They will certainly not be the last.

The lack of a basic, original philosophical study is an obstacle to defining Smoluchowski's views, though not the only one. The philosophical views contained in his writings were also not clearly laid out. *Smoluchowski is not unambiguous in his works, often putting forward theses which he opposes elsewhere. Statements or suggestions can sometimes be found that contradict each other. It is also not possible ascribe them to a specific philosophical system, as his views did not fit into any philosophical school in operation at that time.* This led to conscious

---

[59] M. Stawarz, *Punkt wyjścia filozoficznych rozważań Mariana Smoluchowskiego na temat przypadku i prawdopodobieństwa* (*The starting point of Marian Smoluchowski's philosophical deliberations on the subject of chance and probability*), „Semina Scientiarum" („Seeds of Science') 2008, No. 7.

abuses in the interpretation of his philosophy, *attributing materialistic beliefs and even dialectical materialism to him*. This was a reputation that stuck to Smoluchowski for many years, but—in the light of the philosophical views contained in his works—it is untenable. He can also not be ascribed atheistic beliefs, though in some publications this has occurred despite the fact that he had an indifferent attitude to religious beliefs. The lack of declarativeness in such an extremely personal issue should lead the researcher to reserve judgment.

It should be emphasised once again that Smoluchowski was a physicist whose work is still being used in scientific and industrial research, especially in chemistry, pharmacy, and physics, in case studies, probability and statistics, in advanced studies of stochastic processes, in studies on chaotic dynamics and in descriptions of sedimentation and coagulation processes.

CHAPTER 2

Preliminaries to the philosophy of Marian Smoluchowski

2.1. The beginnings of Smoluchowski's philosophy

As already mentioned, Smoluchowski did not leave any separate publication in which he summarised his philosophical views, but his scientific works contain thoughts and concepts that make a special contribution to the philosophy of science and the philosophy of nature. The fact that he did not present his thoughts in a separate study is not a matter of chance, but a well-thought-out decision related to Smoluchowski's thesis on the transience of philosophical concepts that fail to keep pace with the dynamic changes occurring in science.

> However, Smoluchowski created philosophical concepts which emerged somewhat involuntarily during the physicist's research work. In his essay Marian Smoluchowski – the theory of science and scientific activity, Krzysztof Starzec shows that scientists of Smoluchowski's calibre are not restricted only to their own field and do not limit themselves only to a concise explanation of the studied phenomena. Their thought horizons stretch much farther. They try to find the foundations of the science they practise, its fundamental importance, and often hidden assumptions; they try to stand outside and look at their work from the position of an uninvolved observer in order to better understand science. Considerations of a philosophical nature did not occupy the margins of his scientific activity. On the contrary, Smoluchowski considered reflection on science—the mechanisms of its development, methodology, theory of physical cognition, etc. —to be extremely important[60].

---

[60] See: *Kraków Philosophy of Nature in the Interwar Period*, op. cit., p. 387.

CHAPTER 2

Ernest Rutherford, one of the founders of modern physics, stated quite radically that "all science is either physics or stamp collecting"[61]. This was a thought that many physicists would probably have agreed with in the early years of the twentieth century, even though only twenty or so years earlier the vast majority of them thought that physics was a science with no prospects and that no longer held any secrets. Smoluchowski had a different opinion. On the one hand he was one of the pioneers of the changes taking place in physics, and on the other was a scientist dealing with physics, mathematics, but also with "stamp collecting", as he demonstrated philosophical interests that actually did not emerge by chance. He received his secondary education at the Collegium Theresianum, one of the best middle schools in Central Europe, founded in 1746 by Maria Theresa Habsburg, which grouped a body of outstanding teachers with a philosophical bent. A key figure in shaping Smoluchowski's interests was the physicist and philosopher Alois Höfler (1853–1922), whose student was also Kazimierz Twardowski, the founder of the Lviv-Warsaw school of philosophy.

> *Initially, Smoluchowski's interests were naturally focused on the humanities, but during his studies, under Höfler's influence, he became interested in physics and astronomy. Many years later, in 1915, in a letter written to Höfler, he jokingly reminded him that it was his fault that he became a physicist: "Thanks to you, I learned in high school to revere physics, mathematics and philosophy as the nicest subjects"*[62]. *In 1890, Höfler published a high school philosophy textbook called Propaedeutic logic for secondary schools*[63], *which was also translated into Polish. He went on to become a university lecturer of philosophy in Vienna and Prague, where he collaborated, among others, with Franz Brentano (1838–1917). It was Höfler who developed a philosophical sensitivity in Smoluchowski, which later manifested in his scientific work.*

Another person who initiated the future physicist's philosophical interests was Stanisław Szczepanowski, Smoluchowski's mother's brother. He was an outstanding Galician personality with broad economic, engineering and managerial interests—an oil entrepreneur, a pioneer of the emerging

---

[61] See: J.B. Birks, *Rutherford at Manchester*, London 1962, p. 27.
[62] A. Teske, *Marian Smoluchowski. Życie i twórczość* (*Marian Smoluchowski. Life and Works*), op. cit., pp. 11 and 12.
[63] A. Hoefler, *Logika propedeutyczna dla szkół średnich* (*Propaedeuticlogic for secondaryschools*), trans. Z. Zawirski, Lviv 1927.

oil industry, and an activist for the industrialisation of Galicia. He was also known for his public activity—as a member of the Austrian Parliament and the Diet (Parliament) of Galicia. He was interested in philosophy and, due to his specific and broad studies, was known as the "romantic of positivism". He was distinguished by his uncommon intellect and unique activity to such an extent that he aroused the interest of Polish literary authors. Bolesław Prus and Henryk Sienkiewicz wrote about him, and Stanisław Brzozowski dedicated his book *Philosophy of Polish Romanticism* to him. As an MP, he wrote a book on economics entitled *Poverty of Galicia in Figures and a Programme for the Energetic Development of the Economy of the Country*[64]. This was a programme for the vigorous development of the economy of Galicia, representing not only a compendium of knowledge on Galicia's social and economic condition but also containing propositions for a programme of development for the region.

Höfler and Szczepanowski contributed to arousing Smoluchowski's philosophical preferences. This tendency is confirmed by a statement by his wife, Zofia:

> He considered himself to be a 'romantic', a type of researcher striving to solve many mysteries at once, never finding peace. These characteristics of his spirit explain the fact that his ninety-six published works relate to a very large number of fields. It really was a 'romantic', youthful way of working, very uneconomical. He did not stop, as the 'classicists' did, at broadening one branch, adding new contributions to previous studies, applying the same method, the same experience to a wider range of subjects, but with the extravagance of a researcher who could afford it, he switched from one field to another. Everything interested him and aroused his curiosity. It was probably this tearing through unknown jungles that attracted and enticed him, like mountain peaks inaccessible to people in his time. A romantic! Something too easy, too obvious, held no charm, like easy green hills… compared to rocky peaks and dizzying abysses…[65].

Smoluchowski's distinctive personality was shaped by practising mountaineering during his school and university days, which imbued it with special features that many years later—on the occasion of receiving the 'silver edelweiss'—he summed up: "From what the mountains gave me, I consider

---

[64] S. Szczepanowski, *Nędza Galicji w cyfrach* (*Poverty of Galicia in Figures*), Lviv 1888.
[65] A. Teske, *Marian Smoluchowski. Życie i twórczość* (*Marian Smoluchowski. Life and Works*), op. cit., p. 140.

three things most valuable: the habit of undertaking difficult tasks, the joy of overcoming difficulties, and the ability to beautify everyday life through the most sublime poetry: the poetry of the mountain world"[66]. Smoluchowski was among the pioneers of European mountaineering.

At the University of Vienna, Smoluchowski studied with Josef Stefan (1835–1893), about whom he wrote in a letter to Boltzmann: "In theoretical physics, I owe the first foundations to my venerable teacher Stefan[67]*." Stefan was an outstanding physicist, who went down in the history of science for his work on explaining the radiation of a perfect black body. In 1879, he demonstrated experimentally that the radiation energy of such an object is proportional to the fourth power of its temperature, which led to the resulting law being called the Stefan-Boltzmann law (Boltzmann supplemented the rule with the emission of grey bodies). This was one of the first major achievements in explaining the radiation of a perfect black body. Of significance to Smoluchowski's future work, Stefan also studied the kinetic theory of gases*[68].

Smoluchowski also studied under Franz Serafin Exner (1849–1926), a pioneer in many fields of modern physics. His research interests included: spectroscopy, electrochemistry, issues of electricity in the atmosphere, research related to the theory of colour, and analysis of the structure of meteorites. In 1907, Exner became the rector of the University of Vienna. Among others, his students included Erwin Schrödinger (1887–1961), one of the founders of quantum mechanics. Exner made a name for himself in science, but also influenced Smoluchowski as it was after becoming acquainted with his work in 1900 that the Polish researcher developed the theory of the molecular-kinetic nature of Brownian motion.

Smoluchowski was not a student of Ludwig Boltzmann, but was greatly influenced by him scientifically, as he wrote explicitlyin a letter: "You have exerted the greatest influence on me both through your lectures and especially through your textbooks, which I study with ever renewed

---

[66] Idem, p. 18.
[67] Idem, p. 15.
[68] *Encyclopaedia Britannica*, vol. 40, Poznań 2004, p. 424.

satisfaction and benefit, and I am proud that I can count myself among your students"[69].

He was a member of the Philosophical Society at the University of Vienna and took part in lectures organised by the Society, as mentioned by the authors of the book *Ernst Mach's Vienna 1895–1930: Or Phenomenalism as Philosophy of Science*[70].

2.2. Smoluchowski's philosophy against the backdrop of the main trends in 19th-century philosophy

Smoluchowski's views are not a set of philosophical declarations; there is no clear support for any particular school of thought, nor any defined philosophical positions. Being a physicist, Smoluchowski treated philosophy instrumentally, and conducted philosophical deliberations when needed. By its nature, physics is a science that constantly touches on philosophy, which is why there is a lot of philosophy in his works. The Polish scientist made selective use of the entire spectrum of philosophical concepts functioning at the time, most often not noting which philosophical school was involved, merely practising it. He discussed some philosophical theses—agreeing with some and opposing others. Reading Smoluchowski from a philosophical perspective, it is necessary to uncover the elements of various philosophies contained in his works, as they are usually hidden in comments related to physics, mathematics or methodology. He was not usually concerned with philosophy, which was not the mainstream of his interests, but rather with a useful theoretical tool, helpful in understanding reality, which is mainly perceived through the theories and hypotheses of the exact sciences.

The above reflections prompt a review of the theses of the main trends in nineteenth-century philosophy of nature, both in terms of their presence in Smoluchowski's scientific works and as an analysis of his references to these theses.

---

[69] A. Teske, *Marian Smoluchowski. Życie i twórczość* (*Marian Smoluchowski. Life and Works*), op. cit., p. 140.
[70] *Ernst Mach's Vienna 1895–1930: Or Phenomenalism as Philosophy of Science*, eds. J.T. Blackmore, R. Itagaki, S. Tanaka, Dordrecht 2001, p. 79.

CHAPTER 2

Auguste Comte's (1798-1857) positivism is a philosophical doctrine that has a permanent presence in Smoluchowski's works. *Although the scholar's thinking often surpassed this doctrine's assumptions, Smoluchowski agreed with the basic principles* of positive philosophy, for example, favouring the rejection of metaphysics.

In *La méthode positive en 16 leçons* (The Positive Method in sixteen lessons), Comte wrote:

> It will now be easy for us to determine precisely the exact nature of the positive philosophy. This philosophy considers all phenomena as subject to invariable natural laws. We regard the search after what are called causes, whether primary of final, as absolutely inaccessible and unintelligible. In our positive explanations [...], we do not pretend in any way to disclose the generative causes of phenomena [...], but only to analyse accurately the circumstances of their production and to connect them with one another by normal relations of succession and similarity. (...) The new philosophy differs from the old in its desire to eliminate all searches for first and deliberate causes, considering them sterile[71].

Smoluchowski had an analogous attitude to causality and purposefulness, claiming that in antiquity and in the Middle Ages, natural phenomena were explained on the basis of purposefulness, which had come to be questioned by every naturalist. He wrote: "The means of explaining with the aid of the concept of purposefulness is today removed from natural sciences as a naive anthropomorphism"[72]. The positivism initiated by Comte proclaimed that true knowledge is scientific knowledge, free from any metaphysical elements, which can be obtained only by positive verification of a theory and which is achieved through empirical research. A positive approach to science and philosophy essentially consists of the study of objects accessible to sensory cognition. Only the exact sciences have developed methods of cognition enabling the creation of scientific knowledge. Smoluchowski opposed pseudoscientific speculation and idealism in favour of what is certain, material and possible

---

[71] A. Comte, *La méthode positive en 16 leçons* (*The Positive method positive in sixteen lessons*), translated by W. Wojciechowska, Warsaw 1961, pp. 15 and 301.
[72] M. Smoluchowski, *Poradnik dla samouków. Wskazówki metodyczne dla studiujących poszczególne nauki. Fizyka, Geofizyka, Meteorologia* (*Self-Study Handbook. Methodological guidelines for students of individual sciences. Physics, Geophysics, Meteorology*), vol. I, Warsaw 1917, p. 18.

to research, i.e. empirically knowable through experience. He drew attention to the inductive nature of science and the leading role of the experiment.

The achievements of *Charles Darwin* (1809–1882) played an important role in the process of shaping positivist philosophy. In his work entitled *On the Origin of Species*, Darwin wrote that he could not see any data to support the existence of a plan, let alone a benevolent plan[73]. Smoluchowski was a strong advocate of Darwin's theory; he wrote about it many times in various texts. He maintained that it was mainly thanks to him that science had moved away from the concept of purposefulness in nature, which was a misleading *a priori* assumption, and that it was he who had shown through various examples how seemingly purposeful adaptations are created which are *de facto* the result of natural causes[74].

Smoluchowski did not identify with every thesis of positivist philosophy. *Positivists and Smoluchowski understood science itself differently. The Pole understood science as a set of hypotheses constituting a certain approximation of the laws of nature. His inclination towards pragmatism was evident here. He did not give science the indisputable position it had in positivism or empirio-criticism. He emphasised the difference that separates scientific theories from the real laws of nature and the constant asymptotic path of science striving to know these laws. His epistemological criticism, stemming from Kantian inspiration, did not allow for an acceptance that scientific knowledge could reveal, in a properly justifiable way, the essence of reality*[75].

Smoluchowski cannot be attributed the position proposed by Jean Le Rond d'Alembert and Comte as he had a different understanding of the credibility of science. He wrote about scientific theories as more or less probable hypotheses that could never be fully verified. He believed that the primacy of the experiment in science does not imply that phenomena should be defined only on the basis of experience and that only empirical facts can constitute the real object of cognition. Similarly, he did not agree with philosophy being limited only to its practical nature, as Comte postulated, so that it should be used mainly in the field of everyday life. He shared the view of positivists

---

[73] See: F. Copleston, *A History of Philosophy*, vol. 8, translated by B. Chwedeńczuk, Warsaw 2006, p. 92.
[74] See: M. Smoluchowski, *Self-Study Handbook*, vol. I, op. cit., p. 18.
[75] I. Dąmbska, *On the metascientific views of Władysław Natanson and Marian Smoluchowski*, op. cit., p. 3.

who questioned the metaphysical side of philosophy, but also saw the need for the existence of the philosophy of nature and philosophy of science that he himself practised.

In his article *Marian Smoluchowski's approach to the causality principle in Brownian motion research*, Zenon Roskal notes: "Thanks to the work of scholars such as Smoluchowski, a breakthrough in philosophy was possible, consisting in going beyond the positivist vision of science, but also beyond the barriers that positivist philosophy erected in the way of the understanding and mastery of nature"[76]. The basis of Comte's positivist philosophy was man and his wider society, and science. With this approach, the French philosopher wrote himself into the history books as an advocate of utilitarianism. This was consistent with the later views of pragmatists and close to Smoluchowski.

> *Scientism developed in the second half of the 19th century on the basis of empiricism and positivism. It assumed that science would enable the prediction and management in the desired manner of the natural and social processes occurring in the world. Although Smoluchowski was a supporter of many scientistic convictions, he was undoubtedly not an adherent of scientism. Like positivism, nineteenth-century scientism was somewhat naive in its apotheosis of science, hence direct references to scientism are not found in scientific works. Smoluchowski did not enter what was to him a field of unclear philosophical diversions, remaining more within the realm of methodology and the language of the exact sciences. In the philosophical concept he built, he distanced himself from theories accepted in science and the validity of scientific knowledge. He did not agree with the scientistic claim that the science they preached was consistent 'in itself' with reality*[77]. *This was reflected in his scientific work in the formulation of hypotheses that broke the standards of science accepted and applicable at that time.*

Smoluchowski's sympathy for empirio-criticism—the so-called 'second positivism', a philosophical movement at the turn of the 19th *and* 20th *centuries, the most important proponents of which were* Richard Avenarius (1843–1896)

---

[76] Z.E. Roskal, *Mariana Smoluchowskiego ujęcie zasady przyczynowości w badań ruchów Browna (Marian Smoluchowski's approach to the causality principle in Brownian motion research)*, 'Zagadnienia Filozoficzne w Nauce' ('Philosophical Problems in Science'), 2017, vol. 62, p. 110.

[77] See B. Kotowa, *Scjentyzm jako światopogląd nauki (Scientism as a worldview of science)*, 'Nowa Krytyka' ('New Criticism'), 2004, No. 16, p. 151.

and Ernst Mach (1838–1916)—was *from the outset not uncritical, and over time he firmly distanced himself from their theses.* The aim of empirio-criticism was to exclude all fiction from science, which to Smoluchowski was obvious. The basic assumptions of empirio-criticism stemmed from the roots of Auguste Comte's philosophy. Empirio-critics argued that *unscientific considerations that are not based on 'pure' experiment should be rejected, and as much experimental data as possible should be incorporated to ensure its understanding. They postulated elimination of the tendency to hypostasise reality into objects and their images, claiming that there is only one, monistic, reality. To Smoluchowski, theses posited in this way were too restrictive and he did not accept such an approach as he realised the limitations they impose on theoretical considerations. For example, research on Brownian motion, culminating in scientific evidence confirming atomistic theory, as well as research in the field of kinematics, had proven that scientific inquiries cannot be artificially limited only to the scope of experimental research.*

The fact that empirio-criticism excluded hypothesis from the process of creating scientific theories was unacceptable to Smoluchowski. He argued that the presence of a hypothesis is essential in the development of science, *and that the dynamic progress of science would not be possible without theoretical considerations formulated in a hypothesis and subjected to experimental verification*[78]. The aim of science, he believed, could not be only a description of facts or phenomena, devoid of an attempt to explain them. The description should be formulated in accordance with knowledge, but we should not limit ourselves only to experimentation as knowledge cannot be based solely on empiricism[79].

Avenarius promulgated *a theory on the psychological nature of* perceptions, so-called introjection, i.e. projection of the perceived world into one's own interior, which led him to the conclusion that there are no separately existing physical and mental phenomena, as they are only two aspects of the same experience. Hence his epistemological monism. However, such an assumption consequently led to rejecting the truth of cognitive activities, which Smoluchowski, as a physicist, could not agree with.

---

[78] M. Smoluchowski, *Self-Study Handbook*, vol. I, op. cit., p. 46–50.
[79] Idem, p. 32.

Smoluchowski was a supporter of pragmatism, a philosophical system from the second half of the 19th century, the basic thesis of which was the pragmatic theory of truth. The initiators of this concept were two American philosophers: Charles Sanders Peirce (1839–1914) and William James (1842–1910), who assumed that utility is the criterion for the truth of judgments and concepts, because truth is determined in relation to the goal achieved by knowledge. A belief is true insofar as it contributes to effective action. Smoluchowski shared the assumptions of the pragmatic theory of truth, according to which the truth of theses depends on their practical results[80]. James' theory of truth assumes that "True ideas are those that we can assimilate, validate, corroborate and verify"[81]. The assumption that truth is not a permanent property of adopted theories but may at most only happen to them, and that the most important property of knowledge is its ability to explain, making science not a set of truths about the world but merely a set of explanations enabling us to understand it, was consistent with Smoluchowski's thinking. He advocated such an understanding of truth in science whereby it would be verified by its usefulness in practice and ideas would be validated through their practical consequences. Progress in science is not only about discovering new laws, but also about replacing existing explanations with better ones. He justified this idea by developing the concept of the utility criterion[82].

Smoluchowski shared James's view that science consists of generalities providing an abstract picture of the world, not reality itself, and that beliefs are more dispositions towards important life actions than a mental representation of reality. Although he did not refer directly to James's philosophy in his publications, he shared pragmatism's view that necessity and doubt guide us and make us want to learn something new, which in turn leads to the creation of new convictions and shapes them with the help of rules of conduct.

Some theses of instrumentalism, such as scientific knowledge having no objective reference within actual reality, being more the result of human relations with nature and society, or the instrumentalists' proposal of solving

---

[80] Ibid., p. 51.
[81] W. James, *Pragmatism: A New Name For Some Old Ways Of Thinking. Popular lectures on philosophy*, translated by M. Filipczuk, Warsaw 2004, p. 161.
[82] A discussion on this theory can be found in Chapter 6, 'Smoluchowski's Utility Criterion'.

problems through the formulation of a hypothesis, its logical verification, and then empirical falsification, were used in Smoluchowski's scientific practice. The general attitude of instrumentalist towards knowledge, which assumed that functioning theories and theorems play only instrumental roles and are merely reactions to environmental factors, that they are simply tools enabling people to adapt to the environment, as well as the instrumentalists' thesis that scientific theories should not be assessed by their veracity or falsity as they are only rules of inference, constituting tools to help in adapting to the environment and mastering it, were philosophically unacceptable to Smoluchowski as he saw them as narrowing the perception of the role of science.

He assumed that the criterion of the truthfulness of a hypothesis is not a decisive argument to accept a given thesis as veracity does not exhaust all the prerequisites for a theory's compatibility with nature. Due to the earlier assumption of the uncertainty of knowledge, he introduced a category that provides greater certainty of the reality of scientific theory. This criterion is *usefulness*, as it is more substantive than the category of truthfulness:

> We do not distinguish between true, untrue, more or less probable theories; we distinguish between more or less useful theories. We can talk about their usefulness in three ways. The more useful the theory or hypothesis: 1) the simpler and more illustrative its essence; 2) the larger the area of known phenomena it explains and makes accessible to our mind; finally, 3) the better it turns out to be as a guide in further research. This last role of hypotheses, of predicting things not yet known, is extremely important for a fair assessment of their importance in science[83].

Narrowing science down to an action aimed at both adaptation to the environment and its subordination to man would suggest limiting science to issues falling within the sphere of empiricism, while Smoluchowski's *use* covers a fuller spectrum including all manifestations of theoretical knowledge, which can perform various functions—not only empirical, but also, for example, explanatory, interpretative, and cognitive. The truthfulness or falsehood of theories, as well as all their instrumental functions, are tested, according to Smoluchowski, by the degree of their *use*—the more a theory is used, the

---

[83] See idem, *The importance of the exact sciences in general education*, in: *The Writings of Marian Smoluchowski*, vol. 3, op. cit., p. 194–195.

more it is considered to be consistent with the real laws of nature, which will most likely never be fully known.

Enthusiasm for truth, *fanaticism for veracity*, wrote Smoluchowski, the fanatical pursuit of reliability and truth, is an ethical foundation, corresponding to the exact sciences and strengthened by them. These, in turn, fight against blasphemy and cliché—diseases that affect society and distort language[84]. This statement concerned the issue of ethical orientation in research, but Smoluchowski's attitude to truth in science remained critical, because he believed that physical theories trying to cover material phenomena with a general mechanical view of the world cannot claim absolute truth. They can only be an "image" of the phenomena revealed to us[85]. This idea is close to another philosophical view, which assumed that beneath the surface of observable phenomena is hidden a true, essential reality. To Smoluchowski, essentialism basically meant hiding the essence of the laws of nature and actually ignoring them, so he assumed that functioning hypotheses and scientific theories may turn out to be false and it will never be possible to pronounce definitively the truth of some of them.

Essentialism, a trend initiated by Karl Popper (1902–1994) and aimed at answering the question of what the laws of science are, corresponds to Smoluchowski's views. Following Einstein's suggestion, Popper stated that the laws of science describe the structural properties of the world, which are important properties that "cannot be seen with the naked eye"[86]. He argued that if we want to possess knowledge, we must look for the laws of nature and the regularity present in it. However, we must not assume that there are certain strict regularities; it is enough to realise that our knowledge consists in the search for universal regularities, as if they existed[87].

---

[84] See idem., *The importance of the exact sciences in general education*, in: The *Writings of Marian Smoluchowski*, vol. 3, op. cit., p. 130.

[85] Idem, *O nowszych progressach na polu kinetycznych teoryj materji* (*On more recent progress in the field of the kinetic theory of matter*), in: *The Writings of Marian Smoluchowski*, vol. 1, op. cit., p. 279.

[86] See: S. Wszołek, *Esencjalizm transcendentalny K.R. Poppera* (*K.R. Popper's Transcendental Essentialism*), 'Zagadnienia Filozoficzne w Nauce' ('Philosophical Problems in Science'), 2002, No. 31, p. 127.

[87] Ibid.

This is a thesis close to the beliefs of Smoluchowski, who assumed the existence of such universal regularities, but with the important reservation that we will never know them. Like Smoluchowski, Popper did not believe in the possibility of reaching a final explanation which would need no further modifications, but at the same time he was convinced that we can delve ever deeper into the structures of our world, towards more and more important or deeper properties of it[88]. Although Popper rejected essentialism for methodological reasons, his essentialist views, expressed several decades after Smoluchowski, remain fundamentally consistent with the Polish physicist's theses.

Conventionalism is a trend of French philosophy of science from the beginning of the 20th century, the main proponents of which were Henri Poincaré, Pierre Duhem (1861–1916) and Édouard Le Roy (1870–1954). The concept of science until then, expressed in accumulating judgments about facts and formulating general, inductively justified statements based on them, could not unambiguously and convincingly defend itself against the criticism put forward by David Hume (1711–1776). The research conducted by the Scottish philosopher was aimed at negating the possibility both of making judgements about the world, and about the subject knowing it. In turn, the emergence of non-Euclidean geometries undermined the concept of *a priori* synthetic judgments, to which Immanuel Kant (1724–1804) attributed the axioms of Euclidean geometry. The above-mentioned arguments contributed to a re-assessment of the erstwhile understanding of science. According to Poincaré's position, in the formulation of scientific laws and the description of facts, in addition to individual statements about facts considered to be true, a fundamental role is played by the scientist's decision or the agreement of the scientific community regarding so-called conventions, i.e., judgements introduced into science on a contractual basis.

> Due to the essence of human nature, scientific theorems and theories are conventional in nature, and the attitudes adopted towards the perceived reality are changeable and have a developmental tendency. The reason for their acceptance is not so much their veracity as their cognitive value, considerations of simplicity,

---

[88] Idem, p. 126.

convenience, economy of operation, and even aesthetics[89]. Thoughts acceptant of this point of view often appear in the works of Smoluchowski, who—like Poincaré—believed that changes taking place in science, especially radical ones, are possible thanks to the presence of convention in it[90]. Smoluchowski's inquiries into issues of chance and probability show a clear influence of the French philosopher. Adopting Poincaré's position, the Polish physicist used the concept of chance as a kind of causal link and sought to explain through this relationship the possibility of chance occurring and to apply statistical methods to a world governed by deterministic laws.

Smoluchowski was not a declared conventionalist, and did not agree with some of Poincaré's statements, particularly with the thesis on the independence of the principles of science from experiments, however, the works of the French physicist and mathematician significantly influenced his philosophical beliefs. There are frequent references to Poincaré's theses and arguments, in particular those contained in two articles: Science and Method, and Science and Hypothesis, on which Smoluchowski wrote a separate, exceptionally kind essay entitled Two Books in the Field of 'The Philosophy of Nature'.

In Smoluchowski's philosophy, some elements can be found of neo-Kantianism in line with the Marburg school. Neo-Kantianism arose in the second half of the nineteenth century and constituted a return to Kant's philosophy not only in terms of methodology, but as a reaction to the speculative systems of German idealism and the empirical materialism of the developing natural sciences. Represented by Hermann Cohen (1842–1918), Paul Natorp (1854–1945) and Ernst Cassirer (1874–1945), the neo-Kantianism of the Marburg school contained some assumptions consistent with Smoluchowski's views. For example, he recognised the view of the active role of the intellect in the learning process and focused on logical, epistemological and methodological issues. According to the proponents of this doctrine, the task of philosophy was to determine the logical structure of scientific knowledge and to show the logical conditions occurring between real material relationships. Some analogies to the theses assuming that the practice of philosophy should begin with a critique of knowledge, that knowledge is subjective, that concepts do not describe reality but are a product of the intellect, that only that which the mind constructs is considered knowledge, or that 'knowledge' should be treated as an analysis of how it is possible, can be found in Smoluchowski's writings[91]*. However, he did not agree with all the assumptions of the neo-Kantianists, especially those that as a practising physicist he could not accept, such as proving that knowledge cannot exceed the boundaries of experience and*

---

[89] See: I. Szumilewicz, *Poincaré*, Warsaw 1978, p. 26.
[90] See: M. Smoluchowski, *Dwie książki z dziedziny 'filozofii przyrody'* (*Two books from the field of the 'philosophy of nature*), 'Ateneum Polskie' 1909, vol. IV, pp. 294–295.
[91] See idem, *Self-Study Handbooks*, vol. I, op. cit., pp. 13–14, 16–17 and 50.

that the real object of scientific knowledge is only thought, while the data of the external world have no cognitive value. Smoluchowski had different thoughts on this subject. In conclusion, it should be stated that his connections with neo-Kantianism were sporadic and superficial, but noticeable.

In the second half of the 19th century there was a renaissance of the thought of St. Thomas Aquinas (1225–1274) in the form of Neo-Thomism. Its founders took account of the changes that had taken place in the perception of reality during the several hundred years since the emergence of Thomism. Research on neo-scholastic thought had commenced in many intellectual centres in Europe. The encyclicals Aeterni Patris of Leo XIII (1879) and Pascendi Dominici Gregis of Pius X (1907) confirmed this state of affairs. A key tenet of Neo-Thomism was establishing the relationship between knowledge and faith. According to Neo-Thomists, not only do faith and rational knowledge not contradict and exclude each other, they complement one another. The source of rational knowledge is human reason, and although it is not a perfect tool, it cannot be rejected. The source of faith is revelation, and the truths arrived at on this path are absolute in nature. Neo-Thomism established a strict hierarchy according to which theology was at the peak of all knowledge, philosophy was in the middle of the hierarchical pyramid, and the other sciences formed its base. The boundary of scientific knowledge was to be the world of created things[92], and interference of scientific thought in the sphere of religious dogma was excluded.

The theses of Neo-Thomism did not conform to the general principles of Smoluchowski's scientific thinking. To assume that all things change in accordance with a divine plan is to advocate the existence of purpose in nature. Smoluchowski maintained that belief in the existence of intentional cause in nature is a manifestation of naive anthropomorphism[93]. A thesis assuming theology to be the peak of all available knowledge was unacceptable to the Polish physicist; he was guided by scientific reasoning, and theology went beyond the methodology of science. Smoluchowski did not claim that scientific theories provide a full answer to the questions posed by nature, but he believed that the scientific path is the most appropriate for understanding nature. He wrote: "Who knows whether the human race, bound to the Earth, is not, as a result of its organisation, blind to entire areas of the

---

[92] See: A. Maryniarczyk, *Tomizm*, (Thomism) in: *Powszechna Encyklopedia Filozofii* (Universal Encyclopedia of Philosophy), vol. 9, Lublin 2008, p. 503.
[93] See: M. Smoluchowski, *Self-Study Handbook*, vol. I, op. cit., pp. 18–19.

phenomena of the universe, like a holothuria (sea cucumber), attached to a rock on the sea bed"[94].

His doubts were raised by the limitations to scientific research which had been created in man by evolution. He asked: "Can we even know the phenomena that are really occurring; does some way exist that enables us to study the real world? After all, all the factual material upon which we base our awareness, or rather ideas about the external world, is made up exclusively from our sensory perceptions"[95].

Smoluchowski's philosophical interests were a consequence of questions posed to the exact sciences, primarily physics and mathematics, inspired by relationships occurring in nature, often surpassing the realm in which nature has armed us with cognitive capabilities, as we are only able to achieve localised knowledge of the Universe we live in. As Heller put it: "This field by some miracle emerges—like a commutative centre in a non-cummutative algebra—from the ocean of the Universe, which surpasses the power of our mind an imagination."[96] Smoluchowski argued that we will never transcend certain confines bounded by the possibilities of empirical knowledge.

In his philosophical system, Saint Thomas Aquinas employed Aristotle's (384–322 BC) concept of intentional cause. In the *Summa Contra Gentiles*, he argued that if there were no intentional cause, the same causes would not always have the same results. Intentional causality is the reason all things exist[97]. In Thomism, intentional causes are independent entities which justify the approach to the structure of contingent existence. Smoluchowski regarded intentional cause differently in the extreme. He argued that explaining phenomena related to the functioning of nature is one of the tasks of physics, without reference to intentional cause[98]. Since ancient times, the functioning of nature has been explained through intentional cause. Situations concerning people and animals have been assigned as natural phenomena. Smoluchowski noted that the naivety of such reasoning leads to the ridiculous suggestion

---

[94] Idem, p. 16
[95] Idem, p. 13.
[96] M. Heller, *Time and Causality*, Lublin 2002, s. 42.
[97] See: E. Gilson, *Tomizm: wprowadzenie do filozofii św. Tomasza z Akwinu (Thomism an introduction to the philosophy of St. Thomas Aquinas)*, trans. J. Rybałt, Warsaw 1960, p. 116.
[98] See: M. Smoluchowski, *Self-Study Handbook*, vol. I, op. cit., p. 18–21.

that intentional cause in the functioning of nature essentially entails acting for the benefit of humanity[99]. There are no direct references to Neo-Thomism in Smoluchowski's texts, however reflections on causality and intentionality can frequently be found in them. The conflict of this doctrine's main theses facilitates an insight into Smoluchowski's philosophical views, which place him in opposition to metaphysics.

When writing about Smoluchowski's philosophy, a philosophical doctrine should not be ignored which has had a significant impact on the development of science, especially physics and chemistry, characteristic of which was the phenomenological perception of matter. Classical *thermodynamics, also known as phenomenological thermodynamics*, which was dominant in the nineteenth century, treated solids, liquids and gases as continuous media without molecular structure. A group of scientists, called the energeticists, dealing with macroscopic thermodynamic phenomena, posited that scientific research be based purely on experiment, on the description of facts, and not on explaining phenomena. Smoluchowski was an opponent of the phenomenological approach to the material world, as well as of the philosophical concept of energeticists, which he wrote about many times. This issue is discussed in more detail in section 3.3. 'The Monistic interpretation of nature'.

Another important issue in Smoluchowski's philosophy is his attitude towards materialism, but not because materialist philosophy is of significance to his views. A problem arose in the 1950s, when Smoluchowski, for political reasons, began to be attributed in numerous publications materialistic views that were not supported by his scientific works. As a result of this manipulation, in the following years the philosopher was labelled a materialist. This issue—as an important point requiring examination—has been developed further below (see subsection 2.3. 'Marian Smoluchowski's worldview').

Following publication of the paper *On thermodynamic fluctuations and Brownian motion* in 1906 and after the general acceptance by European academic circles of the kinetic-atomic theory of matter, there occurred a change in Smoluchowski's philosophical mindset, which is perceptible both in the scientific problems he took on and in his publications. This issue is also one of the main topics of this book.

---

[99] Idem, p. 18.

2.3. Marian Smoluchowski's worldview

2.3.1. Marian Smoluchowski—materialist, atheist?

In the hundred years that have passed since Smoluchowski's death, relatively few publications have been released in Poland related to this scientist's achievements, and among them, especially in the 1950s, greater or lesser presumptions appeared as to the physicist's materialist views. Are they really based on source materials? Did Smoluchowski provide grounds for such a conclusion? If so, how serious were the arguments that justified this belief, and if not, what are the origins of such suggestions?

Due to the above-mentioned works, finding reliable answers to these questions should be an important element of any discussions of Smoluchowski's philosophy. The key person in forming this quite controversial interpretation of the Polish physicist's worldview was the philosopher Władysław Krajewski. The marked influence he had on the perception of Smoluchowski's philosophy is evidenced not only by publications presenting his profile and achievements in terms of the materialistic worldview, but also by the generally prevailing belief that "something had to be going on". Krzysztof Starzec's apt conclusion captures the essence of this problem: "It should be noted that the interpretation of Smoluchowski's meta-scientific texts proposed by Krajewski, otherwise quite detailed, was the only one that existed in Polish literature for over twenty years. Therefore, certain ideas contained in it must have penetrated the consciousness of philosophers and scientists and may still be lingering to this day[100]. "

Studying Smoluchowski's achievements, while maintaining the principle of objectivity and critical reference to Krajewski's theses, allows the image he established of Smoluchowski as an atheist and materialist to be undermined. In analysing the topic of Smoluchowski's materialistic worldview, two works should be recalled in which a critical analysis of Krajewski's book was undertaken. The first, entitled *Two interpretations of Marian Smoluchowski's thought*, was written by Krzysztof Starzec, the second is Małgorzata Stawarz's doctoral thesis, *Reconstruction and critical analysis of Marian Smoluchowski's philosophical*

---

[100] K. Starzec, *Dwie interpretacje myśli Mariana Smoluchowskiego* (*Two interpretations of Marian Smoluchowski's thought*), in: *Krakowska filozofia przyrody w okresie międzywojennym* (Kraków philosophy of nature in the interwar period), op. cit., pp. 423–424.

views. Also in the book *I was your opponent [Professor Einstein]…* by Paweł Polak, a number of valuable comments can be found on Krajewski's theory.

The second person who attributed materialistic views to Smoluchowski was Władysław Kapuściński, who claimed that "in very many statements made less '*ex cathedra*', we see Smoluchowski as a naturalist-materialist, deeply convinced of the existence of the external world and its objective laws, sometimes approaching even the position of dialectical materialism"[101]. However, it is Krajewski who is considered the initiator of the theory of Smoluchowski's materialistic worldview. He wrote about it many times over the years, referring to the publications of Soviet scholars and creating a whole range of arguments and suppositions to support his thesis.

2.3.2. General characteristics of materialism

In reference to Krajewski's controversial argument, let us follow the textbook assumptions of this philosophical concept. Materialism argued that the only existent entity is matter, and the whole world is made of matter or is a modification of it. This concept was a variation of so-called ontological monism. Materialism—in its older form called mechanistic materialism—assumed that everything that happens in nature can be explained and reduced to causes. From the epistemological point of view, it assumed that only material reality, existing objectively and independently of the subject perceiving it, constitutes the object of human cognition. Materialism sees consciousness and mental experiences as secondary to matter, consequently reaching the conclusion that every spiritual activity of man is merely a result of the development of nature and society. The thoughts, feelings and emotions that accompany a person, the entirety of human culture and social relations, result from material and economic conditions and remain secondary to matter.

With many varieties of materialism existing, for the purposes of the current study, mechanistic, natural and dialectical materialism will be discussed.

In Marxist terminology, natural materialism is defined as a set of spontaneous, instinctive, unconscious beliefs about the materiality of the world,

---

[101] W. Kapuściński, *Poglądy filozoficzne Mariana Smoluchowskiego* (*Marian Smoluchowski's philosophical views*), op. cit., pp. 201–202.

its objectivity and perceptibility, supposed to characterise the views of most representatives of the natural sciences.

Dialectical materialism was initiated by two nineteenth-century German thinkers—Karl Marx (1818–1883) and Frederick Engels (1820–1895). Applying the principles of Hegel's dialectics, they assumed that all reality is material. The feature distinguishing dialectical materialism from mechanistic or ontological materialism was the assumption that one of the features of matter is its dialectical nature. According to this position, matter is not only subject to the laws of the natural sciences, but also to the general laws of dialectics. Dialectical materialism introduced into the movement of matter the activity of man, which as a being expressing and satisfying its needs, as well as acting consciously and intentionally, affects the changes taking place in nature. Dialectical materialism argued that everything that happens in nature can ultimately be reduced to chemical, physical, and biological phenomena. Matter, having a certain inner wisdom, leads to the formation of ever more complex and perfect forms, and the laws that cause the formation of such structures are called materialist dialectics[102].

A theory that should not be ignored in a discussion of materialism is Darwin's theory of evolution, in which both natural selection and the emergence of random and undirected changeability within a species determine the changes taking place in nature.

2.3.3. The genesis of attributing materialistic views to Smoluchowski

As already mentioned, the publications that appeared in the 1950s suggested that Smoluchowski was either a materialist or an unconscious materialist. These opinions, although controversial and unsupported by reliable evidence, influenced the general perception of the Polish physicist's philosophical views. Smoluchowski's writings do not contain any statements declaring a materialistic perception of the world.

By making some simplifications and distortions of the author's intentions, a thesis was built that cannot be defended. It is debatable whether Smoluchowski can be attributed statements referring to some elements of

---

[102] See: J. Turek, *Materializm* (*Materialism*), in: *Powszechna Encyklopedia Filozofii* (Universal Encyclopedia of Philosophy), vol. 6, Lublin 2005, pp. 913–916.

materialist philosophy, however, the claim that he was a materialist philosopher is an overinterpretation of his views and attempts to inscribe his achievements into the framework of dialectical materialism are absurd. Materialistic views cannot be assigned to the scholar through a suggestion taken out of context, as the views of philosophical pragmatism, positivism, conventionalism or empirio-criticism can often equally well be attribute to him on the basis of the same statement. Research on the structure of matter conducted at the end of the nineteenth century and the promotion of atomic theory do not constitute sufficient proof for declaration of a materialistic worldview. Krzysztof Starzec noted that "The study of the structure of matter, reaching for its internal mechanism, was a manifestation of the materialistic tendency prevailing in 19th-century natural science (the materialism of scientists of the time was, however, generally mechanistic, non-dialectical)"[103].

Marxism in the 1950s was a philosophical system that imposed—through an ideology based on some of the works of Marx, Engels and Lenin—political, economic, social and philosophical views. The Stalinist version was the most radical. During the time of the Soviet Union and in countries with a similar (socialist) political system, including in Poland, it was impossible to separate philosophy from ideology in the publications available. Marxism was always an ideology in the Soviet Union, but it took on its most ideologised form during the Stalinist period. Clarification of the ideological assumptions of the trend called Stalinism ended in 1956 with the 20th Congress of the Communist Party of the Soviet Union[104], although both earlier and later communist ideologues formulated the principles of Marxist theories, creating a new scientific worldview based on dialectical materialism.

The history of constructing the materialist worldview of Marian Smoluchowski in Stalinist times represents a sequence of events characteristic of the era, the individual points of which are determined by the ideology

---

[103] K. Starzec, *Two interpretations of Marian Smoluchowski's thought*, op. cit., p. 13.
[104] The 20th Congress of the Communist Party of the Soviet Union—the first party congress after the death of Stalin, which took place on 14–25 February 1956. The most important event of the congress was Nikita Khrushchev's presentation at the end of the conference, at a closed meeting without the participation of guests, on the cult of the individual and its consequences. The purpose of this speech was to unveil the behind-the-scenes reality of Stalin's rule.

prevailing at a given moment in history, as well as accidental events, and the intentions of the authors.

On May 29, 1952, the front page of the *Trybuna Ludu* (*People's Tribune*) newspaper[105] featured a photo of Marian Smoluchowski with a caption commemorating the 80th anniversary of the scholar's birth. The text under the photo directed the reader to the last page of the newspaper, where Władysław Krajewski's text *The Great Physicist and Materialist Philosopher (On the 80th Birthday of Marian Smoluchowski)* appeared. This article, occupying about a third of a column, described the main events of Marian Smoluchowski's biography, presenting the person and his scientific achievements. From the point of view of a reader in the second decade of the 21st century, it would seem that the 80th anniversary of the great scientist's birth (Smoluchowski was born on May 28, 1872) would be an opportunity to recall the achievements of the Polish physicist, but this is a misguided belief. For a party newspaper of the time, the article was surprisingly extensive, and in addition was reprinted many times, which was neither a common phenomenon nor an accident. The appearance of Krajewski's article became a relatively important event, as its publication initiated a series of texts that were to present Smoluchowski's worldview in terms of the materialist philosophy sanctioned by this scholar.

In the early 1950s, politically neutral topics were not to be found either in the daily party newspapers, or in the so-called organs of the Polish United Workers' Party. A newspaper of that time, and especially a party newspaper, was not an informational device, but primarily a propaganda tool. The articles had ideological references or constituted a tool of overt party indoctrination. Therefore, it is clear that in addition to celebrating the anniversary of the scholar's birth, the article's author had a political agenda. This conviction is reinforced by the history of this non-accidental publication. The same article about Smoluchowski was reprinted the next day (May 30, 1952) by the Kielce newspaper *Słowo Ludu*[106] (*Word of the People*), and then, in July, it was printed in *Nowiny Tygodnia* (*News of the Week*)—a weekly supplement to

---

[105] The journal constituting the official record of the Central Committee of the Polish United Workers' Party (KC PZPR).

[106] The newspaper that at the time was the official publication of the Provincial Committee of the Polish United Workers' Party (KW PZPR).

the *Nowiny Rzeszowskie* (*Rzeszów News*) newspaper[107], and then in November in the magazine *Słowo Tygodnia* (*Weekly Word*)[108]. Essentially the same article, under a slightly altered title: *Marian Smoluchowski as a philosopher and materialist*, was printed in June 1952 in *Widnokrąg* (*Horizon*) magazine[109] and in the supplement to the *Głos Szczeciński* (*Sczczecin Voice*) newspaper entitled *Życie i Kultura*[110] (Life and Culture) and in December 1952 also in *Problemy* (*Problems*) magazine[111].

An article commemorating the anniversary of the birth of a Polish physicist could not appear in several newspapers that are organs of the Central Committee and Provincial Committee of the Polish United Workers' Party without an intended purpose. Knowing the realities of that time, it can be stated that it was undoubtedly a well-thought-out political decision of the then party authorities, which had agreed to such a publication. It is worth asking here what the intentions of the decision-makers were, and consequently of the author of the article.

Some explanation of the circumstances is provided by an article from the magazine *Voprosy Filosofii* (RUS: *Вопросы философии*, ENG: *Problems of Philosophy*). This was a newly created philosophical journal published in the Soviet Union. It appeared under the auspices of the Presidium of the Academy of Sciences of the USSR. In the first issue of 1948, an article appeared entitled *Discussion on the nature of physical cognition* (Дискуссияо природефизического знания), in which Ł.I. Storczak, contributing to a

---

[107] W. Krajewski, *Wielki fyzik i filozof materialista (w 80-lecie urodzin Mariana Smoluchowskiego)* (*The great physicist and materialist philosopher (on the 80th anniversary of Marian Smoluchowski's birth)*), 'Nowiny Tygodnia' ('News of the Week), supplement to 'Nowiny Rzeszowskie' ('Rzeszów News') 1952, No. 23.

[108] Idem, *Wielki fyzik i filozof materialista (w 80-lecie urodzin Mariana Smoluchowskiego) Wielki fyzik i filozof materialista (w 80-lecie urodzin Mariana Smoluchowskiego)* (*The great physicist and materialist philosopher (on the 80th anniversary of Marian Smoluchowski's birth)*), 'Słowo Tygodnia' ('Weekly Word') 1952, No. 14.

[109] Idem, *Marian Smoluchowski jako filozof i materialista*, (*Marian Smoluchowski as a philosopher and materialist*), 'Świat i My' ('The World and Us') 1952, No. 28.

[110] Idem, *Marian Smoluchowski jako filozof i materialista* (*Marian Smoluchowski as a philosopher and materialist*), 'Życie i Kultura' ('Life and Culture'), supplement to the 'Głos Szczeciński' ('Szczecin Voice') newspaper 1952, No. 23.

[111] Idem, *Marian Smoluchowski jako filozof i materialista* (*Marian Smoluchowski as a philosopher and materialist*), 'Problemy' ('Problems') 1952, No. 12.

discussion ongoing in the publication about the nature of physical knowledge and referring to an earlier text by Andrey Andreyevich Markov, wrote: "Smoluchowski's extensive inquiries transpired to be required to finally establish that statistical regularity is a completely new type of regularity, strictly defined by physical conditions… Unfortunately, we still propagate Smoluchowski's wonderful, materialistic ideas very little[112]."

The quarterly *Voprosy Filosofii* was published from 1947 with the exceptional consent of Andrei Aleksandrovich Zhdanov (1896–1948), a member of the political office of the Communist Party of the Soviet Union (CPSU), as the only philosophical magazine in Russia. Previously, philosophical texts had appeared only in the party monthly, *Bolshevik*. The mention of Smoluchowski in *Voprosy Filosofii* appeared a few months before Krajewski's article. In 1951, in the third, May–June issue of the bimonthly *Physics and Chemistry* magazine for teachers, a short article appeared by Armin Teske, entitled *Marian Smoluchowski*. In its final part, Teske included information about Storczak's article in *Voprosy Filosofii* and praise of Smoluchowski by Soviet scholars. For Marian Smoluchowski, this was, in these special times, a unique accolade. He gained the position of a right-thinking scientist in the circles of Soviet science, with a materialistic worldview that could be safely cited. The article in *Voprosy Filosofii* was an inspiration for Władysław Krajewski. The author confirms on multiple occasions that he knew Storczak's assertion, mentioning it both in an article in a party newspaper and in an essay in *Myśl Filozoficzna* (*Philosophical Thought*) (No. 4), which appeared in the second half of 1952. The article *Marian Smoluchowski as a philosopher-materialist*,[113] and the essay about Smoluchowski published in the quarterly, were a further stage in exploiting the personage of the great physicist to promote a new scientific worldview identified with the intentions of Marxist philosophy.

In the article in the *People's Tribune*, Krajewski made quite general statements about Smoluchowski's materialistic views, but in the essay in the *Philosophical Thought* quarterly he posited an unambiguous thesis, counting the Polish physicist among the declared materialists. Krajewski wrote that

---

[112] Ł.I. Storczak, Дискуссия *о природе физического знания*, (*Discussion on the nature of physical knowledge*) op. cit., p. 206.

[113] W. Krajewski, *Marian Smoluchowski jako filozof-materialista* (*Marian Smoluchowski as a philosopher-materialist*), op. cit., pp. 232–248.

"Smoluchowski, as one of the creators of statistical physics, and at the same time a determined materialist, undertook the task of demonstrating the objective nature of the concepts of chance and probability, in order to provide a permanent methodological basis for their application in science"[114]. Embedded into a sentence providing substantive information about scientific research on probability calculus issues is an entirely unrelated comment on Smoluchowski's materialistic views. Not only is the thesis not supported by any source, but several pages earlier Krajewski even contradicted his later statement, writing: "And he, admittedly, does not call himself a materialist"[115]. Elsewhere, the author states that Smoluchowski did not realise he was a materialist[116].

Krajewski does not explain how Smoluchowski could have had a "definitely materialistic" worldview without realising it. The placement within substantive information of such an important claim, unrelated in any way to this information, and supposedly originating from a previously proven thesis, is pure casuistry. The author probably assumed that the truthfulness and unquestionability of the factual context would lend credibility to the unjustified term he used. This is a well-known tactic of psychological suggestion, about which Robert Cialdini has written, for example[117].

Practically the same article, written by the same author, appearing eight times in various newspapers in a relatively short time was not only intended to celebrate the 80th birth anniversary of the great physicist—the anniversary was more of a pretext to use Smoluchowski's authority to propagate the adopted ideology and build a new scientific worldview based on the professed version of so-called Marxist doctrine.

A key publication incorporating the Polish physicist into the circle of Marxist philosophy for many years was Władysław Krajewski's book *Światopogląd Mariana Smoluchowskiego* (*Marian Smoluchowski's Worldview*), published in 1956. This is, to date, the only extensive work analysing the

---

[114] Idem, pp. 243–244.
[115] Idem, pp. 235, 241.
[116] W. Krajewski, *Światopogląd Mariana Smoluchowskiego* (*Marian Smoluchowski's Worldview*), op. cit., p. 126.
[117] See: R. Cialdini, *Wywieranie wpływu na ludzi. Teoria i praktyka* (*Influence: Science and Practice*), trans. B. Wojciszke, Gdańsk 1994, pp. 18–24.

philosophical views of this scientist. The book was a continuation of a plan initiated with publications from 1952, and its aim was to appropriate the philosophical and scientific legacy of Marian Smoluchowski for the needs of party ideology. The communist system needed scientific authorities it could invoke.

Krajewski presented Smoluchowski as a physicist who shared the beliefs of materialist philosophy, and even attributed to him the views of dialectical materialism. Thus, he suggested that the Polish physicist's convictions in the philosophy of science were consistent with Marxist philosophy. Such a reading of Smoluchowski's philosophy was unfounded, being both factually groundless and having no supporting sources.

2.3.4. Alleged Marxism and materialism

Władysław Krajewski's book was written at a specific time. It fitted into the body of publications of the times, and it was no coincidence that its title referred to the Polish physicist's worldview. The newly formed worldview of a scientist-citizen was crucial for communist ideologues. Many books appearing at that time were published with an ideological bent and were aimed at giving credibility to the creation of a new society. There was an attempt to show that more enlightened scientists—consciously or otherwise—had already adopted a new, fundamentally materialistic approach to science. It did not really matter whether the content sought for in these authors' publications and works was consistent with their intentions. Using the dialectical method, they proceeded according to the Hegelian statement: "If the facts do not fit the theory, so much the worse for the facts". Since interpretations often related to the writings of deceased authors, various manipulations were made. An example is a quote from a book by Bolesław Skarżyński (1901–1963), *O Jędrzeju Śniadeckim* (*On Jędrzej Śniadecki*), in which the author wrote about the chemist-philosopher 117 years after his death:

> Standing on the basis of specifically formulated laws of nature, eliminating from his reasoning all idealistic concepts except the concept of the life force, Śniadecki saw in the world around him a struggle of opposites that no one else saw, and he saw specific connections invisible to others. In this approach to natural phenomena, our scholar appears to be a kind of exuberant materialist-dialectician. (…) Today, when the theses of dialectical materialism are becoming more and more

thoroughly assimilated by the general public, we are in a position to properly assess the power of his thoughts[118].

It is not known who the initiator was of the morally dubious practice of coupling deceased authorities with the politics of the day, but it was common. Unfortunately, this comment also applies to Władysław Krajewski's interpretation of Smoluchowski's philosophy.

In 1946, Krajewski graduated from a school of the Polish Workers' Party (PPR, later PZPR) and held a high position in the party apparatus as an ideologist of this organisation. Until 1989, he was one of the main Polish Marxist philosophers with a positivist orientation and was considered the most outstanding expert on Engels' philosophy. In his works, he used the method of dialectical materialism in the field of the philosophy of physics and of other natural sciences[119]. In communist ideology, the dialectical method was a tool enabling alteration of the meanings of words and concepts, facilitating manipulation, and imposing a certain interpretation of the author's intentions. Krajewski uses this method when analysing Smoluchowski's texts.

Leszek Kołakowski writes about this method in *Main Currents of Marxism*:

> Diamat [Russian abbreviation for dialectical materialism – note J.G.] consists of statements of various types. Some of them are common sense clichés and contain nothing specifically Marxist. Others are philosophical creeds, unprovable and unresolvable by scientific means. Still others are simply nonsense. The fourth category includes theorems that are open to interpretation and, depending on which is used, belong to one, two or three of the above categories[120].

An example of a dialectical approach to Smoluchowski's views is Krajewski's considerations contained in a paragraph at the end of the book:

> In summary, it can be said that Smoluchowski took a definitely materialistic and often spontaneously dialectical position in the field of the methodology of physics, and in general philosophical considerations—the materialistic trend flowing from

---

[118] B. Skarżyński, *O Jędrzeju Śniadeckim* (*On Jędrzej Śniadecki*), Warsaw 1955, p. 86.
[119] See: W. Słomski, *Władysław Krajewski*, in: *Polska filozofia powojenna* (*Polish post-war philosophy*), ed. W. Mackiewicz, Warsaw 2001, pp. 537–554.
[120] L. Kołakowski, *Główne nurty Marxizmu* (*Main Currents of Marxism*), vol. 3, Warsaw 2009, p. 157.

his attitude as a whole was obscured by multiple positivist layers resulting from submission to fashionable bourgeois philosophy[121].

Krajewski claims that Smoluchowski was an unconscious materialist, and his positivist convictions were not authentic but resulted from succumbing to bourgeois fashion, while materialism dominated in the depths of his worldview. The fact that the thesis has no basis in factual materials and no proof is to be found of Smoluchowski's philosophical materialism is of no consequence to the method of dialectical materialism. Kołakowski recalls the Hungarian philosopher György Lukács (1885–1971), who, by using the word "dialectical", invalidated all empirical circumstances that may superficially look one way, but dialectically look quite the opposite. In a book about Lenin, Lukács accuses the reformists of having a non-dialectical concept of the majority[122]. It transpires that the majority can be understood in the ordinary sense, but also in the dialectical sense, which is the opposite of the ordinary majority. The Communists did not have a regular majority, but they had a majority in a deeper, dialectical sense[123].

Kołakowski writes in *Main Currents of Marxism* that materialism in Marxism came down to the thesis that the world was not created by a rational being, confirmed by a quote by Engels that materialism is ultimately about creating a world without God[124]. Claims that the world was or was not created by God cannot be empirically proven. There is not and cannot be any scientifically valid evidence of the non-existence of God, writes Kołakowski. A thesis proffered in this way is not a statement of science, but rather a confession of faith[125].

Krajewski adopted a similar path of inference and came to the conclusion that Smoluchowski could be labelled a materialist because he made no statements about God and therefore did not believe in Him, *ergo* he was a materialist. This simplified and tendentious means of arriving at conclusions was the standard in the application of the laws of dialectics, some of which

---

[121] W. Krajewski, Światopogląd *Mariana Smoluchowskiego (Marian Smoluchowski's Worldview)*, op. cit., p. 241.
[122] See: L. Kołakowski, *Główne nurty Marxizmu (Main Currents of Marxism)*, vol. 3, op. cit., p. 308.
[123] Ibid.
[124] Idem, p. 159.
[125] Ibid.

turned out to be common-sense clichés of negligible cognitive or scientific value. Smoluchowski was an outstanding scientist, and his views cannot be reduced to the level at which his scientific worldview is built on the basis of issues lying within the realm of faith, which for Smoluchowski belonged to a completely different order of knowledge than science.

Not only would the Polish physicist not have described himself as a materialist, but he would most likely not have agreed with the definition of materialism proposed by Marxist ideologues. Kołakowski claims that the main thesis of Marxist materialism takes the form of an unprovable profession of faith. The statement that "the world is inherently material" loses meaning when matter is defined as it was by Lenin, divorced from its physical properties, and assuming its existence to be independent of consciousness, i.e., objective. The extant world is material, therefore independent of consciousness, which is a fundamentally false thesis, since certain phenomena in the world—also according to the theories of Marxism and Leninism— depend on consciousness. Moreover, religious notions—God, angels, devils—are also, according to these beliefs, independent of consciousness[126].

2.3.5. Smoluchowski's philosophical predilections

Smoluchowski's philosophy is in opposition to Krajewski's thesis, which attributes materialism to the scholar. In terms of the possibility of comprehensive and clear knowledge of the world of phenomena *available to us*, Smoluchowski's views were close to pragmatism; he believed that the scientific perception of the world was not consistent with the real state of affairs. On the question of whether we were allowed to believe in it, he stated that this was a personal matter that had nothing to do with physics but was a matter of faith. By nature, people cannot remain in a state of constant uncertainty, do not like doubt or the systematic mental effort that critical thinking requires, and prefer to rely on an indisputable and safe faith in reality. Daily existence prompts us to create certain cognitive frameworks within which our mind must function[127]. Is this a thought that a materialist could have formulated?

---

[126] Idem, p. 158–159.
[127] See: M. Smoluchowski, *Pisma Marjana Smoluchowskiego (The Writings of Marian Smoluchowski)*, vol. 3, op. cit., pp. 165–166.

In his book, Krajewski also raises social and economic issues, considered from a Marxist perspective, which have nothing to do with Smoluchowski's philosophy and were most probably included with a view to reinforcing the suggestions in the discourse about materialism. These are attempts to influence the reader's perception, so that the arguments cited are seen through the prism of certain associations and from a specific perspective. They were intended to persuade the reader of the author's arguments, which do not relate to rational premises, but instead lead the thought on a specific path, suggesting ideas, concepts and attitudes that the reader is to adopt as his or her own over time.

Two facts are significant for an assessment of the book's scientific value. Firstly, Krajewski was one of the main dogmatists for whom Marxist doctrine together with dialectical materialism constituted the zenith of the development of human thought. Secondly, the book was produced during the most restrictive period of Marxism, when indoctrination affected almost every aspect of life and science, in particular philosophy. This was the end of the Stalinist era—admittedly after the death of Stalin, but before the so-called 'thaw', which in Poland was to take place only after October 1956. Hence this book cannot be seen as an objective scientific study as it is burdened with the ideological Marxist attitude of the Stalinist period.

A statement by Krajewski contained in the introduction to the monograph, that Smoluchowski's views were by no means consistent, creating fertile ground for the drawing of divergent conclusions[128], cannot excuse his interpretative abuses. The inference based on quotations from Smoluchowski's works, intended to prove his materialism, is biased and unreliable. The quotations cited and analysed are often over-interpreted and sometimes even manipulated.

In semantic terms, Krajewski's remarks bear the hallmark of the times in which his text appeared. The author summarises the understanding of the issue of chance and probability as follows:

> A consistent scientific solution to the issue of chance and probability, as well as to other philosophical issues, is possible only from the standpoint of dialectical materialism. The foundations of such a solution can be found in the works of the classics of Marxism. Using and critically processing the valuable content of Hegelian

---

[128] See: W. Krajewski, *Światopogląd Mariana Smoluchowskiego* (*Marian Smoluchowski's Worldview*), op. cit., p. 11.

dialectics, Marx and Engels raised the whole of philosophical materialism—including determinism—to a new, higher developmental stage[129].

Introducing the physicist's texts into such an unscientific discourse, full of ideological references and arguments based on worldview, is a reprehensible practice that has little in common with science. Krajewski confirms the lack of substantiveness in his argument:

> As far as an interpretation of the concept of probability and statistical regularities is concerned, the classicists of Marxism did not directly address these issues. However, some 'supporting points' for the solution of these issues can be found in their works. In addition to the concepts of chance already discussed, it should be noted that Marx included in *Das Kapital* the issue of economic rights operating on the capitalist market, in particular the law of value[130].

The above quotes indicate at least a lack of consistency in the argumentation. First of all, the author expresses an unfounded conviction that a scientific solution to the issue of chance and probability is possible only from the standpoint of dialectical materialism, only to conclude a moment later that the classicists of Marxism did not directly deal with interpretation of the concept of probability and statistical regularities, and that only in Marx's *Das Kapital* can unspecified economic topics relate to the issue of chance, in some way not specified by the author. The argument that Smoluchowski's theories of chance and probability are related to dialectical materialism is jarring and unconvincing, all the more so their presentation as an exemplification of the Marxist theory of necessity and randomness.

Marxist philosophy much more often referred to fields such as sociology or economics than to scientific subjects such as physics or chemistry, which it could not cope with due to their rapid development. Marx, Engels and Lenin never commented authoritatively on physics or mathematics, and therefore the Marxist classics could not be referred to in these fields. It was different with economics, philosophy or sociology, where appropriate patterns of thinking were developed and accepted. Perhaps this is why Krajewski repeatedly

---

[129] Idem, p. 108.
[130] Idem, p. 111.

invokes in the text comments and comparisons that are highly unconvincing in the field of physics and the philosophy of nature[131]:

> We know that Marxism means in practice a broad social practice, above all production activity, industry. Such practice is, as the classics of Marxism show, the basis and purpose of theoretical knowledge and the highest criterion of truth. Could Smoluchowski have shared the belief that the practice of production is the basis for the development of physics?

And another example: "Marx's analysis of economic laws may offer some help in the philosophical analysis of the nature of statistical laws studied by physics"[132]. Undoubtedly, it was not the classicists of Marxism who were the authorities in physics, and it is not socio-economic issues that should be the main pathway to arriving at the truths of physics.

Characteristic of Krajewski's ideological thinking at that time is his opinion on the philosophy of two great physicists: Niels Bohr (1885–1962) and Werner Heisenberg (1901–1976): "So for example, today we negatively evaluate the philosophical views of outstanding physicists—Bohr or Heisenberg—primarily because they occupy a retrograde position in the methodology of physics, inhibiting the development of science (indeterminism, subjectivism, the absolutisation of quantum mechanics)"[133].

Opinions built on the basis of ideological interpretations binding at the time do not lend the author's thoughts credibility. Without doubt, if Krajewski had written the book several decades later, his assessment of Bohr and Heisenberg would have been completely different. In the essay *Marian Smoluchowski—A Forerunner of the Chaos Theory*, published in 2001, annexed to the collection *Polish Philosophers of Science and Nature in the 20th Century*, Krajewski presents Marian Smoluchowski and his scientific successes, portraying him as a pioneer of chaos theory and not mentioning his materialist worldview. Compared to previous publications, the article is factual and

---

[131] Idem, p. 231–232.
[132] Idem, p. 112.
[133] Idem, 235.

contains no ideological references. Circumstances bring to mind Marx's famous dictum that it is "being that determines consciousness"[134].

In the 1950s and 1960s, writers invoked the classics of Marxism, sometimes completely detached from the subject of the publication. When the subject matter precluded such an invocation in the work's contents, Marxists were supported, for example, in the preface. It was required at that time, and for some works even essential in order to be published. Every article and book was subject to censorship, and citation of the appropriate authority determined the work's publication. Most often, the intentions of the authors were clear and legible, and it was mainly about satisfying censorship requirements. Unfortunately, this cannot be said about Krajewski's book, which is a serious and deliberate attempt to distort Smoluchowski's philosophical thought.

In 1963, another book was published by this author, entitled *Szkice filozoficzne (Philosophical Sketches)*. It is a collection of papers previously published in various journals and after being edited, collected in one volume. Some essays contain references to Marian Smoluchowski's philosophy of science. A subtle change can be detected in them that has taken place in the author's philosophical thought process. It is true that he continues to analyse from the standpoint of Marxist philosophy, but not as orthodoxly as in the mid-1950s.

Smoluchowski cannot be pigeonholed into a single narrow system of perceiving the world, much less into the system of materialistic views. It cannot be said that he was a materialist if he relativises every theory and every definition. In his publications, he virtually never refers directly to materialism, because this issue did not represent a scientific problem to him. The dilemma he contested with was the energy monism of Wilhelm Ostwald (1853–1832) and Ernst Mach and, after 1906, scientific problems related to probability theory and coagulation. The mention in which he indirectly expresses an opinion about materialism illustrates his attitude to this philosophy: it is dangerous to trust too much in the reality of physical hypotheses; perhaps those who believe only in the real existence of matter (in connection with

---

[134] See W. Krajewski, *Marian Smoluchowski—A Forerunner of the Chaos Theory*, in: *Polish Philosophers of Science and Nature in the 20th Century*, 'Poznańskie Studia z Filozofii Nauk i Nauk Humanistycznych' ('Poznań Studies in Philosophy of the Humanities'), 2001, No. 74, pp. 185–188.

the immutability of mass) will one day seem as naive as those who worship energy in a similar way today[135]. Smoluchowski treats the theoretical position according to which matter is the only material of reality[136] not as a scientific approach, but as an ideological thesis, and from his point of view—naive.

2.3.6. Substantive analysis of Władysław Krajewski's arguments

Although the era of raising materialism to the rank of an imperative idea in science is over, relics of the past still appear. One of them is the labels of materialist and atheist which have clung to Smoluchowski and often appear in publications devoted to him. Władysław Krajewski contributed to this situation, however the intention of these deliberations is not to analyse this author's publications as it is not them but Smoluchowski's philosophy that is the subject under consideration. However, it is worth presenting one of the theses that appeared in Krajewski's publications devoted to Smoluchowski. Krajewski analyses the scholar's views, following the issues of his basic physical research and their resulting conclusions, hypotheses and theories. His approach to the Polish philosopher's texts is particular as it is burdened *a priori* with the thesis of Smoluchowski's materialistic worldview.

In *Dwie interpretacje myśli Mariana Smoluchowskiego* (*Two interpretations of Marian Smoluchowski's thought*) Krzysztof Starzec described this situation quite tactfully:

> Having been written in the 1950s, the book has a tone that fully conforms to the political correctness of those times. (…) Its fundamental distortion was not so much the conclusion that Smoluchowski was ultimately to be a materialist, but something that could be called an overinterpretation, i.e. the selection of certain extracts from the physicist's texts that actually seem to testify to a materialist attitude rather than a view of philosophy as a whole, or accusing him of contradictions because he was not consistent in materialist thinking[137].

---

[135] See: M. Smoluchowski, *Kilka uwag o analogiach fizycznych, zwłaszcza w teoriach prądów elektrycznych, prądów cieplnych i zjawiskach dyfuzji* (*Several observations on physical analogies, especially in theories of electrical currents, heat currents and diffusion phenomena*), in: *Pisma Marjana Smoluchowskiego* (*The Writings of Marian Smoluchowski*), vol. 3, op. cit., p. 242.
[136] See: J. Turek, *Materializm* (*Materialism*), op. cit., p. 913.
[137] K. Starzec, *Two interpretations of Marian Smoluchowski's thought*, op. cit., p. 423.

Meanwhile, Krajewski, quoting Smoluchowski, not only overinterprets, changing the meaning of his thought, but all too often makes unjustified redactions to the quoted texts, changing their meaning and the essence of the physicist's statement, so that Smoluchowski's remarks constitute an argument for the adopted thesis.

Krajewski writes:

> Ignorance of Marxist dialectics particularly weighed on the erroneous understanding by physicists of the relation of relative truth to absolute truth, on the misunderstanding that the relativity of truth does not contradict its objectivity, that every relative truth contains a grain of absolute truth[138].

Continuing this thought, he refers to Smoluchowski:

> Smoluchowski's biggest problems are the issue of truth, the problem of the relationship between the claims of science and objective reality. We can already see that he is convinced of the objectivity of the laws of physics; he views the issue of the objectivity of scientific theories differently. Smoluchowski realises that no physical theory "can claim absolute truth". But what conclusion does he draw from this? He asserts that scientific theories should not aspire to adequately recreate reality at all… Our scientist stumbles here upon the same issue upon which many other physicists have already stumbled: the relation of relative truth to absolute truth[139].

Marxists defined truth as the conformity of human representations with objective reality, which was a reference to the Aristotelian tradition of comprehending truth. Marxist philosophy additionally assumed that absolute truth arises from the sum of relative truths, which by virtue of their very relativity do not have to become biased.

The reflections Smoluchowksi conducted on the essence of the issue of truth in science are included in the book *Marian Smoluchowski's Worldview*, but Krajewski's analysis serves as an example of biased interpretation of the physicist's views in the spirit of the truths of Marxist ideology.

The Polish scientist spells out his thesis in the article *On thermodynamic fluctuations and Brownian motion*, writing:

---

[138] W. Krajewski, *Marian Smoluchowski's Worldview*, op. cit., p. 206.
[139] Idem, p. 213.

> If we care about the accuracy of terminology, we should not talk at all about the "truth" of any physical theory, or even about its probability, but about its greater or lesser *usefulness*, and we must abandon the concept of *experimentum crucis* as a meaningless experience of earlier, more naive times, when science was not yet, as it is today, imbued with the scepticism of the theory of knowledge[140].

Smoluchowski's position was contrary to the assumptions of dialectical materialism, because it was not rational in the sense that it assumed that laws could not be fully known through reason. Rationalism was one of the foundations of dialectical Marxism. According to dialectical Marxism, the laws (concerning matter) are an inalienable trait of matter, rather than merely hypotheses that we try to create about it. Krajewski limited himself to a biased interpretation of an extract of Smoluchowski's statement, adjusting it to the assumed ideological thesis. Such an act is neither objective nor scientific.

In his book, Krajewski also addresses a statement by Smoluchowski regarding the issue of hypotheses in science:

> Smoluchowski, like the vast majority of naturalists, takes a materialist position on these fundamental issues. He has no doubt that reality exists objectively, regardless of the mind of the perceiver. He often mentions, as something self-evident, that physics deals with the "external world". At one point, emphasising the huge role of hypotheses in science, he notes somewhat ironically that "the claim of the existence of some external world outside us is also a hypothesis, except that all those endowed with sound senses accept it". We see here a certain tribute paid to positivist fashion (recognition of the external world as a hypothesis), but also a clear declaration of the author's position: one must be out of one's mind to become a subjective idealist[141].

Krajewski selectively quotes part of Smoluchowski's statement narrowed down for his own needs, attaching a commentary that ends with conclusions aimed at presenting the scientist's philosophical beliefs in an appropriate light. The same quotation, rendered more fully, shows that the author of the passage would not agree with Krajewski's comment:

> As Boltzmann and Poincaré correctly point out, every theorem of physics contains some hypotheses and without them we cannot talk about issues of nature at all.

---

[140] M. Smoluchowski, *O fluktuacjach termodynamicznych i ruchach Browna (On thermodynamic fluctuations and Brownian motion)*, in: *Pisma Marjana Smoluchowskiego (The Writings of Marian Smoluchowski)*, vol. 2, op. cit., p. 268.

[141] W. Krajewski, *Marian Smoluchowski's Worldview*, op. cit., p. 183.

After all, according to what we have said before, the claim of the existence of some external world outside of us is also a hypothesis, except that all those endowed with "sound senses" accept it. Any induction-based generalisation, any statement referring to the external world that goes beyond the scope of our direct sensory perceptions, is a hypothesis, and various hypotheses differ only in the degree of probability and how accustomed to them we are[142].

This undoubtedly deep thought of Smoluchowski's illustrates the fact that all physical theories, even theories accepted as true, are to a greater or lesser extent only hypotheses. This is a thought that is repeated many times in Smoluchowski's writings. Krajewski manipulates the scholar's text, completely changing the meaning of his statement to achieve a "further argument" to justify his claim of Smoluchowski's materialism. This deep epistemological thought of the Polish physicist is reduced to a philosophical dispute over the reality of the physical world, in which the fashions prevailing in science play an important role.

From time to time, Krajewski provides summaries in the text intended to assure the reader of the soundness of his arguments. For example, he writes:

> Smoluchowski's dialectic, like his materialism, was not consistent; in some cases, as we have seen, he still reasoned in a metaphysical way. However, his methodological approach as a whole justifies the conclusion that Smoluchowski rose above the old mechanistic materialism and in many cases approached dialectical materialism within the realm of his specialisation[143].

Smoluchowski's overall scientific attitude, including his methodological one, does not entitle similar conclusions to be drawn.

However, it is difficult to expect an objective assessment from the author of *Marian Smoluchowski's Worldview* if the assumptions for the research methods of physics are found in Vladimir Lenin's *Materialism and Empirio-Criticism*, a book that is not a scientific work but merely an ideological textbook of Marxism.

---

[142] M. Smoluchowski, *Poradnik dla samouków. Wskazówki metodyczne dla studiujących poszczególne nauki. Fizyka, Geofizyka, Meteorologia (Self-Study Handbook. Methodological guidelines for students of individual sciences. Physics, Geophysics, Meteorology)*, vol. II, Warsaw 1917, p. 48.
[143] W. Krajewski, *Marian Smoluchowski's Worldview*, op. cit., p. 203.

According to Lenin: "Physics was moving towards dialectical materialism 'not directly, but by zigzags, not consciously but instinctively, not clearly perceiving its 'final goal,' but drawing closer to it gropingly, hesitatingly, and sometimes even with its back turned to it[144]."

In a situation where the starting point for an assessment of Smoluchowski's writings are assumptions of dialectical materialism, a reliable analysis of philosophical views is not possible, hence—according to Krajewski—on the Marxist path of the development of science, Smoluchowski was often unable to avoid obstacles.

Let us present Smoluchowski's quoted statement in its entirety:

> The kinetic theory of matter, which attempts to encompass material phenomena in a general mechanical view of the world, *cannot claim absolute truth* [emphasis added – J.G.], just as with all other physical theories. Like energy theory and electrical theory (still very undeveloped today), and perhaps other theories that will arise in the future, it can only be an "image" of the phenomena revealed to us; an image whose advantage over the aggregate of individual facts lies in the fact that: 1) it includes them in a systematic whole, built on as many assumptions as possible; 2) it leads us by theoretical deduction to the discovery of previously unknown phenomena[145].

The tone of this argument undoubtedly differs from that suggested by Krajewski. Smoluchowski assumes an extraordinary complexity of the laws of nature, which results in theories being able at most to reveal fragments of the truths and principles of nature's functioning, and perhaps we will never fathom the absolute truth. At the turn of the 19th and 20th centuries, few physicists and philosophers were aware of the complexity of the laws of nature, hence this thought testifies to an in-depth understanding of reality, while the Marxist explanation simplified and flattened the author's argument, changing the essence of Smoluchowski's thought.

Let us look at quite an important issue of materialism, which was determinism. Krajewski put it simply: "The starting point of the materialist approach to the issues of chance and probability must, of course, be the recognition of

---

[144] Ibid.
[145] M. Smoluchowski, *On more recent progress in the field of the kinetic theory of matter*, in: *The Writings of Marian Smoluchowski*, vol. 1, op. cit., p. 279.

determinism[146]." In many statements, Smoluchowski talked more about indeterminism in physics. He argued, for example, that kinetic theory assumes the existence of indeterminism at the level of microscopic material phenomena[147].

Despite the physicist's unambiguous statements, Krajewski constructed his own interpretation of them, claiming that although Smoluchowski writes in his two works "about indeterminism" entering physics with the application of probability calculus[148], his approach results from the fact that: "he did not realise, did not at all understand, that he was fighting for materialism against positivism"[149]. Continuing this thought, he cites as an example the paper *On thermodynamic fluctuations and Brownian motion* and claims that in writing in it about "indeterminism" in kinetic theory, in contrast to deterministic thermodynamics, Smoluchowski gave the term a different meaning. As he goes on to write, this is due to terminological clumsiness, which is the result of Smoluchowski's "philosophical awkwardness". He concludes his deliberations with the statement that "From behind Smoluchowski's philosophical awkwardness, a healthy, materialistic and dialectical tendency breaks through: the desire to go beyond Laplace's classical determinism and create some broader concept of determinism, corresponding to the needs of modern science"[150].

However, Smoluchowski's text presents a different research perspective, because the physicist was not interested in the philosophical aspect of the defined terms. He wrote: "We use the word 'indeterminism' to indicate that the course of a phenomenon depends on circumstances that are never accessible to direct experimental control, i.e. on the coordinates and velocities of all atoms; however, we assume that they are subject to these values in the correct way. Therefore, it is not proper indeterminism in the philosophical sense of the word"[151].

His discourse was rooted in research conducted in physics and was exact science by its nature. Krajewski gave it a specific philosophical connotation, without entering into the essence of the issue raised by Smoluchowski and thus distorting the author's intentions.

---

[146] W. Krajewski, *Marian Smoluchowski's Worldview*, op. cit., p. 123.
[147] See: M. *On thermodynamic fluctuations and Brownian motion*, op. cit., p. 271.
[148] W. Krajewski, *Marian Smoluchowski's Worldview*, op. cit., p. 124.
[149] Idem, p. 126.
[150] Idem, p. 126–127.
[151] M. Smoluchowski, *On thermodynamic fluctuations and Brownian motion*, op. cit., p. 189.

This situation is confirmed by Smoluchowski's next statement:

> Studies on fluctuations—have even more momentous general significance, because they are directly related to the fundamental feature of kinetic theory, which, in contrast to the thermodynamic view, emphasises a certain indeterminism of macroscopic material phenomena, entailing the introduction of the concepts of chance and probability into the field of physics and expressing itself in this form: a tool of the statistical method of reasoning[152].

Krajewski's commentary is an analysis of the text from the position of dialectical materialism, about which Kołakowski wrote, and in which, with the help of the word 'dialectical', it is possible to invalidate all empirical circumstances that in the common perception may look a certain way, but dialectically look quite the opposite.

Krajewski uses a similar method when analysing the paper *O pojęciu przypadku i pochodzeniu praw fizyki opartych na prawdopodobieństwie* (*On the concept of chance and the origin of the laws of physics based on probability*). In terms of Smoluchowski's research on the issue of chance and the origin of statistical laws in physics, he shows that the researcher abandons the erroneous terminology of previous years and clearly declares his deterministic position. According to Krajewski, Smoluchowski's works are aimed at demonstrating that the concept and genesis of chance can be precisely defined, including when standing firmly on the ground of determinism[153].

However, in the above-mentioned work, Smoluchowski writes that if chance is treated as a negation of regularity, the resulting contradictions are certainly insurmountable. Such a concept of chance is incompatible with the determinism prevailing in science at the time[154].

So why does Krajewski, in the face of Smoluchowski's unambiguous statements, invariably assign him a deterministic position? In addition to the obvious ideological motives that enable Krajewski to treat the discussed issue dialectically, there is another aspect to this matter. In some extracts can be seen a rather ambivalent approach by Smoluchowski to determinism in terms

---

[152] Idem, p. 188–189.
[153] See: W. Krajewski, *Marian Smoluchowski's Worldview*, op. cit., p. 127.
[154] See: M. Smoluchowski, *On the concept of chance and the origin of laws of physics based on probability*, op. cit., p. 29.

of chance and probability. By no means does this imply the possibility of assigning him a deterministic position, but Smoluchowski permits a certain type of determinism in the indeterminism perceived in chance, necessary in his opinion for the mathematical calculation of physical probability. The pragmatism of a scientist whose mind is engaged in the sciences does not allow excessive importance to be attached to philosophical terms when they stand in the way of achieving a stated aim in physics.

This opinion of Smoluchowski's is illustrated by a sequence of quotations from the paper *On the concept of chance and the origin of the laws of physics based on probability*:

> we assume that while there is a valid causal relationship between the relevant cause and effect, the nature of this relationship is unknowable to us, because it is a complex phenomenon. In this sense, chance should be described as "a partial cause unknown to us"[155].
>
> (…) strict natural science is interested in objective or 'mathematical' probability, that is, the relative frequency of the occurrence of marked random events[156].
>
> (…) the effect of unknown partial causes can be calculated with reference to the 'law of large numbers'[157].
>
> The concept of objective probability can, in a completely analogous way, be applied to all such not entirely determined phenomena ('accidental' in the sense explained earlier) in which the same type of elementary phenomenon is constantly repeated over time[158].
>
> (…) the concept of probability, in the usual sense of the correct frequency of random events, has a strictly objective meaning, that the concept and genesis of chance can be precisely specified, although based on determinism, and that the law of large numbers results from this not as a mystical principle and not as a purely experimental empirical law, but as a very simple mathematical consequence of this special form, which in such cases assumes a causal relationship[159].

---

[155] Ibid.
[156] Idem, p. 30.
[157] Idem, p. 31.
[158] Idem, pp. 48–49.
[159] Idem, p. 51.

Krajewski exploited the natural attitude of a physicist who does not identify with the terms 'deterministic' or 'indeterministic' philosophically, let alone ideologically, because he is interested in their substantive meaning and practical application in physics. The leap that Krajewski makes several times, moving from the level of physics to the level of philosophy and vice versa, causes terminological confusion and allows the physicist to be accused of terminological and philosophical awkwardness[160]. Krajewski is either unable to appreciate the fact, or he does it intentionally. I do not think that we are dealing here with Smoluchowski's philosophical awkwardness, but on the contrary, these are pioneering steps, culminating in success, to incorporate probabilistics into research in physics. Writing Marxist intentions into the course of the physicist's reasoning creates a philosophical contrivance with which Smoluchowski would not agree. Krajewski analysed the text from the position of the philosophy of dialectical materialism, causing a distortion of Smoluchowski's statements and intentions related to the subject of physics.

A quote is significant from the essay *On the concept of chance and the origin of the laws of physics based on probability*, intended to support Krajewski's thesis in the context of his commentary. Smoluchowski wrote:

> It is therefore clear, as far as their application in theoretical Physics is concerned, that all theories of probability conceiving of chance as an "unknown partial cause", should be considered in advance to be insufficient. A given event's probability useful for Physics may depend only on the conditions affecting the occurrence of the event, but cannot depend on the degree of our knowledge![161]

Krajewski commented on this, arguing that "This is an extremely clear, materialistic approach to the problem: the probability of an event is determined by the set of conditions in which the event occurs, and thus by objective factors independent of our consciousness"[162]. The cited conclusion is an example of the biased interpretation of Smoluchowski's text.

Smoluchowski's opinion is a conclusion resulting from reflection on the issue of coincidence, which he carries out a little earlier:

---

[160] W. Krajewski, *Marian Smoluchowski's Worldview*, op. cit., p. 126.
[161] M. Smoluchowski, *On the concept of chance and the origin of laws of physics based on probability*, op. cit., p. 31.
[162] W. Krajewski, *Marian Smoluchowski's Worldview*, op. cit., p. 134.

> If chance is treated, as it is popularly, as a negation of regularity, then these contradictions[163] are certainly completely insurmountable. But such a conception of chance cannot be reconciled with the determinism prevailing in today's science. Therefore, we generally used to explain the thing in such a way that we assume there is indeed a valid causal relationship between the relevant cause and effect, but that the nature of the relationship is unknowable to us when the phenomenon is too complex; hence the apparent break from regularity. In this sense, chance should be described as "a partial cause unknown to us"[164].

Both quotes convey a certain opinion of Smoluchowski regarding the analysis of chance, which has nothing to do with materialism, but it is worth noting that the comment on determinism emphasises the author's position against materialism. Typically, Krajewski shortened Smoluchowski's statement, twisting his opinion to suit the needs of his thesis.

Another example of demonstrating the Polish physicist's materialism is the analysis of Smoluchowski's statements referring to the hypothesis of the 'heat death' of the universe. In the chapter *Criticism of the theory of the 'heat death' of the universe*, the researcher writes:

> This hypothesis leads to idealistic, fideistic consequences. These consequences were brought to light by Engels in *Dialectics of Nature* (first published in the USSR in 1925). Indeed, if energy tends only to dissipate, if heat tends inevitably to equalise temperature levels, then where did the universe's variation in these levels, the concentration of energy in small areas (stars), come from? Since it cannot arise in a natural way—it had to be given to matter from the outside, had therefore to be the work of supernatural forces[165].

There are no grounds to associate the concepts of the 'heat death' of the universe with an idealistic or materialistic belief. It is a scientific hypothesis, not a philosophical or ideological premise. Another issue is whether Smoluchowski

---

[163] Smoluchowski had previously asked two questions: how is it that accidental causes have the right effects and how can the right causes have an accidental effect? He identifies these questions as potential contradictions.

[164] See: M. Smoluchowski, *On the concept of chance and the origin of laws of physics based on probability*, op. cit., p. 29.

[165] W. Krajewski, *Marian Smoluchowski's Worldview*, op. cit., p. 92.

declared himself an opponent of 'heat death'. Neither in the essay *Lord Kelvin*[166] and in the paper *On thermodynamic fluctuations and Brownian motion*[167] are there any such unambiguous declarations, as Krajewski himself states, writing that "the author was not convinced of the falsity of the view presented".[168]

In the study *On more recent advances in the field of kinetic theories of matter*, Smoluchowski states:

> Significantly, according to the data of the kinetic theory, there must come a moment when, for example, in 1 mm^3 of gas, all molecules will have equal velocities, that is, this part of the gas will 'by itself' change the movement of heat into a progressive movement, so that the law of entropy will be violated. But this is an extremely unlikely case, meaning it may happen once over an extremely long period; after this moment has passed, the system will for a long time again strive for a 'probable' state, i.e. entropy (= disorder!) will increase[169].

Smoluchowski did not adopt an unequivocal position on the issue of 'heat death', perhaps because he did not have one. It is natural that a scholar might distance himself from the latest hypotheses, especially those that are difficult to falsify.

As an argument to support his thesis, Krajewski recalls a statement by the Soviet scholar Kliment Timiryazev (1843–1920), who in his book *Kineticheskaya teoriyamaterii (The Kinetic Theory of Matter)*, published in 1954, demonstrates that Smoluchowski's works dealt a "devastating blow" to the theory of heat death[170], without, however, citing unambiguous arguments in support of this thesis.

The Polish author gave a similar treatment to the phenomenon of fluctuation, which at one time was one of the main subjects of Smoluchowski's research, "there is no doubt—and all Marxist authors agree on this—that

---

[166] M. Smoluchowski, *Lord Kelvin*, in: *Pisma Marjana Smoluchowskiego (The Writings of Marian Smoluchowski)*, vol. 3, op. cit., p. 5.

[167] Ibid., *O fluktuacjach termodynamicznych i ruchach Browna (On thermodynamic fluctuations and Brownian motion)*, op. cit., pp. 273–274.

[168] W. Krajewski, *Marian Smoluchowski's Worldview*, op. cit., p. 96.

[169] M. Smoluchowski, *On more recent progressin the field of the kinetic theory of matter*, op. cit., pp. 285–286.

[170] W. Krajewski, *Marian Smoluchowski's Worldview*, op. cit., p. 98.

the fluctuation hypothesis played a remarkably progressive role in its time, as its blade was turned against idealism, against fideism. Boltzmann and Smoluchowski were among the few physicists at that time who firmly rejected the idealistic theory of 'heat death'[171]."

He goes on to cite Boltzmann's fluctuation hypothesis: "The eternally existing universe is—generally speaking—in a state of thermodynamic equilibrium. However, here and there, there occur inevitable fluctuations of various sizes, and therefore unlikely states. (...) Of course, any fluctuation must disappear, giving way again to a state of equilibrium—however, similar fluctuations must inevitably arise elsewhere[172]."

In the working methodology used by Krajewski, certain specific meanings are apparent, which makes them the so-called key, a characteristic mental shortcut that triggers specific associations. Throughout the post-war years, such a mental shortcut was, for example, 'electrification', associated in publications with a widespread civilizational leap, another example was the term 'kulak', used in the 1950s to refer to a wealthy peasant who inhibited progress and opposed the social changes brought about by the new system. The code-like acceptance of such terms meant that they could be used safely.

Krajewski attempts to use these tried and tested mechanisms of communist propaganda in his work. In order to achieve his goal, i.e. to prove Smoluchowski's materialism, he follows this philosophy. Hence attempts to imbue terms such as determinism, the heat death of the universe, or the theory of fluctuations, with specific connotations. In practice, however, it sometimes happened that the terms used in support of the proclaimed theory at the same time contradicted it. This is the case with the theory of fluctuation, which was originally intended to be a key argument against idealism, after which it transpired to not entirely fit into Marxist ideology. Its intended effective use fell apart, as Krajewski says in the final part of his argument: "We can therefore see that Smoluchowski did not consider the fluctuation hypothesis to be a dogma, foreseeing the possibility of other

---

[171] Idem, p. 102.
[172] Idem, p. 95.

solutions"[173]. Moreover, something Krajewski does not write about is that the theory of fluctuation, through the elements of indeterminism contained in it, constitutes an argument against Smoluchowski's supposed materialism.

The theory of fluctuation was not directed against idealism, and much less constituted an argument against the theory of heat death, but resolved a significant physical dispute, as Smoluchowski writes:

> This is the main significance of fluctuational phenomena, that they decisively resolve the old dispute in thermodynamics—to its detriment. They prove that macroscopic thermodynamic parameters are not sufficient to accurately determine the state of the material system and that the indeterminism of microscopic molecular phenomena is not a product of the imagination of theoretical physicists, that it indeed manifests itself, in a perceptible way, in these random fluctuations[174].

Indeterminism, as Krajewski also noted, was at odds with the determinism proclaimed by Marxists in their materialist theory.

Smoluchowski's involvement in the seemingly scientific discourse on materialism, and indeed in the Marxist ideological discussion, created the appearance that the Polish physicist was engaged in materialistic deliberations on fluctuation, so the reader could get the impression that Smoluchowski had joined in materialistic inquiries. And although the Polish scholar had been dead for several decades, through his manipulation, Krajewski presented him as an active participant in this dispute.

2.3.7. Smoluchowski's silence about God

To consider Krajewski's suggestion on the subject of Smoluchowski's atheism, after analysing the physicist's work, it is difficult to find any references to his religious views. Explicitly expressed religious declarations, reflections on God or on faith are nowhere to be found in Smoluchowski's publications. From certain philosophical inquiries, the nature of his religious views can be presumed or—more generally—a guess can be made as to the direction in which his thoughts were moving. Therefore, in the context of the materialistic

---

[173] Idem, p. 102.
[174] M. Smoluchowski, *On thermodynamic fluctuations and Brownian motion*, op. cit., p. 273–274.

views attributed to Smoluchowski, *the suggestion of his atheistic outlook* should be considered erroneous. *It is difficult to find religious declarations in his writings, but it is also impossible to find any evidence there of an unambiguously atheistic attitude. Agnostic statements, appearing in the context of deliberations on the existence of the external world and reaching for materialistic arguments, are made from the position of a naturalist, never having a metaphysical context and not addressing the question of the existence of God.* This is an issue that Smoluchowski does not directly address anywhere. There are however statements, although not many of them, regarding his attitude towards religion.

Smoluchowski biographer Armin Teske, in the book *Marian Smoluchowski. Życie i twórczość* (*Marian Smoluchowski. Life and Works*) also addresses the subject sparingly, though unambiguously: "to prevent erroneous conclusions, it should be highlighted that Smoluchowski was far removed from any mystical institutions. He was also indifferent to religious beliefs; quoting the words of his family: 'He did not believe in the afterlife'"[175]. Indifference to religious beliefs or disbelief in the afterlife demonstrate that religious thought was not very close to Smoluchowski, especially at the exoteric level. However, this does not imply a lack of esoteric sensitivity, and certainly does not constitute evidence of his atheistic attitude. It is not possible to clearly define Smoluchowski's spiritual attitude, although from his philosophical deliberations it can be presumed that he did not believe in a personal God.

His philosophical convictions were built through reflection on the exact sciences, especially on physics and mathematics, hence his philosophy did not allude to theology or religion, but had a secular nature, based on the paradigms of science. *Charles Darwin's* work *On the Origin of Species* played an important role in the process of shaping his philosophy. Smoluchowski was a strong advocate of Darwin's theory and wrote about it many times in various texts. He maintained that it was mainly thanks to him that science had moved away from the concept of purposefulness in nature, which was a misleading *a priori* assumption, and that it was he who had shown through various examples how seemingly purposeful adaptations occur which are *de*

---

[175] A. Teske, *Marian Smoluchowski. Life and Works*, op. cit., p. 35.

facto the result of natural causes[176]. Smoluchowski argued that if we remove associations such as, for example, *aim* or *action*, that are aroused in us when we use them to apply them to the concept of *force*, we will understand that thought conceived of in this way is an anthropomorphic element, alien to inanimate nature. This means that if we say that there exists a force of universal gravity, we state only the regular existence of a certain property of the common movements of all bodies[177].

Another of his statement is significant:

> Today, no naturalist accepts purposefulness in nature, because nature cannot be personified as if it were a thinking and planning being. Purposefulness would be comprehendible only from the point of view of those who accept the hypothesis that nature was created by an intelligent being, with a plan and a goal arranged in advance. In inanimate nature we see no traces of purposefulness, and in the field of living nature we have learned from Darwin to do without this concept[178].

Smoluchowski distances himself from Neo-Thomistic concepts and does not believe that the sensory world was created by a personal God, with some preconceived plan. *He leaves no illusions; his approach to religion is highly sceptical. However, there are also no statements to be found questioning the existence of God. Despite his great scientific distance to metaphysics, it cannot be said that he had no understanding of the esoteric conception of spirituality or the questioning of any relationship between man and God, because Smoluchowski remains silent on this sphere of human cognition.*

This lack of any statement does not testify to the author's atheistic attitude but is rather a manifestation of scientific agnosticism. Smoluchowski was a physicist for whom—in terms of scientific reflection—the existence of a supernatural force was not unambiguous and obvious. In the context of his scientific research, he saw no need to ask questions about God, like in the famous anecdote about Pierre Simon de Laplace (1749–1827), who, when asked by Napoleon where there was

---

[176] See: M. Smoluchowski, *Self-Study Handbook*, vol. I, op. cit., p. 18.
[177] Idem, p. 21.
[178] M. Smoluchowski, *Self-Study Handbook*, vol. II, op. cit., p. 18.

room for God in his system, replied to the emperor: "Sire, I had no need of that hypothesis". Being a person sensitive to spiritual experiences, and having an open mind, reaching beyond the discourse of the exact sciences, Smoluchowski could not restrict the known world only to scientific reality, which he talked about many times. Most probably, wishing to be reliable in his statements, he remained silent about God. As the philosopher wrote: "What you cannot talk about, you must be silent about"[179]. *He did not find in himself that indisputable truth to which he could refer and which would convince him that "science requires faith, and religion requires truth".*

Krajewski tried to interpret some of Smoluchowski's statements on the basis of materialism in order to demonstrate his unequivocally anti-religious attitude. Paweł Polak points out that as part of his argument, this author cited, among others, examples in which Smoluchowski was supposed to refer critically to people from church circles. In support of his thesis, he invoked arguments devoid of any scientific value, such as Smoluchowski's opposition to granting Archbishop Józef Bilczewski an honorary doctorate from the Faculty of Philosophy of the University of Lviv, or his support for the position of rector—as Krajewski stated—of the atheist Benedykt Dybowski[180]. Smoluchowski's position on both issues had no religious basis. He expressed his opinion according to his own conviction and attitude towards these people. Elsewhere, Krajewski argued even more strongly that Smoluchowski was "an atheist, criticising teleology, vitalism and creationism"[181]. Apart from the Polish scholar's personal position on this issue (which he has *de facto* nowhere publicly expressed), the very attempt to reconstruct some of his statements from the perspective of religious faith, opposed to the fundamentally atheistic materialistic philosophy, is abuse enough[182].

---

[179] L. Wittgenstein, *Tractatus logico-philosophicus*, trans. B. Wolniewicz, Warsaw 2004, p. 83.
[180] P. Polak, *Byłem pana przeciwnikiem [profesorze Einstein]… (I was your opponent [Professor Einstein]…)*, Kraków 2012, p. 87.
[181] W. Krajewski, *Marian Smoluchowski's Worldview*, op. cit., p. 238.
[182] M. Stawarz, *Rekonstrukcja i krytyczna analiza poglądów filozoficznych Mariana Smoluchowskiego (Reconstruction and critical analysis of the philosophical views of Marian Smoluchowski)*, doctoral dissertation, supervisor dr hab. Paweł Polak, Kraków 2016, pp. 42–43.

> *Faith and science are two different realms of thought that function in different cognitive spaces. Smoluchowski focused his attention as much as possible on the study of available physical reality. His work does not include deliberations that go beyond the sphere of science, which, however, as already noted, cannot constitute evidence of his atheistic worldview.*

2.4. Summary

The above considerations are aimed at revising the image of Marian Smoluchowski's worldview common in publications. A researcher of the exact sciences, especially a physicist, for obvious reasons is inclined to a materialistic treatment of reality, especially in the epistemological sense, because the material on which he works at the turn of the 19th and 20th centuries exhibits primarily a mechanistic nature. However, the research and empirical treatment of reality should not be equated with a philosophical orientation. For Smoluchowski, although he proved the existence of atoms, the world was a decidedly more complex structure, for which science, its hypotheses and theories, merely represent accepted conventions enabling the physicist to function in the scientific space. They did not translate into a literal understanding of reality. In this context, attributing views of materialist philosophy to Smoluchowski constitutes a complete lack of understanding of his philosophy of science and is an unjustified narrowing of the horizons of the scholar's thinking about the reality around us. Smoluchowski would not agree to such an instrumental treatment of his philosophical views.

With his publications, Krajewski fulfilled his obligation to the political expectations of the time, as evidenced by the fact that in the 1960s and later, when politicians' expectations towards writers had changed, the topic of materialism was marginalised or omitted in his publications regarding Smoluchowski.

Smoluchowski was a physicist building a scientific reality, so it is impossible not to perceive certain shades of materialism in his scientific methodology. Materialism is also visible in his epistemological treatment of reality. However, his attempts at an ontological perception of nature are devoid of aspects of materialist philosophy.

According to Paweł Polak, Smoluchowski was critical of the materialist interpretation of the construction of reality:

The electromagnetic concept of mass and its confirmation by Kaufmann were proof for him of the shallowness and fallacy of materialism. The way he talked critically of the materialists and their dogmatic understanding of matter indicates that in no way can Smoluchowski be considered a materialist (even an unconscious one), as Władysław Krajewski and Władysław Kapuściński wanted to do during the Stalinist period[183].

Starzec believes that the fundamental distortion of Krajewski's book was not so much the conclusion that Smoluchowski was ultimately a materialist, but something that could be called an overinterpretation, i.e. the selection of certain extracts from the physicist's texts that actually seem to testify to a materialist attitude, rather than a view of his philosophy as a whole, or accusing him of contradictions because he was not consistent in his materialist thinking[184].

Krajewski "Adopts (…) a position in advance, in the light of which he interprets Smoluchowski's philosophy. Its intention is not only to present the scientific theories and methodological concept of the physicist, but also (and perhaps above all) to judge these, and thereby—to show to what extent they are similar to, and to what extent they differ from, the only true Marxist philosophy"[185].

In *Zarys najnowszych postępów fizyki* (*An outline of the latest advances in physics*) Smoluchowski wrote that

> the concept of mechanical mass and the associated—or rather confused—metaphysical concept of matter, the tangible "substrate of physical phenomena", which some philosophers considered to be the only thing that really exists, is presented in a strange light from the point of view of today's physics: the concept evaporates as if it were a dream, an apparition, and we get the impression that the true basis for the phenomena of nature is electricity, and that electrical forces have only deceived us, deluding us that there is a material world, guided by the unbreakable laws of mechanics[186].

---

[183] P. Polak, *I was your opponent [Professor Einstein]…*, op. cit., p. 99.
[184] K. Starzec, *Two interpretations of Marian Smoluchowski's thought*, op. cit., p. 423.
[185] Ibid.
[186] M. Smoluchowski, *Zarys najnowszych postępów fizyki* (*Outline of the latest advances in physics*), op. cit., p. 43.

Smoluchowski's statement confirms the thesis that the Polish physicist's thinking could not have been limited to relatively narrow aspects of materialist philosophy. Like quite a large number of scientists at the turn of the 19th and 20th centuries, he sought answers to fundamental questions about the structure of reality, which is why he also reflected on matter.

Starzec states:

> Therefore, it seems imperative to try to re-valuate the philosophy of scientists, which was somehow contorted by later interpreters; a philosophy that addressed extremely important issues for science, such as the process of knowledge itself, its subject and object, or research methodology, which in the interwar period were the subject of lively discussions within the scientific community, many of whom provided fruitful thought on this topic[187].

The need to re-assess Smoluchowski's views stems from the necessity to objectify the interpretation of his opinions, but primarily in terms of their value and importance for science.

---

[187] K. Starzec, *Dwie interpretacje myśli Mariana Smoluchowskiego* (*Two interpretations of Marian Smoluchowski's thought*), op. cit., p. 424.

Smoluchowski's approach to the methodology and epistemology of science

3.1. Fields and issues of physics

3.1.1. The subject of physics

Smoluchowski's reflections on the methodology of physics, embedded in the contemporary philosophy of science, *de facto* went beyond the framework of its functioning at the time. His methodological concepts are not inferior to the standards of the modern methodology of physics. The 'scientific method', being a research procedure determining the rules of operation, is based on assumptions made about the subject, as well as the purpose of the research[188].

When studying the works of the Polish physicist, it is impossible to ignore the fact that Smoluchowski repeatedly referred to a work that is a classic of the methodology of natural research: *A Preliminary Discourse on the Study of Natural Philosophy* by John F.W. Herschel (1792–1871).

In the *Self-Study Handbook*, in the essay *Subject, task, method and the division of physics*, and in the work *Fields and issues of today's physics*, both the references to Herschel's thought and the characteristic methodology of the English scientist are so obvious that it is surprising this work is not cited in the footnotes. Although some of Herschel's theories and hypotheses have not survived the test of time, and Smoluchowski's methodology far exceeds the issues dealt with by the English researcher, the fact remains that his methodological concepts are evident in the works of the Polish scholar. The eighty-year

---

[188] See: Z. Hajduk, *Filozofia przyrody. Filozofia przyrodoznawstwa. Metakosmologia* (*Philosophy of Nature. Philosophy of natural science. Metacosmolgy*), Lublin 2007, p. 95.

interval separating *A Preliminary Discourse on the Study of Natural Philosophy* from Smoluchowski's works, as well as progress in the methodology of the exact sciences, mean that the works of the Polish physicist are characterised by a much higher scientific level, a wider range of methods used, and a more complete definition of the tasks of physics.

In trying to define the subject of physics, Smoluchowski poses the question of whether it can be assumed that physics (along with chemistry) is the science of the phenomena of inanimate nature as well as of phenomena common to both animate and inanimate nature. He argues that in this spirit, astronomy, geophysics, meteorology, geology and other sciences would also have to be included which undoubtedly are ultimately based on physics, but are at the same time independent[189]. The subject of astronomy, geophysics, meteorology or geology is nature (excluding living phenomena), so their subject is the same as that of physics. However, their tasks are different—in physics the general laws of nature are studied, while the task of the other sciences is to study individual events. Both the physicist and the astronomer deal with the motions of the planets around the Sun. However, they are of interest to physics as an example in the field of mechanics and as proof of the existence of general gravity, while the data determining the orbit of Mars or Earth are individual data, and the orientation of the orbits of the planets in relation to the constellation of stars is immaterial to physics[190].

Physics studies the rules of the movement of gases and air currents resulting from pressure differences. However, recording what winds blow in different parts of the globe and explaining these phenomena is the remit of meteorology[191]. Creative work in the field of physics involves discovering phenomena, as well as creating new theories to explain known phenomena. The physicist deals with the repetitive, general features of the laws of nature, and an individual event is a mere example of a general law. Such research gives science direction; it brings with it new intellectual trends and determines the progress of human thought[192].

---

[189] See: M. Smoluchowski, *The subject and task of physics*, in: *The Writings of Marian Smoluchowski*, vol. 3, op. cit., pp. 194–195.
[190] Idem., p. 157.
[191] Idem., s. 157–158.
[192] Idem., p. 102.

Physics does not deal with discovering the actual structure of the external world, and does not definitively resolve issues of matter, space or time. It also does not answer metaphysical questions, such as what things are in essence, beneath the surface of empirical phenomena, because according to Smoluchowski these issues are inaccessible to physics. Reflecting on the subject and tasks of physics, he notes that its main subject, in the general sense of the word, should be considered not individual facts, but rather the permanent phenomena of nature. Determining the tasks of physics is more problematic, because considering this question involves complex philosophical issues; they may seem to many an empty, sophistic play on words, but their explanation—or at least an attempt at it—is necessary because these tasks are directly related to beliefs about the results of scientific research and the place that physics occupies in people's view of the world[193].

Physics examines and formulates theorems about the phenomena occurring and empirical data obtained, since only this is available to scientific inquiry. Scientific research methodologies, and in particular the methods of conducting research in the physics contemporary to Smoluchowski, based on specific philosophical assumptions, frequently constituted the content of the Polish physicist's scientific works.

He devoted part of the discourse he conducted in various places to the methodology of the deductive and inductive drawing of conclusions from experiments and research. According to Smoluchowski, the historic breakthrough that occurred in the methodology of scientific research at the turn of the 16th and 17th centuries caused a dynamic development of science: "The astonishing, unprecedented development of technology against the background of scientific advances in physics and chemistry is undoubtedly the main feature of the last 150 years. It resulted in an innumerable range of inventions that we have become accustomed to as if they had existed for centuries, from matches, gas and paraffin to aeroplanes, wireless telegraphy, X-rays and submarines[194]."

On several dozen pages of the Self-Study Handbook, Smoluchowski analyses 150 years of the development of science, focusing mainly on physics. He begins his deliberations with an analysis of the essence of experiment,

---

[193] Idem., p. 161.
[194] Idem., *The importance of the pure sciences in general education*, op. cit., pp. 286–294.

and then enquires into the importance of methodological differences in the experiments conducted. Chance has always played an important role in the making of scientific discoveries, but it would not have been so significant had it not awakened in researchers the observational sense of experimenters, which enabled them to notice seemingly imperceptible or insignificant phenomena.

It is important to distinguish between a scientific, accidental observation and a planned experiment. Smoluchowski mentions this idea several times in his work, however it comes from the work of John Herschel, who analysed the experiment as an important and ultimate source of knowledge about nature and its laws. He thought it could be achieved in two ways. Firstly by perceiving facts as they occur, without any attempt to influence them—he called this observation. Secondly, "by putting in action causes and agents over which we have control (…), and noticing what effects take place"—he called this an experiment[195].

3.1.2. Reflection on Aristotle's philosophy of nature

Smoluchowski argued that both the era of ancient Greece and the later medieval period were devoid of an important element of modern science, namely the experiment. In these epochs, philosophical speculation dominated, at best supported by deduction. The experiment happened accidentally, as for example to Archimedes. Smoluchowski wrote: "Today, we cannot understand at all the aversion of the Greeks and scholastic academics to experiments, this attraction to groundless speculations and their application to the natural sciences. This trend seems to us to be a kind of madness, inhibiting the development of these sciences, especially due to the authority of Aristotle, throughout the entire course of antiquity and the Middle Ages"[196].

Herschel expresses a similar sentiment, writing that the fundamental error of Greek philosophy was the belief that the same method that proved so effective in mathematical research could be applied to natural research. Therefore, research on nature was neglected, and humble and careful inquiry into the facts was held in utter contempt as unworthy of the *a priori* position a

---

[195] J.F.W. Herschel, *A Preliminary Discourse on the Study of Natural Philosophy*, London, 1830 p. 76.
[196] M. Smoluchowski, *Subject, task, method and the division of physics*, op. cit., pp. 177–178.

true philosopher should adopt[197]. Ancient and medieval philosophers believed the laws of physical phenomena could be learned using the deductive method: they proceeded from a general statement related to the essence of the world, which to them seemed obvious, and by logically deductive inference they drew conclusions about various phenomena of nature. They were not at all discouraged by the fact that each philosopher arrived at different conclusions depending on their starting point[198].

In The Copernican Revolution, the American physicist and philosopher Thomas Kuhn (1922–1996) refers to the general views of the Stagirite, suggesting they arose through speculative reasoning. It was reasoning based on logic and deduction, introducing Aristotle's characteristic method used in the investigation of the truths of observed nature: "Aristotle was able to express in an abstract and consistent manner many spontaneous perceptions of the universe that had existed for centuries before he gave them a logical verbal rationale. In many cases they are just the perceptions that, since the seventeenth century, elementary scientific education has increasingly banished from the Western mind"[199].

Aristotle's system was burdened with many limitations, causing, as Kuhn writes, that "the organic realm has a conceptual priority, and the behaviour of clouds, fires, and stones tends to be explained in terms of the internal drives and desires that move men and, presumably, animals"[200].

Remnants of Aristotle's own thinking in relation to the laws of physics can be seen in his intellectual speculations:

> the movements of the simple natural bodies (fire, earth and so on) show not only that there is such a thing as place, but also that it has a certain power. For unless prevented from doing so, each of them moves to its own place, which may be either above or below where it was. Above and below and the other four directions [i.e. right, left, forwards and backwards] are the parts or forms of place (...). But in nature, each of these directions is independent of our position. 'Above' is not just any random direction, but where fire and anything light move towards. Likewise,

---

[197] See: J.F.W. Herschel, *A Preliminary Discourse on the Study of Natural* Philosophy, op. cit., p. 107.
[198] See M. Smoluchowski, *Self-Study Handbook*, vol. I, op. cit., pp. 30–31.
[199] T. Kuhn, *The Copernican Revolution: Planetary astronomy in the development of Western Thought*, Cambridge, Massachusetts and London, England, 1992, p. 96.
[200] Ibid.

'down' is not just any random direction, but where things with weight and earthy things move towards. So their powers as well as their positions make these places different[201].

The essential flaw in these analyses was the lack of experiments in support of the conclusions. However, Kuhn believes it is impossible not to perceive the breakthrough achieved under the Stagirite's influence in the "methodology" of thinking about nature and in creating the foundations of physics.
He writes:

> Aristotle's world view was not the only one created in antiquity, nor was it the only one that gained adherents. But Aristotle's was far nearer to many primitive conceptions of the world than its ancient competitors and it corresponded more closely with the evidence of unaided sense perception. That is another reason why it was so immensely influential, particularly during the late Middle Ages[202].

In the book Medieval and Early Modern Science, Alistair Cameron Crombie (1915–1966) *notes that Aristotle's* Physics, *translated in the 12th century from Arabic to Latin and presenting scientific concepts in a common sense way, with faulty views on physics contained in it, turned out to be a work constituting a closed rational system explaining the universe on the basis of natural causes. Aristotle's system was more than natural sciences in the 20th century sense*[203]. *Crombie asserts that:*

> "[i]n the course of the twelfth century a wonderful concept appeared, which immediately enabled the development of science; this was the idea of a rational explanation in the form of formal or geometric evidence, according to which a particular fact is explained when it can be deducted from a general principle. This occurred due to a gradual familiarity with Aristotle's logic and Greek and Arabic mathematics[204].

---

[201] Aristotle, *Physics*, trans. Robin Waterfield, Ed. David Bostock, Oxford University Press, 1999, pp. 78–79.
[202] T. Kuhn, *The Copernican Revolution*, op. cit., p. 99.
[203] A.C. Crombie, *Nauka średniowieczna i początki nauki nowożytnej, t. 1, Nauka w średniowieczu w okresie V–XIII w. (Medieval and Early Modern Science, vol. 1, Science in the Middle Ages)*, trans. S. Łypacewicz, Warsaw 1960, p. 77.
[204] Idem, p. 11.

Smoluchowski considered the deductive research method used by the Stagirite to be correct and effective but only in certain areas of study since—he argued—such a way of thinking hindered the development of science. Although it was supported by the authority of many philosophers, according to the Polish scholar, the indisputable position of Aristotle's philosophy had slowed the development of science.

Did the common sense method of perceiving natural phenomena, attributed to Aristotle, called by Smoluchowski "a madness holding back the development of those sciences, especially as a result of Aristotle's authority throughout the whole course of antiquity and the Middle Ages"[205], really impede that development or was it rather an essential link making modern science possible, a stimulus fertilising the mind to think more astutely?

In the *Middle Ages there existed schools of thought based on the Aristotelian conception of the universe, inspired not by metaphysics but by the mathematical, physical, astronomical and medical aspect of the Stagirite's science. The English Franciscan Roger Bacon* (1214–1292) of Oxford was an opponent of the speculative resolution of philosophical problems. He considered the experimental research of phenomena with the use of mathematics to be important. Interestingly, he believed—unlike most scientists—that the speed of light was finite. He foresaw the emergence of many inventions, such as ships without oars, carts driving without horses, and flying machines. He was much less interested in the metaphysical views of the Stagirite.

In 14th-century Oxford and later in several other European university centres, there worked Calculators who used Aristotelian logic and physics in their calculations. They divided motion into uniform and uniformly accelerating and were able to determine the mean velocity of motion and the law of free-fall. They even attempted to calculate some physical properties like heat, colour and the density of light. Their research was certainly not speculative in nature. The ideas of the Oxford Calculators did not spread to Europe however.

In the opinion of 19th-century historians of science, the founder of the scientific method was considered to be Francis Bacon (1561–1626), who freed scientific inquiry from speculative reasoning. According to Herschel, there was no natural science in the scientific sense of the term prior to Bacon's

---

[205] M. Smoluchowski, *Self-Study Handbook*, vol. I, op. cit., p. 31.

Novum Organum. He also pointed out that the principle of induction gained wide publicity thanks to Bacon and that it is to him that we owe the development of the idea that the whole of natural science consists of a series of inductive generalisations that start from the establishment of individual facts and lead to universal laws[206].

Smoluchowski wrote in the *Handbook* that Bacon formulated the principle of the inductive method, which introduced a gradual generalisation of knowledge instead of immediately compiling general statements based on sensory data, adding at the same time:

> Bacon's approach to the principles of induction was even flawed in many respects since Bacon conceived of induction as too mechanical an act, occurring with participation of speculative reason. But despite this, the historical significance of his input was enormous; only subsequently did unfounded, fantastic speculation gradually have to give way to experimental, empirical research[207].

Without referring directly to Herschel, the Polish physicist develops his thought from the A Preliminary Discourse on the Study of Natural Philosophy.

The method of induction was applied before Bacon. Humanity has been using it unconsciously in everyday life since time immemorial. On the basis of everyday experience, we develop views that create our beliefs regarding the future. There are known cases of the use of induction in scientific research in ancient times, as Smoluchowski writes. Ptolemy (100–168) made systematic measurements of the angle of incidence of the sun's rays and tried to establish the relationship of the results achieved to the angle of refraction of light rays[208]. The use of experiments can be seen in the research of Leonardo da Vinci, who called the experiment "a question posed to nature[209]. Some of Galileo's (1564–1642) later studies show the application of induction.

However, it is thanks to Bacon that induction became a scientific method consciously used to draw conclusions from experiments and that the experiment became recognised as the basic tool of science. Smoluchowski argued that as a consequence of applying the new method, a change occurred in

---

[206] See J.F.W. Herschel, *A Preliminary Discourse on the Study of Natural Philosophy*, op. cit., pp. 103–104.
[207] M. Smoluchowski, *Self-Study Handbook*, vol. I, op. cit., p. 31.
[208] Idem, p. 32.
[209] Idem, p. 37.

the way scientific research was conducted. He argued that knowledge is also required of how to ask questions the right way. Performing experiments without a plan and without proper preparation is just worthless fun[210].

Today, similar considerations are interesting generally only from the point of view of the history of the methodology of science and show how the model of experimentation at the time was shaped. At that time, methodological thinking was not obvious. Smoluchowski notes that research conducted by some scientists in a way considered the only scientific one—i.e. when a researcher, with no preconceived theories or accepted concepts, simply wants to classify the known facts—rarely yields valuable results. Progress in science occurred when we started approaching the subject of research with a certain specific plan, for example by following a theory, even the simplest one, which needs to be checked, or by examining the predicted relationships between numbers[211].

In the context of the above facts, can Aristotle's achievements be summarised with the thesis that his *Physics* halted the development of science for many years? Could his philosophical concepts have so bound the minds of philosophers and thinkers that for nearly 1,900 years they were unable to go beyond the framework of perceiving nature imposed by the Stagirite? Referring to Aristotle's *Physics*, Smoluchowski makes the accusation that were it not for this school of thought, focusing on philosophical speculation, "some Bacon" would most probably have appeared in European culture and science much earlier, and modern science would not have emerged only in the 17th century.

However, Ptolemy is not accused of holding back the development of science, and especially astronomy, for nearly 1,600 years by building a geocentric system and with this assumption contributing to a philosophy according to which the Earth is at the centre of the Universe. It cannot be denied that his theory placed man at the centre of the world for a long time, which influenced the development of theology, philosophy, and especially the philosophy of nature. Euclid is also not charged with stalling the development for many years of other geometric systems in which the postulate of straight parallelism does not hold, by describing in *The Elements* a geometric system based on axioms.

Leaving aside these issues, it should be added that Smoluchowski could not have known whether "some Bacon" had not appeared earlier and, for

---

[210] Idem, p. 38.
[211] Idem, p. 52.

example, in the medieval scholasticism so despised by the 19th century, simply gone unnoticed.

Let this thesis be exemplified by the previously described Calculators or the theorems of the forerunner to Nicolaus Copernicus (1473–1543)—Nicolaus of Cusa[212] *(1401–1464). The latter argued that the Earth is not the centre of the Universe and that it is smaller than the Sun and larger than the Moon (which—he argued—can be observed during a solar eclipse) and rotates on its axis every 24 hours*[213]. *He had an extraordinary record of achievements in the field of mathematics, for example research on the length of a circle. Another argument for the thesis of the Earth's rotation are the achievements of Nicolaus Copernicus, which are evidence of overcoming medieval tendencies, especially in mathematics and astronomy, as well as a new direction of thinking.*

Referring to Josef Pieper, in his book Fizyka (Physics) *in* Quaestiones super octo libros 'Physicorum' Aristotelis John of Głogów, *Stanisław Bafia writes that until the 12th century, and in the Augustinian school of 13th-century philosophy and theology, mathematics was an introduction to metaphysics, or—in line with the terminology of the time—to theology. However, in the 14th century, under the influence of Ockham's views, mathematics became an indispensable tool of the science of nature. This heralded a new science*[214].

Not until the late renaissance *is a systemic change visible in the approach to natural philosophy. The Dominican Domingo de Soto (1494–1560), a Spanish theologian from Salamanca considered a forerunner of modern physics, is a figure characteristic of that change. In 1545, thanks to observation, he determined that a body in free-fall undergoes constant acceleration. This was 14 years before the birth of Galileo, but it was he who turned this observation into a scientific fact. Galileo made use of de Soto's work but went much further, proving the independence of Earth's acceleration from mass in the free-fall of bodies*[215].

Both Aristotle's philosophy and his physics were still present in the renaissance era. What then influenced the revolutionary transformation of natural philosophy into science despite the clear continued presence of Aristotle? It may have been a change in the system of

---

[212] See: Nicholas of Cusa, *O oświeconej niewiedzy* (*On Enlightened Ignorance*), trans. I. Kania, Kraków 1997.
[213] See J.M.R. Morales, *Kościół i nauka. Konflikt czy współpraca?* (*The Church and Science. Conflict or cooperation?*), trans. S. Jędrusiak, Kraków 2003, p. 82.
[214] See: S. Bafia, *Fizyka in Quaestiones super octo libros 'physicorum' Aristotelis Jana z Głogowa*, Kraków 2013, p. 9.
[215] See: J.M.R. Morales, *The Church and Science, op. cit., p. 83*.

thinking, which became ever more creative, while there were transformations in perception and the way of thinking about reality.

The scientific and philosophical thought that had been forming since ancient Greece needed time to mature in its reflection and take on the characteristics of a scientific project, based in the 20th century on induction and hypothesis. The change went deep and required a new methodology for studying nature. It is possible that European thought, which been developing for a long time—until Galileo—under the influence of the Stagirite's philosophy, matured, by no means in opposition to his eight books, so that a model of scientific reasoning would eventually develop. Without Greek philosophy and the philosophy of Aristotle, that process would probably have taken longer as it is significant that the scientific model we know has so far only emerged in European culture, the foundation of which is Greek philosophy.

The Stagirite's contribution to building the foundations of the science of the future, and especially to the creation of the scientific method, is key. He is responsible for developing methods of the logical construction of thought, the creation of rules of inference, of definition and classification. It was Aristotle who made us aware that concepts arise through abstraction. This Greek philosopher gave importance to the ability to make observations and ask questions. His bivalent logic has left its mark on all science (as exemplified by modern computer programs based on the binary system).

In his philosophical and scientific views, Smoluchowski, like Galileo, opposes speculative analysis of nature and treating Aristotelianism as an intellectual basis in the study of natural science. He does not see (at least he does not write about) the philosophical and logical potential created by the Stagirite, enabling the later development of Bacon's philosophy. In fundamental logical issues and rules of inference, Aristotle's achievements form the basis of scientific thinking, creating the foundations of reasoning enabling the development of science.

In the Middle Ages, Aristotle distinguished three basic cognitive activities: distinguishing (Latin: *distinctio*), logical division (Latin: *divisio*) and defining (Latin: *definitio*). These represented the pillars on which in time the scientific method could be built.

In early Christianity, philosophy was treated as a pagan thought, and in the Middle Ages—despite the use of elements of ancient philosophy and the incorporation of Aristotle's philosophy into the circle of medieval thought—research was guided by other priorities. The attention of the main minds of those times was primarily focused on theological inquiries, and these represent the basis of their considerations. This state of affairs persisted until the late Renaissance, when a change of interests occurred.

Aristotle's *Physics* could not effectively inhibit the scope or pace of changes in philosophical deliberation over such a long period. Paraphrasing Smoluchowski's *utility* concept, it should be stated that knowledge in the field of natural philosophy did not make significant progress because the *utility* of the previous theories created sufficiently strong pressure not to pursue this change.

Aristotle's philosophy constituted another level in building the methodology of learning about nature. It opened the way for logical thought, inference, definition, classification, abstraction, observation, and the asking of questions, or everything that we owe to Aristotle and which modern science uses. However, for quite a long time, these important attributes of the methodology of science were not used, because it was not useful—thought was focused on theological goals.

The situation in mathematics, seen in the Middle Ages as an exemplary science, was slightly different, as *Crombie writes. Earlier in Arabia, and in* the 12th and 13th centuries in medieval Europe, mathematics developed on the basis of Aristotle's logic among others, and though relegated, as noted above, to a role subordinate to theology, it was one of the historical factors that significantly influenced the development of that discipline.

3.2. Methods of physics

3.2.1. The experiment in the history of science

According to Smoluchowski, progress in science accelerates when research is approached by following a theory, even the simplest one, or even by studying a numerical relationship. However, the key factors for the growth of progress transpired to be: rejection of teleological thinking in science, an increase in the role of inductive reasoning, and an emphasis on creating hypotheses.

It is assumed that inductive reasoning in science has been functioning since the time of Galileo and that teleological thinking was questioned in the time of Isaac Newton (1643–1727), but the creation of scientific hypotheses was still contested at the end of the 19th century. The extremely slow evolution of change in the methodology of science accelerated only in the 20th century. This evolution was unique and can be compared only with the rate of change in philosophy that took place in the period between pre-Socratics and classical Greek philosophy. It remains unknown whether the current expansion will last as long as it did in antiquity.

Neither speculative thinking nor Aristotle's physics prevented progress in science as they represented only successive steps in its development. Aristotle's physics, like any scientific theory, is only a tool—more or less useful—for learning about reality. Tools are used according to their *usability*. It is hard to say unequivocally how far Aristotle's speculative thinking is removed from the methodology of building science from hypotheses and theories, such as M-theory or the theory of cyclic universes currently being constructed in astrophysics and physics, which are also not based on experiment. So it is impossible to agree with Smoluchowski's radically negative assessment of Aristotle's physics.

Apart from the Polish physicist's unambiguous attitude towards the experiment in science, a separate issue is the reliability of data obtained from it. Smoluchowski outlines the procedures used, depending on the purpose of the research, which enable the elimination of errors being superimposed on the results of experiments. These are essentially random or systematic errors. In trying to determine quantities such as the speed of light, the magnitude of an electron's elementary charge, or the gravitational constant, we proceed to obtain the most precise quantities possible. According to Smoluchowski, in order to obtain them, a very large number of measurements should be made under the same conditions and average values should be calculated from the results obtained, which will allow accidental errors to be avoided. Calculations made using probability calculus show that in this way random errors are eliminated with accuracy increasing by the square root of the number of measurements. According to Smoluchowski, removing systematic errors in the process of conducting an experiment is important, but also difficult. If the sources of these errors can be predicted, they are eliminated by first calculating a correction to the data obtained. The way to avoid errors in experimental research is to systematically change the conditions of experiments and check whether these changes affect the results of measurements or whether they are independent of these changes[216].

---

[216] See: M. Smoluchowski, *Self-Study Handbook*, vol. I, op. cit., p. 42.

Addressing the problem of avoiding errors in experimental research in *A Preliminary Discourse on the Study of Natural Philosophy*, Herschel described the situation as follows:

> Though we admit the necessary existence of numerical error in every observation, we can always assign a limit that such error cannot possibly exceed; and the extent of this latitude of error of observation is less in proportion to the perfection of the instrumental means we possess, and the care bestowed on their employment. In the greater part of modern measurements it is, in point of fact, extremely minute, and may be still further diminished, almost to any required extent, by repeating the measurements a great number of times, and under a great variety of circumstances and taking a mean of the results, when errors of opposite kinds will, at length, compensate each other[217].

The means of conducting research depends on the method we decide on. In the case of using the induction method, it is indispensable to repeat the experiment several times. If the result of the experiment can be quantified, we call it quantitative research, and otherwise—qualitative research. Qualitative experiments are usually the first step to performing quantitative studies. The latter are the proper goal of experimental physics, because we always strive to learn the laws of nature as accurately and quantitatively as possible[218].

In contrast to passive observation, the experiment has the advantage that it can be induced it and its course even influenced, while the repeatability of the research enables the appropriate quantity of required data to be obtained. As a result, the results obtained in this way enable use of the inductive method. Of course, there is a certain range of phenomena that we cannot influence, such as the movements of planets, Earth or meteorological phenomena. A separate issue is unexpected phenomena, such as the discovery by Hans Christian Ørsted (1777–1851) in 1820 of the influence of electric current on a magnetic needle, or the discovery of X-rays in 1895. However, as Smoluchowski writes, it would be wrong to suppose that this is a common genesis of scientific discoveries.

Usually, discoveries are the result of planned, persistent experimental or theoretical work. Smoluchowski states:

---

[217] J.F.W. Herschel, *A Preliminary Discourse on the Study of Natural Philosophy*, op. cit., p. 127.
[218] See: M. Smoluchowski, *Self-Study Handbook*, vol. I, op. cit., p. 38.

> The foundation for induction, as well as the test for deduction, is experience in the broader sense of the word (*Erfahrung, l 'expérience*). Depending on whether we cause the phenomena that concern us intentionally or whether they occur without our participation, we distinguish experience in the stricter sense (*Versuch, une expérience*), or experiment, from observation[219].

The Polish physicist places the experiment in the context of functioning theories. It is important for the effectiveness of the experiments carried out that they fit into the context of current theoretical physics. Thus, while opposed to theoretical speculation not based on experiment, he also believed that an experiment not based on theoretical considerations would not be of full value. The importance of experimental work depends not only on the accuracy of measurements, but mainly on whether the result is an isolated fact, whether more general conclusions can be drawn from it, or whether it determines the value of a given theory. What is important is the theoretical significance of the experiment; its relationship to theoretical physics. An experimenter who wants to work fruitfully must constantly follow the development of theoretical physics[220].

Analysing the methods of conducting experiments in the *Self-Study Handbook*, Smoluchowski does not divide observations into direct and indirect. He also does not explicitly address this issue, and from an analysis of his arguments it should be presumed that this division is irrelevant to the evidential value of an experiment[221].

A key element in the process of conducting an experiment has become inference, made on the basis of data obtained from the experiment and consisting in arriving at specific formulae—new ones, or those confirming formulae already held to be true. Similar deliberations were conducted eighty years earlier by Herschel, who wrote:

> The only facts which can ever become useful as grounds for physical enquiry are those which happen uniformly and invariably under the same circumstances. This is evident: for if they have not this character, they cannot be included in laws. They

---

[219] Idem, p. 35–36.
[220] See: M. Smoluchowski, *Kierunki i zagadnienia fizyki dzisiejszej, odczyt wygłoszony na kursach dla nauczycieli we Lwowie 12 marca 1913 r.* (*Themes and issues in today's physics, a lecture given on courses for teachers in Lviv, March 12, 1913*) w: *The Writings of Marian Smoluchowski*, vol. 3, op. cit., p. 207.
[221] Idem., *Self-Study Handbook*, vol. I, op. cit., pp. 37–39.

want that universality which fits them to enter as elementary particles into the constitution of those universal axioms which we aim at discovering[222].

The research methodology of Smoluchowski's physics is much more modern than *A Preliminary Discourse on the Study of Natural Philosophy*. It is an expression of the next stage of the development of science, especially of physics. By emphasising the essence of chance and the use of statistics in physics, and being a forerunner in the introduction of probability theory in the study of the exact sciences, Smoluchowski steered the methodology of research in physics onto new paths.

3.2.2. The induction method

Reliable inference is inference based on deduction. This is achieved when the conclusion follows logically from the adopted premises (only a true conclusion can be reached from true premises). Since the truth of the premises guarantees the truthfulness of the conclusion, this means that the reasoning is contained within its assumptions and does not require the creation of new statements or concepts; it is merely a simple drawing of conclusions. Assuming that the set of premises does not contain false statements, correctly conducted deductive reasoning leads us to conclusions that must be true and that cannot be reasonably questioned. A classic example of using the method of deduction in the process of inference is the building of mathematical and logical theories. This way of thinking was initiated by the Greeks in the third century BC and is subject to constant improvement with the development of mathematics.

According to Smoluchowski, a much more efficient method, especially in physics, although not as sure as deductionism, is the inductive method, which is based on inference, in which generalisations are derived from specifics. The truthfulness of consequences is based on the truthfulness of their reasoning. Despite Hume's reservations about its verifiability, it is, as the history of science has shown, a very effective method. It is distinguished from deduction by its efficacy, particularly in experimental research. Smoluchowski had no doubt about the effectiveness of induction when he wrote:

---

[222] J.F.W. Herschel, *A Preliminary Discourse on the Study of Natural Philosophy*, op. cit., p. 116.

> Since the subject of physics is the phenomena of the world around us or, more strictly speaking, our impressions about it, of which *a priori* we have no intuitive awareness, it is clear that the basic method of research must be the inductive method, also characteristic of other natural sciences: consisting in gathering experimental material, classifying this material, and abstracting general rules from it[223].

As the researcher notes, the use of induction in physics does not exclude the use of deduction. Empirical sciences, being based on mathematics and logic, necessarily employ deductive methods. For a scientist, it is obvious that reliable inference is deductive inference, which occurs when the conclusion follows logically from the adopted premises, so physics cannot be deprived of deductive analysis.

According to Smoluchowski, the distinguishing feature of physics in the natural sciences, in addition to the use of deduction and the wide use of mathematics, is the ability to use the deductive method in the aspect of deduction based on a hypothetical footing, with experimental testing of the conclusions. Deduction in physics has a different meaning than in mathematics, where the fundamental theorems of science are obvious and certain. We build our conclusions on them, with confidence in the durability of the foundations. In physics, on the other hand, the fundamental laws are the unknowns that are being sought, so we adopt conditional assumptions in relation to them, a 'hypothesis', and deductively derive from it all the special consequences that can be experimentally tested. And based on whether they align with the known facts, whether they can be verified, we infer the truthfulness of the hypothetical assumption through experiment[224].

The history of science, and especially the history of scientific discoveries, is replete with instances of the deductive testing of inductive hypotheses. Smoluchowski cites examples confirming this fact: the discovery of the planet Neptune by Urbain Le Verrier (1811–1877) and John C. Adams (1819–1892), the existence and position of which were predicted mathematically on the basis of observations of discrepancies in the orbit of the planet Uranus. It was hypothesised that some of the observed deviations in Uranus's orbit may have been connected with the gravitational influence of another celestial

---

[223] See: M. Smoluchowski, *Subject, task, method and the division of physics*, op. cit., p. 177.
[224] See idem., *Self-Study Handbooks*, vol. I, op. cit., pp. 33–34.

body unknown at the time. This celestial body was discovered by Johann Gottfried Galle (1812–1910) using a telescope—the hypothesis was confirmed.

The situation was similar with the hypothesis of Christian Doppler (1803–1853), who assumed that as a result of an observer approaching or moving away from a source of radiation, a change must occur in the frequency of the waves emitted. For example, in the case of sound, a change can be perceived in the pitch of the tone, and in the case of light, there is a change in the colour seen as a result of a shift of the spectral lines towards the red or blue end of the spectrum. It was not until 1901 that Aristarch Biełopolski (1854–1934), and in 1906 Johannes Stark (1874–1957), could actually confirm experimentally that movement of the light source causes such a phenomenon. Another famous example is the hypothesis of James C. Maxwell (1831–1879) in 1864 regarding electromagnetic waves. The existence of this type of radiation was experimentally confirmed by Heinrich Hertz (1857–1894) in 1887[225].

The methodology of scientific research did not prefer only one of the methods. Smoluchowski saw the advantages of both induction and deduction, believing that each of these methods should be used when it is more effective. In certain cases, the advantages of Baconian induction should be combined with the cautious deduction of Hume. According to Smoluchowski, both of these methods are useful, however, it is induction—creating and giving impetus to the development of science—that is preferred in physics, while the deductive method, used to verify the hypotheses formulated, helps in creating new theorems and new scientific theories.

Bacon considered the method of eliminative induction to be the best way of arriving at knowledge, in which inference proceeds from the detailed to the general. Knowledge is practical in nature and the experiment plays a large role in establishing facts. In order for the induction method to be used effectively, it is necessary to first indicate the sources of deformation of both the intellect and of knowledge, as he demonstrated in *Novum Organum*[226]. It is worth mentioning here that Hume, who was considered an opponent of

---

[225] Idem, p. 34–35.
[226] See: F. Bacon, *Novum Organum*, New York 1902, p. 83.

the inductive method, claimed that we use it in reasoning out of habit and that there is a serious problem with justifying the use of induction[227].

The inventor of the inductive method, as Aristotle writes in *Metaphysics*, is Socrates. Popper mentions this in *The Growth of Scientific Knowledge*.[228] Ultimately, it was Aristotle who was the first to use deduction as a research method; he also discussed the method of inductive reasoning in the form of a simplified complete enumerative induction, in which the inference, as defining a general regularity, is recognised on the basis of a finite number of sentences stating the occurrence of this general regularity.

Popper argued that "the belief that we can start with pure observation alone, without anything in the nature of a theory is absurd"[229]. In the considerations being conducted, two aspects of the problem should be distinguished: the analysis of the tool itself, which is the principle of induction, and issues verifying the falsifiability of conclusions resulting from induction. A similar idea was put forward forty-five years earlier by Smoluchowski, according to whom progress in science occurred when we began to approach the subject of research with a specific plan[230].

The Polish physicist raises this issue in his essay *Subject, task, method and the division of physics*, writing:

> In physics, fundamental laws are what we are looking for; we only make conditional assumptions about them, we adopt some 'hypothesis' and derive, in a deductive way, all the consequences that lend themselves to experimental testing. From the result, if they agree with the known facts or can be verified by means of deliberate experiments, we conclude the converse about the truth of the hypothetical assumption. So it is a deduction on a hypothetical basis, with experimental testing of the conclusions[231].

In her essay *On the metascientific views of Władysław Natanson and Marian Smoluchowski*, Izydora Dąmbska directly quotes Smoluchowski's thought

---

[227] More on Hume's position on induction in subsection 6.5.1 'Hume's inductive scepticism'.
[228] See K. Popper, *Conjectures and Refutations: The Growth of Scientific Knowledge* Routledge, London and New York, 2014 p. 16
[229] Idem, p. 61.
[230] See M. Smoluchowski, *Self-Study Handbook*, vol. I, op. cit., p. 52.
[231] Idem, *Subject, task, method and the division of physics*, op. cit., p. 180.

about giving both types of reasoning, inductive and deductive, equal weight in physical research—asking questions of nature comes down to formulating a working hypothesis, followed by deductive inference and then confronting the result of this deduction experimentally. Smoluchowski saw the physicist's work—Dąmbska continues—as an activity involving the constant creation of new theories explaining previously observed phenomena, and subjecting these theories to an experimental test. He believed that any experimental research, in order to be effective and enable learning about a given field of phenomena, should be preceded by theoretical consideration[232].

The example he gives is a specific illustration of the possibility of using the inductive method in everyday life, at the same time testifying to Smoluchowski's sense of humour and intellectual distance from scientific disputes:

> However, the most interesting example of induction is given by the principles of thermodynamics. From the point of view of logic, can the position be justified of the Paris Academy, which in 1755 decided not to accept any works related to a perpetual motion project, because it considered the possibility of such an invention to be ruled out in advance? On what basis did it think so? For the sole reason that, despite their most strenuous efforts, no one has succeeded in solving this problem? It proceeded not according to the principles of deductive logic, but according to induction, and to this day we agree with it[233].

3.2.3. Reflections on the mathematics of nature

Smoluchowski's findings on the use of probability calculus in assessing the reliability of the inductive method seem in retrospect to be accurate and essentially consistent with the views expressed several decades later by Popper. As happened many times, Smoluchowski was generally insusceptible to the views imposed by the academic community. The belief that the value of a hypothesis can be determined using probability calculus was, as the Polish scientist writes, quite widespread at that time. In the first half of the 20th century, this way of thinking was presented by Hans Reichenbach (1891–1953).

---

[232] See: I. Dąmbska, *On the metascientific views of Władysław Natanson and Marian Smoluchowski*, op. cit., p. 7.
[233] Idem, p. 33.

Popper was also an opponent of the probabilistic theory of induction, believing that the inductive method leads to a preference for specific theories, which does not translate into probability calculated by probability calculus. For example, it may turn out that a theory will be accepted that has a lower probability of veracity, but is more desirable[234].

In order to make an objective assessment of a given theory, Popper introduced the so-called idea of the degree of corroboration. He intended to thereby show that any probabilistic theory of induction is ludicrous. By the degree of corroboration he meant a concise account of the state (at a given time, t) of critical discussion of a theory, taking into account the way in which it solves its problems, the degree of its verifiability, the severity of the tests it has undergone, and how it has survived those tests. The degree of corroboration is an account of the theory's performance to date[235]. It was intended as a means of expressing preferences for the truthfulness of a given inductive theory, but it did not constitute a verification of the theory since even a high degree of corroboration does not result in a hypothesis being recognised as true. The confirmation obtained is not indisputable proof of the truthfulness of a theory, merely the degree of its credibility, an indication that the theory under scrutiny has no other competing against it and that there are no arguments that would contradict it.

Corroboration is relative, meaning that its degree may determine the greater credibility of theory A over theory B, but always in relation to time t. Preferring one theory does imply probabilistic relationships, and therefore it is impossible to determine its reliability using probability calculus. Popper writes that a theory can be called more desirable, more "probable", but these are only words until we allow ourselves to be convinced by them[236]. The concept of Popper's corroboration substantively justifies Smoluchowski's reservations over the use of probability theory in assessing the likelihood of hypotheses being true.

To make decisions when formulating a hypothesis is to create new but uncertain knowledge. This knowledge can be used to search for new

---

[234] See: K.R. Popper, *Wiedza obiektywna: ewolucyjna teoria epistemologiczna* (*Objective Knowledge: evolutionary epistemological theory*), trans. A. Chmielewski, Warsaw 1992, pp. 30–32.
[235] Idem, p. 31.
[236] Idem, p. 32.

decision-making rules that minimise the number of erroneous decisions or the losses incurred because of them. Optimal decision-making enables deductive reasoning. Inductive inference can be introduced into the rules of deductive logic. An example of decision-making in conditions of high uncertainty is the way weather forecasts are made. Forecasts, given in the form of categorical statements defining a type of weather (for example, tomorrow will be sunny or, on the contrary, rainy), were forecasts with a high risk of error. Preparing a weather forecast in a form that includes, for example, the statement that a 20 % chance of rain is expected tomorrow makes it possible to make more accurate decisions, for example, whether to leave the house with or without an umbrella. Deductive inference allows the selection of a subset of premises to prove a thesis, while inductive inference for different data subsets can lead to different, often contradictory conclusions— hence it is necessary to use all available data[237].

In 1917, during the Conference of the Society of High School Teachers, Smoluchowski gave a lecture entitled *The importance of science in general education*. In a few concise words, he captured, as if in a lens, the basic idea of the functioning of inductive-deductive reasoning in science. He noted that in physics, chemistry or astronomy, each step gives classic examples of either inductive reasoning, i.e. generalising inference on the basis of empirical material and expressing it in a strict law, or deductive reasoning, arriving at special conclusions from these general laws by means of strict mathematical calculations.

A long series of facts from the history of these sciences, such as the evolution of our knowledge of planetary motions, Kepler's laws and Newton's law of gravitation, the discovery of Neptune, knowledge of the laws of light refraction, the development of fundamental concepts in the field of heat and so on, is found in all textbooks and can be cited as a classic example of scientific reasoning[238].

Smoluchowski notes that the inductive nature of physics has the consequence that no hypothesis can ever be proven in physics in the sense in which

---

[237] See: C. Domański, *Statystyka bliżej biznesu*, w: *XIX Ogólnopolska Konferencja Dydaktyczna „Nauczanie przedmiotów ilościowych a potrzeby rynku pracy"* (*Statistics closer to business*, in: *XIX National Didactic Conference 'Teaching quantitative subjects and the needs of the labour market'*), Łódź 2010.

[238] See: M. Smoluchowski, *The importance of science in general education*, op. cit., p. 128.

the word "proof" is used in mathematics. In the interests of the accuracy of terminology, we cannot say that any physical theory is true, or even judge its degree of probability. We can talk only about its greater or lesser *usefulness*, and we must abandon the concept of *experimentum crucis* as devoid of intrinsic meaning, used when science was not yet, as it is today, permeated with scepticism about the theory of knowledge. Nonetheless, as the Polish physicist claims, it still happens that such systematic evidence is found for the validity of a given theory, such astonishing confirmations of conclusions that initially aroused distrust, that it seems that trust in this theory has gained a tangible basis, although science warns against blind faith[239].

This is an important philosophical thought of Smoluchowski which forms part of a still unsettled dispute in philosophy, running since the time of Hume through Bertrand Russell (1872–1970) to Popper, Kuhn and Rudolf Carnap (1891–1970), and concerning the solution of the problem posed in scientific research by uncertain knowledge, i.e. knowledge obtained in empirical research through non-deductive inferences.

Hume similarly distanced himself from science, claiming that our knowledge is exposed as something that has the nature of an opinion, moreover, an opinion that cannot be defended by reason—it is therefore irrational faith.

Today, many accepted theories have passed through a sieve of repeated attempts at falsification, not thereby increasing the probability of their veracity. In turn, a change has repeatedly been witnessed not only in the theories themselves, but also in the paradigms on which these theories were built. Popper proposed a criterion for the relationship of a new theory to the old theory being amended. He wrote:

> The new theory, although it has to explain what the old theory explained, corrects the old theory, so that it actually contradicts the old theory: it contains the old theory, but only as an approximation. Thus I pointed out that Newton's theory contradicts both Kepler's and Galileo's theories—although it explains them—owing to the fact that it contains them as approximations; and similarly, Einstein's theory contradicts Newton's theory, which it likewise explains and contains as an approximation[240].

---

[239] See idem: M. Smoluchowski, *On thermodynamic fluctuations and Brownian motion*, op. cit., p. 268.
[240] K. R. Popper. *Objective Knowledge: An Evolutionary Approach*, Edition 2, Oxford, 1979, p. 24

CHAPTER 3

In his methodological considerations, Smoluchowski argued that the typical task of inductive experimental research in physics should not be to determine one quantity, but to find a relationship between different quantities[241]. As an example, he cited research carried out by Henri Victor Regnault (1810–1878), who continued the work of Russian chemist Dmitri Konovalov (1856–1929) on the relationship between the pressure of vapour rising above a mixture of water and alcohol and the temperature of this mixture and its composition. The research covered two cases—the dependence of vapour pressure on temperature and the dependence on the temperature and composition of the liquid. In the first case, the vapour pressure depends on one factor, which is the temperature of the mixture, in the second on two factors—the temperature and the composition of the liquid. The procedure for conducting this type of research, as Gustav Kirchhoff (1824–1887) put it, is to gradually simplify the description—from the raw material to the most perfect and simplest form, which is a mathematical formula. The first step is to cleanse the raw measurement results of random errors, the second involves presenting the results in tabular or graphic form, and the third is to express these results in the form of a mathematical formula.

According to Smoluchowski, these are three successive levels of the mental processing of raw material, or—according to Kirchhoff's principle—a gradual simplification of the system, in which the most perfect and simplest description will be a mathematical formula[242].

For phenomenalists such as Kirchhoff and Mach, the proposed method is a key aspect of theoretical work, primarily involving finding mathematical equations that fit observations, without even building models[243]. In Smoluchowski's view, however, it contains a certain inconvenience as the use of a mathematical formula for a large number of observations means its form is usually intricate and, moreover, excessive inaccuracies appear. On the other hand, excessively precise measurements may also constitute an obstacle to uncovering fundamental laws[244]. However, the key issue is the ability to draw conclusions based on research for which data are obtained in two different

---

[241] See: M. Smoluchowski, *Self-Study Handbook*, vol. I, op. cit., p. 42.
[242] Idem, p. 43.
[243] R. Piotrowski, *Demon Maxwella. Dzieje i filozofia pewnego eksperymentu* (*Maxwell's Demon. The history and philosophy of a certain experiment*), Warsaw 2011, p. 80.
[244] See: M. Smoluchowski, *Self-Study Handbook*, vol. I, op. cit., pp. 42–43.

ways from the same experiment, which undoubtedly makes the quality of the conclusions drawn more credible. Smoluchowski adds that a new field of applied mathematics, called nomography, which deals with the theory and construction of nomograms for individual equations or systems of algebraic equations, has become helpful in conducting similar research. As the age of steam transitioned into the age of electricity and the atom, Smoluchowski's innovative thought was noticed by few people and accepted by even fewer.

In the *Self-Study Handbook*, Smoluchowski recalls a thought of the Königsberg philosopher: "Kant said, with some exaggeration, that in every field of knowledge there is only as much science as there is mathematics". Summing up Kant's thoughts with use of the phrase "with some exaggeration" is no accident. In his commentary, Smoluchowski applies the meaning contained in Kant's statement to the exact sciences, in particular to the understanding of mathematics in nature as approaching phenomena quantitatively. He believed that the most important activity of a researcher practising experimental physics is measurement. Progress in each field is closely related to the quantitative approach to a phenomenon, i.e. the introduction of the concept of quantity and the carrying out of appropriate measurements. The most important part of quantitative experiments is measurement and it is this that constitutes the main task of a researcher dealing with experimental physics, not demonstrative experiments, which are colloquially called experimentation[245]. Seeing Kant's thought from the point of view of a physicist, Smoluchowski argued that mathematics should be understood as a way of seeing nature in the category of space and time, i.e. in terms of the possibility of measurement in space and time. However, contained in Kant's statement is also another idea that explains the mathematics of nature slightly differently—philosophically.

A statement by Michał Heller in the book *Matematyczność przyrody* (*The Mathematicality of Nature*) indicates such an understanding:

> the world is a book, and science is trying to decipher the information that this book contains. The carrier of the information must be some kind of structure; without structure, there is no information. It turns out that the structure of the universe is strangely similar to those structures that mathematics studies. The similarity is so astonishing that some thinkers are inclined to treat it as a kind of

---

[245] Idem, p. 38.

singularity. To emphasise this, they say that nature is mathematical or that it is made of mathematics[246].

However, this is not as obvious as Heller and Józef Życiński (1948–2011) suggest. Nima Arkani-Hamed, an American physicist of Azerbaijani origin, claims that we use standard mathematical constructions and other abstractions because they work, but they are not at all inevitable; if they were, we could arrive at them on the basis of pure logic, and yet it cannot be proven that any mathematical structure provides a true description of nature, because mathematical proofs concern only the structures themselves, rather than their relationship to reality[247]. How mathematics relates to reality is a mystery that haunted philosophers long before scientists came along, and we are none the wiser today[248]. Eugene Wigner, in reference to Gottfried Wilhelm Leibniz (1646–1716), posed the question: "Why can we describe the world mathematically at all?"[249]. Why is the universe governed by simple, elegant mathematical laws, when there is no reason for it?

We do not know the answers to these questions and perhaps we will never know them, which does not change the fact that, according to Sabine Hossenfelder, physicists use a lot of mathematics and are very proud that it works so well. However, she further states that physics is not mathematics, and the development of a theory needs experimental data as guidance[250]. This idea is the author's reference to the situation that has prevailed in physics for over forty years (more on this subject in subsection 6.4.2. 'The utility criterion and social demand').

Returning to Smoluchowski, it should be noted that Kant's dictum can be understood more fully and deeply, because the mathematicality of nature is not only its receptivity to counting and measuring, but also the possession of

---

[246] M. Heller, J. Życiński, *Matematyczność przyrody* (*The Mathematicality of Nature*), Kraków 2010, p. 8.

[247] See. S. Hossenfelder, *Zagubione w matematyce. Fizyka w pułapce piękna* (*Lost in Math: How Beauty Leads Physics Astray*), trans. T. Miller, Kraków 2019, p. 110.

[248] Idem., p. 80.

[249] E.P. Wigner, *Niepojęta efektywność matematyki w naukach przyrodniczych* (*The unreasonable effectiveness of mathematics in the natural science*), trans. J. Dembek, 'Zagadnienia Filozoficzne w Nauce' ('Philosophical Problems in Science'), 1991, No. XIII, pp. 5–18.

[250] See: S. Hossenfelder, *Lost in Math*, op. cit., p. 32.

encrypted information about the structure of the universe. It is not certain whether Smoluchowski's understanding of "measuring" concerned only the physical performance of this activity for the purposes of deductive verification of hypotheses, or whether it also included the study of the carrier of encrypted information forming the structure of the universe.

In describing the essential aspects of conducting physical research, Zygmunt Hajduk in a way repeats Smoluchowski's formulations:

> Despite the changes that have occurred in the development of natural physics and cosmology, one feature remains constantly present in it. It is the quantitative approach to the world. It is believed that mathematical models correspond objectively to a given order, which is more real than the perceptually given world. Since the time of Galileo, physics has taken into account only the quantitative aspects of things. They are considered primary qualities and— as objective— are contrasted with subjective, sensory-perceptible qualities[251].

The tendency to quantify reality, a factor in physics since the time of Galileo Galilei, required consolidation, which undoubtedly happened over time, and Smoluchowski's unequivocal advocacy of the quantitative study of reality is a classic example of this. The fact that mathematics manifests itself in nature in such a diverse and unexpected way, and at the same time the mathematical description gives the impression of being the most in line with the truth, does not solve the issue that the mystery of science remains the constant search for the path to full knowledge, and the inductive method is reasoning focused on creativity and exploring other, new solutions.

3.3. The explanatory function of physical theories

According to the traditional view, Smoluchowski writes in his essay *Subject, task, method and the division of physics*, the task of both physics and the natural sciences, is twofold: learning about phenomena and explaining them[252]. On the other hand, in A Preliminary Discourse on the Study of Natural Philosophy, Herschel defined natural science as knowledge of causes and

---

[251] Z. Hajduk, *The philosophy of nature*, op. cit., p. 58.
[252] See: M. Smoluchowski, *Subject, task, method and the division of physics*, op. cit., p. 162.

their effects as well as of the laws of nature[253]. Smoluchowski's definition is only eighty years younger than Herschel's, but it can be seen that in their approaches to science these are two contrasting visions, as if from different eras. The epistemic and exploratory function of science, outlined by Smoluchowski, is an opening to a range of philosophical problems but, Smoluchowski writes, the first part of this seemingly innocent statement raises serious doubts on closer reflection. Can we even know phenomena? Is there a way to explore the real world? The questions remain unanswered, since all the factual material on which we base our knowledge or concepts about the external world is made up exclusively of our sensory impressions[254].

The interpretation, explanation and description of things, facts, and events without evaluating them is the fundamental task of physics. However, learning the empirical side of the reality available to us is fragmentary knowledge of the laws binding in nature. We must realise that our goal is not to learn the essence of a thing hidden behind appearances; the task of physics is, as far as possible, to thoroughly and clearly know the world of phenomena available to us. It is about examining these phenomena as thoroughly as possible and linking them into a whole that is comprehensible to our mind[255]. Despite the differences between theories of scientific explanation, their common point is the assumption that a given empirical phenomenon is explained by other empirical phenomena that stand in a certain characteristic relationship to the phenomenon being explained.

Smoluchowski considers the basic goal of physics the most thorough investigation of these phenomena, in particular the detection of their regularities, linking them into an understandable whole and presenting them in the simplest way possible, in the form of strict functions reflecting the relations of real quantities. The author highlights that the advantage of the educated view over the naive one lies primarily not in the fact that the former is true and the latter erroneous, but in the fact that the former is definitely more accurate and more comprehensible[256].

---

[253] See: J.F.W. Herschel, *A Preliminary Discourse on the Study of Natural Philosophy*, op. cit., p. 20.
[254] See: M. Smoluchowski, *Subject, task, method and the division of physics*, op. cit., p. 162.
[255] Idem, p. 165.
[256] Ibid.

The naive view is usually the view of the majority and even the greatest scholar is guided by it in everyday life, as it is the result of direct, everyday experiences and facilitates existence. The perception of reality by the senses in a 'pure' way is deceptive and leads us astray, because—of course—our senses can be faulty and for this reason alone we can succumb to optical, thermal, acoustic or other illusions[257].

As early as infancy—Smoluchowski noted—a child becomes accustomed to associating certain sensory impressions: muscular sensations, pains, thirst, hunger and so on, with its own self, with what in time it calls "me" and which later may be divided into "my body" and "my mental world", while other sensory impressions, forming entirely separate chains of association, it involuntarily relates to a separate whole, from which in time it forms the notion of the external world. The boundary between these two types of sensory impression is very clear and even a completely uneducated, utterly naive person differentiates perfectly between what is related to their own body and what is the external world, apparently entirely separate and independent of the person. They ascribe independent existence to this external world and imagine it to be essentially the way it appears to them. They believe, for example that there exists gold, which is yellow, shiny, relatively heavy—they can see and touch it; they believe that soft green objects exist of different shapes, called leaves; they believe that luminous objects exist—stars and the Sun, it does not occur to them that they be illusions of their own senses[258].

Awareness of the fact that subjective perception is burdened by the limitations of our senses and is very often also hampered by their faulty functioning is the reason why we strive to eliminate all anthropomorphic features from our view of the world. In this context, Smoluchowski writes: "The theorem of the intelligent person that light is a transverse wave phenomenon is superior to the statement of the naive person: light is that which illuminates."[259] This statement is therefore a step, as the scholar notes, away from anthropomorphising our perception. It is the task of physics that through research science discovers the reality hidden beneath the illusion of our senses.

---

[257] Idem, p. 162.
[258] Ibid.
[259] Idem, *Self-Study Handbook*, vol. I, op. cit., p. 17.

If the perception of reality is burdened with the imperfection of our senses, and anthropomorphisation affects the laws and hypotheses created, then can we believe in them?—Smoluchowski asks. As he asserts, this is a personal matter having nothing to do with physics. As a rule, people do not like to have doubts, or make the systematic mental effort that critical thinking requires; they prefer to rely on a safe and undisputed faith in reality.[260] Everyday existence prompts us to create certain frameworks within which or mind will function.

It would be ridiculous to demand that an astronomer keep repeating in his mind: I cannot know anything about the real universe, meanwhile I stick to Copernicus's system since it is the simplest. It would be similar if the chemist kept reminding himself: I do not know whether atoms exist; all I know is that everything happens as if matter is really made up of atoms. Over time—somewhat involuntarily—a certain faith is created in these theories, which have so often proven to be infallible signposts. If we really need to believe something, it is obviously the system that is the latest fruit of science[261].

Let us beware, Smoluchowski recommends, of the stubborn conservatism usually associated with such faith. If a theory turns out to be flawed in the face of scientific progress, let us not hesitate to replace it with another and let us not becry the bankruptcy of science; science does not at all demand that we believe fully in the reality of a given view of the world[262].

This would seem to be a declaration of quite a controversial message. Smoluchowski, as a scientist-physicist, argues that science does not represent certain knowledge, based on unwaveringly stable foundations, as many other scientists have argued, but rather the opposite—it demands trust, and even faith, in currently accepted laws. However, as is the way of faith, scepticism is essential in relation to the subject of faith, enabling the creation of new hypotheses and laws.

Smoluchowski pointed out that we should not see theories of physics as the permanent contents of science but rather as an instrument of research. Let no one be offended by the audacity of new theories. Even seemingly bizarre ideas are received with enthusiasm if they point the way for new research. This does not at all mean that "uncritical fantasists" have currently

---

[260] Idem, p. 165.
[261] Idem, p. 166.
[262] Idem, pp. 165–166.

won. Let he who is not schooled in strict mathematical thinking, who is not accustomed to precision in experimental work or in logical reasoning, who does not possess a thorough awareness of the whole field of physics, stay far away from creative scientific work; physics remains, as it was, the prototype of strict natural science[263]. With a light heart, as Smoluchowski points out, sanctified traditional dogmas and hitherto unshakable principles should be destroyed if they even seem to us to be inappropriate.

3.4. The role of analogy in the methodology of physics

In the article *Several observations on physical analogies, especially in theories of electrical currents, heat currents and diffusion phenomena*, Smoluchowski addresses the issue of applying the inference method through analogy in physics. He does this through the example of theories that arose as a result of the use of analogous reasoning.

The oldest of these is the theory of heat conduction—formulated by Jean Fourier (1768–1830) in 1822. This theory was published in the famous work *Théorie analytique de la chaleur (The Analytical Theory of Heat)*. It inspired George Ohm (1789–1854) to adopt similar assumptions for electric current in 1827, and also helped Adolf Fick (1829–1901) in 1855 to develop the theory of diffusion[264]. Smoluchowski wrote: "The analogy in this case lies in the assumption that the electric current, heat current or current of matter penetrating other matter is proportional to the decrease in potential or temperature, or concentration, and is revealed in the calculations regarding the steady state"[265].

Inference by analogy is one of the basic and most primary forms of thinking that can be studied in many scientific disciplines[266]. According to Mieczysław Krąpec (1921–2008), analogy is the relational unity of what is complex. It is an appropriate ratio of proportions, similarities between what is fundamentally

---

[263] Idem, *Themes and issues in today's physics*, op. cit., pp. 221–222.
[264] See, idem, *Several observations on physical analogies, especially in theories of electric currents, heat currents and diffusion phenomena* in: *The Writings of Marian Smoluchowski*, vol. 3, op. cit., p. 237.
[265] Ibid.
[266] See: A. Biela, *Analogia w nauce (Analogy in science)*, Warsaw 1989, p. 121.

different[267]. However, Adam Biela in *Analogy in Science* states that "we understand 'analogy' as a kind of similarity. Similarity consists in a correspondence of the relations constituting compared domains, objects or phenomena."[268]

It can also be added that: "The specificity of analogous knowledge is that we embrace in it what is common to many different things, and what is unifying, for what is multiply complex, while maintaining awareness of the diversity and uniqueness of things. Analogy is a method of learning many things at the same time based on the perceived similarities between them (or in them)"[269].

It consists in exploiting the similarity occurring between certain circumstances in physics. It takes place between such phenomena in which similar or identical relationships prevail, despite the fact that the elements included in both physical processes are fundamentally different.

This parallelism of features and the correspondence between them allows for the functioning of *per analogiam* reasoning, which has many times made it possible to make momentous scientific discoveries. Reasoning by analogy was the engine of building new scientific hypotheses, creating new solutions to old problems, drawing preliminary conclusions—this is to an extent due to the heuristic function attributed to this reasoning in science. Analogies, which often occur in various fields of theoretical physics, create a rewarding and quite frequently used topic for research. For the researcher, investigating them is an excellent means of logically deepening the meanings of various concepts and supplementing theories by adjusting deductions borrowed from other fields of physics; they are also an invaluable aid in the didactic use of the subject[270].

It should be stipulated that analogy in physics is reasoning of a different type than the reasoning by analogy carried out in non-mathematical sciences. Typically, in the colloquial sense, analogy means a significant similarity of certain things, external similarity of phenomena, situations or features. This is probabilistic reasoning by analogy, found in non-mathematical sciences, where on the basis of n singular propositions it is inferred with a certain

---

[267] See: M.A. Krąpiec, *Analogia (Analogy)*, in: *Powszechna Encyklopedia Filozofii (The Universal Encyclopedia of Philosophy)*, vol. 1, Lublin 2000, p. 210.

[268] A. Biela, *Analogy in Science: From a Psychological Perspective*, Frankfurt am Main, Berlin, New York, Paris, 1991 p. 17, p. 14.

[269] See: *Analogia w filozofii (Analogy in philosophy)*, ed. A. Maryniarczyk, K. Stępień, P. Skrzydlewski, Lublin 2005, pp. 7–8.

[270] Idem, p. 237–238.

probability that the next singular proposition $n + 1$ will occur. However, this has nothing to do with the reasoning by analogy used in science or physics.

Physical analogy is defined as the similarity between the laws describing various natural phenomena, but such similarity can be established in different ways; for example, completely different phenomena can be described by the same laws, or in other words—an 'unknown' phenomenon can be treated as analogous to another 'known' phenomenon.

> *Dutch philosopher Henk de Regt in his article* Philosophy and the Kinetic Theory of Gases, *giving the above examples, alludes to the process of building analogies in the physics of Maxwell, who uses analogy, treating an 'unknown' phenomenon as an order operating in an analogical way to another 'known' phenomenon. For example, he explains a certain electromagnetic phenomenon, understood as a mechanical process, using the well-known ether model, according to which ether, being a substance, is a transmission medium essential for the propagation of electromagnetic forces.*

In another example, stemming from early work on kinetic theory, gases were treated as systems of elastic spheres. If such an analogous explanation seems consistent with empirical facts, a physical analogy is established. For Maxwell, physical analogies, at least in the latter sense, differ from hypotheses in that they cannot be interpreted realistically. In other words, according to Maxwell, physical analogies and hypotheses do not differ ontologically, but only in the epistemological sense.

Maxwell's physical analogies can have three different effects: first, they can have a heuristic value—when physical analogies are established between phenomena of two different branches of science, there may be an opportunity to create new discoveries in both branches. Secondly—analogies can provide visualisations or illustrations of phenomena, and thirdly—the establishment of analogies constitutes scientific knowledge.

> *According to Maxwell, the discovery that a given phenomenon is to some extent analogous to another phenomenon constitutes true knowledge of this phenomenon. However, the analogy should not be interpreted as real recognition of a formal relationship; the analogy between two systems of ideas leads to in-depth knowledge of both of them, unlike as a result of studying each system separately*[271].

---

[271] See: H.W. de Regt, *Philosophy and the Kinetic Theory of Gases*, 'The British Journal for the Philosophy of Science' 1996, no. pp. 35–36

In the article *Several observations on physical analogies, especially in theories of electric currents, heat currents and diffusion phenomena*, Smoluchowski sees an analogy in the common assumption that both electric current and heat current or a current of matter penetrating other matter, which occurs in the diffusion of gases, is proportional to the decrease in potential or temperature, or the decrease in concentration. These phenomena are subject to a common equation: $\Delta^2 U = 0$. The analogy of the phenomena that suggests itself to the researcher consists in referring the equation $\Delta^2 U = 0$, where U expresses the distribution of electrical potential in the environment of a sphere with a radius α, to another understanding of the same equation, where U denotes body temperature. In this way, we obtain a corresponding case from the theory of heat conductivity—when we substitute U with the concentration of the solution, we get the concentration distribution caused by diffusion[272]. This is a type of analogy in which completely different phenomena can be described by the same laws.

Smoluchowski distinguishes between the often cited examples of scientific discoveries in works using analogy, made by the method of scientific inference used in physics, and discoveries made using analogy in the colloquial sense. As examples of the latter, he cites the discoveries of James Watt (1736–1819), who, observing steam escaping from a kettle, came up with the idea of building a steam engine, or Ernest Rutherford, who created a planetary model of the structure of an atom based on the Solar System.

The way the human mind functions makes analogous reasoning natural reasoning, enabling a person to function in the reality that surrounds them. It is obvious that discoveries are made as a result of analogous associations. However, in physics there is a different type of reasoning—based on inference resulting from analogy.

Analogy and its effectiveness in creating new hypotheses and new theories constitute a specific methodological value. Science evolves in such a way that of course we naturally attribute greater importance to a simpler analogy than one that has a more limited scope of validity of common features, although

---

[272] See: M. Smoluchowski, *Several observations on physical analogies, especially in theories of electric currents, heat currents and diffusion phenomena*, in: *The Writings of Marian Smoluchowski*, vol. 3, op. cit., p. 237–238.

this one is often not completely useless—just as today, in many cases, we imagine electricity or heat in the form of a 'fluid'[273].

Anyone who naively believes in the reality of physical theories does not understand how a researcher can perceive the same object from different, even opposite points of view, for example as Hermann von Helmholtz (1821–1894) did in developing first the theory of the mechanical absorption of light, and then the electrical theory. However, this becomes completely understandable if we consider any physical theory only as an analogy. In this sense, it can be said that all analogies in physics are only formal analogies, and that the task of theoretical physics generally involves the search for such analogies, which makes it possible to encompass the widest possible range of physical phenomena[274]—such as in the relationship noticed by Smoluchowski between the phenomena of voice, light and electrical vibrations, between which he sought analogies, and which in the twentieth century became the basis of the theory of electromagnetism.

His conviction about the inspiring function of analogy is complemented by the important belief that our knowledge, consisting of hypotheses and theories, is only an analogy to the reality we know.

3.5. Convention and experience

3.5.1. The concept of simplicity of scientific theories

Scientists such as Smoluchowski, who study the fundaments of nature, cannot limit themselves to exploring only their own field. Rather than explaining only the phenomena under study, they should have the ability to relate their hypotheses and theories to a broader, philosophical understanding of reality. The theoretical underpinning of the research conducted should be the search for the foundations of the science practised and determining their assumptions. When developing a methodology for his research in physics, Smoluchowski used considerations from the fields of epistemology, the philosophy of science and the philosophy of nature.

This is not a practice used in building a research methodologies, because while the philosophy of science was and remains a central pillar of science,

---

[273] Idem., p. 241.
[274] Idem, p. 142.

the philosophy of nature did not and does not have a strong position there. Many scholars believe that the philosophy of nature does not contribute to the development of science and is doomed to systematic extinction because it is being displaced by the exact sciences. It is worth noting that this tendency has not been present since the turn of the 19th and 20th centuries, but it has appeared in modern times.

During his inaugural lecture in Jena in 1789, the German poet and philosopher Friedrich von Schiller (1759–1805), when asked what we call the philosophy of nature and why it is practised, replied that "the updated philosophy of nature is a field that uses the results of the natural sciences and the humanities. In addition, it reflects the basics of contemporary natural science"[275]. Schiller's 18th-century voice is testament to the endless dispute over the status of natural philosophy.

Since the modern period, the exact sciences have taken over subsequent fields of interest which historically fell within the spectrum of attention of natural philosophy. The development of the exact sciences, extremely dynamic, based on extensive empirical research and the mathematical calculations embedded in the natural sciences, distances them from fanciful philosophical speculations. However, the situation is not so clear-cut, because in tandem with the development of the exact sciences, new areas of knowledge arise, in which questions posed about the phenomena under study become ineffective and even meaningless, and it is impossible to construct sensible answers within the framework of their subject matter. The same questions make sense when they become the subject of natural philosophy.

The developing natural sciences imply philosophical assumptions that are often very diverse and even contradictory in the scope of their research. The forum of discussion becomes the philosophy of nature, the subject of which exceeds the scopes of research in individual sciences. Spectacular contemporary examples of such discourses are problems that can be listed endlessly, for example the issues of determinism, causality, deterministic chaos, the participation of the subject of cognition in the act of observing macro-objects or the issue of so-called quantum logics, as a special case of non-classical logic, also called philosophical logic. Within the field of relativity, these will be, for example, the issues of the structure of space-time, the nature of time, the

---

[275] See: Z. Hajduk, *The philosophy of nature*, op. cit., p. 18.

nature of space, in natural cosmology—the issues of the genesis, evolution and structure of the Universe or the significance of the so-called anthropic principle, connecting the nature of the cosmos with the fact of the presence in it of intelligent beings. In the field of the biological sciences, on the basis of the theory of evolution, the problem is being considered of the emergence and spatial spread of specimens of Homo sapiens.[276]

The scope of natural philosophy, along with the development of the exact sciences, is not only not limited, but the opposite phenomenon is experienced—the constant expansion of the subject of philosophy. The structural element of research programmes are hypotheses about the supposed image of the world, which are verbalised in qualitative language. Such hypotheses are formulated using a specific philosophy of nature, for example in classical mechanics or electrodynamics, and then belong to both science and the philosophy of science[277].

In Smoluchowski's time, the problem of the relationship between philosophy and science was as topical as it is today. In the publication *Two books from the field of the philosophy of nature*, he writes about the antagonism between naturalists and philosophers, which—according to the author—is to a great extent unnatural. Apart from the fact, however significant, that in English physics is to this day called *natural philosophy*, Smoluchowski states further, it should be noted that the area of research of these sciences is common, especially in terms of methodological issues, reflections on the basic concepts and principles of mathematics and physics which create a point of contact between these sciences and metaphysics (especially the theory of knowledge), logic and psychology. The main difference is the point of view and the way things are treated. Philosophers despise naturalists appropriating what the former consider their exclusive scope of research, accusing their opponents of superficial judgement and lack of qualifications to create precise intellectual arguments, while naturalists shun everything that smacks of philosophy, considering it synonymous with fantastic, speculative systems, a futile struggle for words, and ostentatious scholarship[278].

---

[276] Ibid.
[277] Idem, p. 20.
[278] See: M. Smoluchowski, *Dwie książki z dziedziny „filozofii przyrody"* (*Two books from the field of 'philosophy of nature'*), Lviv 1909, vol. 4, p. 291.

Smoluchowski was of the opinion that stereotypes should be broken, especially among representatives of the exact sciences, and that the antagonisms arising between science and philosophy should also be discarded. He saw links between philosophy and the mathematical and natural sciences, and especially physics. He argued that in-depth philosophical education is a cardinal condition for general education. He valued philosophical propaedeutics highly, as a subject that should be taught in every secondary school. He advocated the skilful encouragement of philosophical reflection, especially through debate in the form of a Socratic discussion of basic psychological phenomena and their relation to physiological phenomena and physical motive. It is important to pay attention to the forms of logical inference, to analyse the essence of definitions, evidence, hypotheses, theory, to be interested in the problems of the theory of knowledge. Above all, in Smoluchowski's view, it is about stimulating philosophical reflection[279].

The interaction of philosophy and the exact sciences is mutually inspiring. An example, as Smoluchowski writes, is the extremely interesting issue of searching for the simplest possible patterns in physics. Why does this happen? The compelling answer is: because the fundamental laws of nature must be simple. Some philosophers considered this argument to be a dogma testifying to the unity and simplicity of nature. A sceptic may just as rightly consider it an empty, baseless statement, especially if he does not believe in the objective reality of the laws proclaimed by physics. He may even cite the historical development of science, in which various laws, initially adopted with faith in their simplicity, had to be replaced by incomparably more complex models as observations were refined[280].

Adopting the position of Kirchhoff and Mach, we seek simplicity, because we want to describe physical phenomena as simply as possible. Not merely because only such an approach meets the needs of science. But also because only in the simplest form is the human mind able to accept and understand these laws. The scientific system is therefore useful insofar as it makes it easier for our mind to grasp the whole and strives to achieve economy of thought[281].

---

[279] See idem., *The importance of science in general education*, op. cit., p. 129.
[280] Idem, *Subject, task, method and the division of physics*, op. cit., pp. 189–190.
[281] See idem, *Self-Study Handbook*, vol. II, op. cit., p. 45.

This thought of Smoluchowski's prompts the belief that the simplicity of definitions and physical laws stems not only from nature's own tendency towards simplicity, which its philosophers have attributed to it, but from the human capacity to understand of these laws. The cognitive abilities of the human mind, enabling it to understand increasingly abstract science, have changed over time and are now undoubtedly different than in ancient times, though the tendency to write laws as simply as possible has always remained the same.

Based on historical and intellectual experience, writes Smoluchowski, three reasons have led to the premise being accepted that the laws of nature should be as simple as possible. The first, unconscious for centuries, concerns the capacity of the human mind, for which the simplicity of laws facilitated an understanding of the world. The human mind most readily assimilates laws in their simplest possible form. As knowledge developed, theories took on more complex forms, but they always strove for the greatest possible simplicity—so that the state of knowledge would enable the economy of thought to be maximised. The second reason is the influence of idealistic, especially Platonic, philosophy on the perception of nature. Through the building up of an idealistic, 'perfect' reality, belief in the superiority of perfect forms over imperfect ones was strengthened. Perfect forms were ideal constructs of philosophical thought, which could be aspired to by imperfectly ex-pressed (i.e. presented in the simplest form) laws of physics. The third reason is our perspective on the perception of reality, as in many cases the simplicity of the laws of physics is due to the fact that we see only a very small fraction of nature[282].

It should be emphasised that this perfection has always been understood in the context of anthropomorphic aesthetics. We seek simplicity, because we want to describe physical phenomena as simply as possible. Consciously or otherwise, we divide the complex whole of natural phenomena into its component parts in such a way that their laws take the simplest form possible, because this is the form that meets the needs of science. After all, the scientific system is only useful to the extent that it helps our mind grasp the whole, insofar as it satisfies our desire for the economy of thought. This would correspond to the so-called nominalist view, according to which the

---

[282] Idem, p. 45–47.

laws of nature would be a subjective product of our mind, not something objective, existing in nature[283].

Smoluchowski emphasises two arguments that recur in his work. We expect a certain simplicity in the laws of physics, because they seem more objective and are not derived from historical and intellectual experiences, but result only from our cognitive predispositions, and we see only a very small part of nature[284].

Science, in particular physics, studies nature only in a certain narrow range of perception, studies a certain fragment of it; the scientist does not have the capacity to observe nature in its entirety. The reason for this is our perception of reality, which limits us to taking in only a small fragment of nature. It is similar to the perception of part of a circle—the larger the circle's radius, the stronger the suggestion that the perceived arc is a section of a straight line[285]. The development of science has shown that on the one hand Smoluchowski underestimated his own method of developing knowledge based on creating hypotheses, and on the other was quite right, as modern theories are not simple.

Discussion on the simplicity of theories is still ongoing in physics. An interesting perspective, to some extent summarising the problem, is the formalisation of the concept of simplicity using Kolmogorov's complexity regarding the problem of optimal information coding. A theory achieves appropriate simplicity when the optimum amount of information is accumulated in the shortest possible statement. Therefore, it needs not only to be concisely written, but also contain an appropriate wealth of terms used[286].

Ancient and medieval philosophers believed that the creation of laws of physical phenomena should be guided by the idea of simplicity. This led them to put forward theses about various phenomena of nature which often led scientific knowledge astray. Smoluchowski gives as an example the fact that it seemed obvious to scientists at the time that the planets had to move in circles, because this is the simplest and most perfect curve.[287] This idea

---

[283] Idem, p. 45–46
[284] Idem, p. 46.
[285] Idem, p. 47.
[286] See: W.P. Grygiel, *Stephena Hawkinga i Rogera Penrose'a spór o rzeczywistość* (*Stephen Hawking and Roger Penrose's dispute over reality*), Kraków 2014, p. 37.
[287] See: M. Smoluchowski, *Self-Study Handbook*, vol. II, op. cit., p. 45.

endured until modern times, when Johannes Kepler (1571–1630) found out that his model based on circular orbits was wrong and introduced ellipses instead. At that time, one of his critics was Galileo, who believed that "only circular motion can naturally suit bodies that are integral parts of the universe as constituted in the best arrangement" – writes Sabine Hossenfelder in the book *Lost in Math: How Beauty Leads Physics Astray*[288].

The problem of the simplicity of a theory in science will be analysed in more detail in subsection 6.1. 'The principle of economy of thought'.

3.5.2. Poincaré's conventionalism

Smoluchowski's methodologies and epistemologies were greatly influenced by Jules Henri Poincaré's philosophy. His area of study is a critical analysis of the foundations of science and an attempt to find answers to the questions posed at that time regarding where the basic concepts of the exact sciences come from and what the value is of scientific hypotheses. The problems raised were part of the discourse of new theories of science of the late 19th and early 20th centuries, in which inviolable dogmas were challenged, at the same time creating new perspectives for development. Important discoveries were made and new theories created that started a revolution in physics and chemistry, prompting a revision of the basic concepts of mass, matter, time and space. In the quest for answers to such questions, a thorough, critical analysis of the foundations of science was needed. It was necessary to ask where the basic concepts of the exact sciences come from, to state which laws can be trusted, which should be considered hypothetical, and what value scientific hypotheses have. These and similar issues were the subject of Poincaré's work.

Smoluchowski was interested in the author himself, a brilliant scholar of mathematics, astronomy and theoretical physics. The penetrating mind of this great scientist, his mathematical and logical way of thinking and self-criticism intrigued the Polish physicist. Poincaré not only worked in the field of the exact sciences but also dealt with philosophical considerations related to these sciences. Hence the works of the eminent academic enjoyed huge popularity. This is how Poincaré was described by Marian Smoluchowski in

---

[288] S. Hossenfelder, *Lost in Math*, op. cit., p. 35.

an article reviewing two books by the French scientist: *Science and Hypothesis* and *The Value of Science*[289].

Smoluchowski's flattering opinions were not only a positive assessment of Poincaré's scientific works, but—very importantly—led the Polish researcher to adopt the French scholar's scientific and philosophical views. Smoluchowski and Poincaré share many common, and at the same time scientifically important, positions on issues of methodology, the philosophy of science, and the philosophy of nature.

Poincaré repeatedly emphasised the valuable role of empirical research and the importance of experiments, which are essential for the formulation of laws[290]. Smoluchowski also argued that confirmation of a scientific hypothesis should occur through scientific experiments, as the surest way for the law created to become a binding principle[291]. Both scientists argued that the laws introduced into science should be characterised by simplicity and convenience in their application and that the number of these laws, due to the empirical nature of science, should be as limited as possible[292]. They also took a similar position on the laws already in force in science. They believed that they should be subject to constant reappraisal, which in turn would lead to their modification. They thereby showed that they are always only a certain approximation of real principles[293].

Both Smoluchowski and Poincaré believed that the real value of science is not theories about the essence of things, but what it teaches us about the relationships between things. Relationships between things are a knowable reality, and dressed in a changing robe of transient theories, they are something permanent, which we learn more and more accurately[294]. They argued that recognised principles are only accepted conventions that can be changed

---

[289] See: M. Smoluchowski, *Two books from the field of 'philosophy of nature'*, op. cit., pp. 291–292.
[290] See: H. Poincaré, *Nauka i hypoteza (Science and Hypothesis)*, trans. M.H. Horwitz, Warsaw 1908, p. 117; I. Szumilewicz, op. cit., p. 25.
[291] See: M. Smoluchowski, *Self-Study Handbook*, vol. II, op. cit., p. 35–47.
[292] See: H. Poincaré, *Nauka i metoda (Science and Method)*, trans. M.H. Horwitz, Warsaw 1911, pp. 11–12.
[293] See: A. Lubomirski, *Henri Poincarégo filozofia geometrii (Henri Poincaré's philosophy of geometry)*, Wrocław 1974, pp. 21, 27 and 28; M. Smoluchowski, *Self-Study Handbook*, vol. II, op. cit., pp. 16–18.
[294] See: M. Smoluchowski, *Two books from the field of 'philosophy of nature'*, op. cit., p. 294.

the same way they were established, i.e. by an appropriate body of scholars. They realised that in the light of scientific discoveries, there existed a high probability of thermodynamics and mechanics being replaced in the near future by other principles.

Both scholars had a similar approach to scientific knowledge in terms of understanding it as a category of objective epistemological truth. Poincaré wrote: "Sometimes truth frightens us. And in fact we know that it is sometimes deceptive, that it is a phantom never showing itself for a moment except to ceaselessly flee, that it must be pursued further and ever further without ever being attained."[295].

Smoluchowski also did not categorise the laws of physics as objective truth, but saw them merely in terms of their degree of *utility*. Poincaré believed that most theories have an extremely short life span, but that is usually the fate of theories that aim to teach us what things are. However, there is something in these theories that remains constant; if the theory contains any element of truth, then it is often found that it can be discovered again in a new form in other theories that come in to replace the outgoing one[296].

Smoluchowski had similar thoughts, for example in relation to thermodynamics, which, within certain parameters, remained a testable theory. He believed that it creates excellent tools for researching average and macroscopic events, but where there are contradictions with kinetic theory, in the microscopic details of phenomena, in accidental deviations from the normal course of things, the superiority of kinetics is undoubtedly established[297].

Both Poincaré and Smoluchowski appreciated the importance of the hypothesis in science, which—according to both—constitutes an important element in building scientific theories; Smoluchowski, in particular, commented on this topic many times[298]. Their attitude to the *ad hoc* hypothesis was similar, which they approached with great caution, especially Poincaré. The *ad hoc* hypothesis, argued the Frenchman, formulated temporarily for phenomena against which science remains powerless, posited to save certain

---

[295] H. Poincaré, *The Value of Science*, Cosimo, New York, 2007, p. 11.
[296] See: A. Lubomirski, *Henri Poincaré's philosophy of geometry*, op. cit., p. 26.
[297] See: M. Smoluchowski, *Themes and issues in today's physics*, op. cit., p. 215.
[298] See idem, *Self-Study Handbook*, vol. I, op. cit., p. 47; H. Poincaré, *Science and Hypothesis*, op. cit., pp. 125–127.

previously adopted principles, according to some scientists does rescues the principle, but makes it useless for science.

Poincaré defined *ad hoc* hypotheses as those that have no experimentally testable consequences. The author gives a specific example of a calorimetric experiment with radium, carried out by Maria Skłodowska-Curie, the result of which was inexplicable to the science of the time and, more importantly, questioned the principle of the conservation of energy. In an attempt to save the situation and maintain the accepted principle, various more or less confused hypotheses—so-called *ad hoc* hypotheses—began to be formulated. One of these was the assumption that radium is only an intermediary that accumulates within itself a radiation alien to its nature, which permeates space in all directions and—with the exception of radium—passes through all bodies without changing their properties or acting on them in any way. Radium itself was supposed to absorb some of the energy and then return it in various other forms[299].

It is not possible now (without additional research that will not be carried out in this work) to assess to what extent Poincaré's philosophical concepts were directly adopted by Smoluchowski, and which merely confirmed his own thoughts. This is not the most important aspect of research on Smoluchowski's philosophy of science. The main principles that formed the basic framework of Poincaré's conventionalism were essentially close to Smoluchowski's.

Inspired by the dynamic development of mathematics and physics in the second half of the 19th century, Poincaré created a philosophical concept referred to as conventionalism. Intensive progress in both these fields resulted in the philosophy of nature becoming anachronistic in the understanding of many scientists, for over three centuries remaining in its concepts at the stage of a mechanistic understanding of physics and nature, which resulted in different perspectives on the perception of reality by philosophers and naturalists and to them treating things differently[300].

Earlier philosophical trends related to views on the structure and function of science, shaping relationships between the experiment and the laws

---

[299] See: H. Poincaré, *The Value of Science*, op. cit., p. 133.
[300] See: M. Smoluchowski, *Two books from the field of 'philosophy of nature'*, op. cit., p. 293.

of science and seeking criteria of truth, could essentially be reduced to two doctrines: rationalism and empiricism. However, the scientific revolution in physics exposed the weaknesses of both, especially in theories concerning the development of science.

Like Poincaré, Smoluchowski had a critical attitude towards speculative philosophy. He was impressed by the French philosopher's concept of science and wrote that in his research he showed a thorough and critical analysis of the foundations of science. He studied the origin of the basic concepts of science, analysed which rules can be trusted, what must be considered hypothetical and what the value is of scientific hypotheses[301].

The Polish physicist also appreciated Poincaré's illustration of the epistemological distance separating the various achievements of science. In his published works, he referred directly and indirectly to the scientific and philosophical achievements of the French scientist[302]. Following publication of Poincaré's books *Science and Hypothesis* and *The Value of Science* in 1904, elements of conventionalism began to play an important role in Smoluchowski's philosophy of nature, methodology and philosophy of science.

Poincaré's reflections on the relationship between theory and experiment are interesting, especially in the context of Smoluchowski's research, as they show how much the French philosopher's views influenced the Pole's concept of science. According to Poincaré, a physicist using the scientific tools available to him to experimentally test binding laws and hypotheses, never addresses an isolated law or hypothesis, but always the full set of these laws. The experiment conducted is usually so complex that it undoubtedly involves a combination of multiple conditions and laws occurring in it. Some of these laws—having been repeatedly verified and confirmed experimentally—are not questioned, and this fact justifies the need to introduce conventions into science. A law that has been sufficiently confirmed experimentally may remain in the body of other laws, and is then subjected to constant testing, which will undoubtedly end with an indication that the law is only approximate

---

[301] Idem, p. 296.
[302] Idem., pp. 291–296; see also M. Smoluchowski, *On the concept of chance and the origin of laws of physics based on probability*, op. cit., pp. 32, 35 and 49; idem., *Notes on the concept of chance in physical phenomena*, in: *The writings of Marian Smoluchowski*, vol. 3, op. cit., p. 75.

or that it can be elevated to the dignity of a principle, accepted as true on the basis of agreement[303].

Smoluchowski essentially shared this belief. In the introduction to the *Self-Study Handbook*, he wrote that the basis for testing theories and hypotheses obtained by induction and deduction is experience in the broader sense of the word, i.e. through both passive observation and experiment[304]. In his essay *Two books from the field of the 'philosophy of nature'*, he stated that experiments that falsify various dogmas, fundamental theories that had been considered inviolable for centuries, have become the basis for further hypotheses that are bolder than the previous theories and have even revised the basic concepts of mass, matter, time and space[305].

In his article Pewna interpretacja konwencjonalizmu Poincarégo (A certain interpretation of Poincaré's conventionalism) *Wiesław Wójcik addresses the described relationship between laws of physics and the experiment as follows:*

In the course of the development of science, the law of physics evolves. A sequence of laws is created: f1, f2, f3... These laws are further agreements with the experiment. An experiment can confirm the law totally, but nearly always there is a certain inconsistency between the law of physics and the experiment. This inconsistency, increasing over time, requires certain modifications—after such modification, the law is considered experimentally confirmed. A given law often remains unaltered and only the laws accompanying it are modified, which still changes its meaning[306].

Thus, a law of physics, f1, that has been formulated on the basis of certain experimental data, has experimental confirmation thanks to the group concept. Modification reconciling a given law of physics with new experiments is possible thanks to the second component of the basic structures of knowledge (i.e., thanks to the principle of a priori induction), which shows that a law's compliance with experience can be reasonably maintained. This means that a clear mind comes with an a priori ability to realise the possibility of making modifications an infinite number of times. And so a principle of physics is created—the convention. Subsequent laws of physics have emerged on the basis of convention (due both to experiment and to mathematical formalism), which have established the necessity for

---

[303] I. Szumilewicz, *Poincaré*, op. cit., p. 37.
[304] See: M. Smoluchowski, *Self-Study Handbook*, vol. I, op. cit., p. 37–38.
[305] See idem., *Two books from the field of 'philosophy of nature'*, op. cit., p. 291.
[306] W. Wójcik, *Pewna interpretacja konwencjonalizmu Poincarégo*, (*A certain interpretation of Poincaré's conventionalism*) 'Kwartalnik Filozoficzny' ('Philosophical Quarterly'), 1993, vol. 3, p. 28.

compliance of a given law with experience. The principle that the conventional decision to reconcile a given law with experience is always considered possible and reasonable is also in itself a convention[307].

Poincaré did not treat the condition of a law or theory's compliance with experiment in a dogmatic way, because he thought that the experiment is inherently imperfect, hence inaccurate, while law and theory are exact. A law relates to an infinite number of experiments, while the scientist always deals with a finite number of them.

Marian Smoluchowski's approach to the compatibility with experiment of the laws applicable in science, and particularly in physics, was more unequivocal. In his opinion, a lack of compliance constituted a stimulus for testing adopted principles and possibly correcting or amending them. Even basic conventions, such as for example the principle of energy conservation, which are key and constitute pillars of science in their deepest philosophical understanding, are subject to constant reappraisal. This is a subtle but quite significant divergence between these two great scholars in their philosophy of the approach to science.

An important contribution by Poincaré to the creation of research methodology in the exact sciences was to dispel the myth of the 'pure experiment' as well as the myth of so-called crude facts. In the book *The Value of Science*, Poincaré points out that, contrary to the views prevailing at the time, a scientist does not create facts, as Édouard Le Roy claimed, but at most creates scientific facts, though subject to the restrictions imposed on him by crude facts. According to Poincaré, a scientific fact is nothing more than a crude fact translated into more convenient language. However, a scientist cannot freely create scientific facts, because the crude fact imposes restrictions on him[308].

The above discussion seems trivial more than a hundred years on, but was of great importance at the time. During a period when research methodology was being formed and scientific language established, these issues were controversial and quite frequently discussed. Poincaré believed that although statements about facts are entangled in conventions, they are not conventions themselves as both crude and scientific facts can be attributed the value of

---

[307] See: I. Szumilewicz, *Poincaré*, op. cit., pp. 22–24.
[308] See: H. Poincaré, *The Value of Science*, op. cit., pp. 141–142.

truthfulness or falsehood, while conventions do not possess such a property. A convention is neither true nor false, but more or less convenient[309].

It should be emphasised that in choosing a convention, according to Poincaré, the so-called *convenience* criterion is used, determining which convention is best for describing the laws of science, i.e. it is the most *convenient*. It is significant that this criterion is found in Smoluchowski's philosophical concept. However, the Polish scientist went further, exchanging *convenience* for *utility*, treating this criterion as a degree of verification of the truth of accepted theories and assuming that the theory that is the most *useful* is the closest to the truth. Smoluchowski's *utility* criterion is undoubtedly more *useful* than the convenience criterion.

Smoluchowski notes that, for Poincaré, the real value of science is not theories about the essence of things (e.g. the essence of light or electricity), but what science teaches us about the relationships between things. They are the knowable reality, and dressed in a changing robe of transient theories are something permanent which we get to know more and more accurately[310]. When analysing the famous fifth axiom of Euclid in terms of non-Euclidean geometry, the French philosopher stated: "What then are we to think of the question: Is Euclidean geometry true? It has no meaning. We might as well ask if the metric system is true, and if the old weights and measures are false"[311].

Geometric theories, writes Smoluchowski, Euclidean and non-Euclidean, constitute for Poincaré adopted scientific conventions created through the acceptance of certain axioms, laws and principles. Their veracity is not unambiguously defined and any of them may be binding[312]. It is significant how in Smoluchowski's philosophy these statements translate into a general attitude towards science, especially physics and the theories applicable therein. The Polish physicist notes that Poincaré introduced a pioneering interpretation of the understanding of the subject of knowledge. The *novum* lay in the perception of the subject of knowledge as collective and group-based, i.e., relating to a specific body of scientists building the laws and principles of science as a result of the observed processes and phenomena available at a given time.

---

[309] See: I. Szumilewicz, *Poincaré*, op. cit., p. 35.
[310] See: M. Smoluchowski, *Two books from the field of 'philosophy of nature'*, op. cit., p. 294.
[311] Henri Poincaré, *Science and Hypothesis*, New York, 1905, p. 59
[312] Idem, M. Smoluchowski, *Two books from the field of 'philosophy of nature'*, op. cit., p. 293.

According to Smoluchowski, this collective subject of knowledge constantly subjects all the achievements of modern science to a process of change, which is tested in a permanent process by successive falsified hypotheses.

In an attempt to answer the question of what science is, Poincaré argued that it is essentially a classification, a way of juxtaposing facts, related to each other by a hidden similarity. The objectivity of science is expressed in the relations perceived in the permanent nodes that connect the external objects we deal with[313]. Smoluchowski expresses this idea quite similarly, arguing that the task of physics is, as far as possible, to thoroughly and clearly know the world of phenomena intelligible to us, i.e., *making the most thorough examination possible of these phenomena and combining them into a whole that is understandable to our mind*[314].

> *In response to questions raised about the value of the subject of science—whether science acquaints us with the true essence things, or whether, perhaps, it acquaints us with the true relations of things—Poincaré answers that not only does science not acquaint us with the essence of things, but there is no possibility that anything could acquaint us with it, and that even if some god knew this essence, he would not find words to express it. He argued that not only could we not guess the answer, but even if it were given to us, we would not be able to understand it, and it is even doubtful whether we understand the question itself. Therefore, attempts to include concepts such as heat, electricity or life in scientific theory are doomed to failure from the outset*[315].

Smoluchowski's position was not as radical as the French philosopher's, but he also claimed it is impossible to know the essence of things and that only subsequent theories built on the basis of hypotheses can bring us closer to the truth.

Poincaré published *The Value of Science* in 1905, and in the same year a young and unknown Einstein announced the special theory of relativity. The conventionalism promoted by Poincaré contributed to the relegation of Newtonian laws from inviolable pillars of physics to principles forming the conventions adopted by a body of scientists. In *The Value of Science* he wrote:

> Perhaps, too, we shall have to construct an entirely new mechanics that we only succeed in catching a glimpse of, where, inertia increasing with the velocity, the

---

[313] See: H. Poincaré, *The Value of Science*, op. cit., pp. 171, 172.
[314] See: M. Smoluchowski, *Self-Study Handbook*, vol. I, op. cit., p. 17.
[315] See: H. Poincaré, *The Value of Science*, op. cit., p. 172.

velocity of light would become an impassable limit. The ordinary mechanics, more simple, would remain a first approximation since it would be true for velocities not too great—so that the old dynamics would still be found under the new[316].

He was one step away from a breakthrough in physics, but it was Einstein who made this change, taking physics to whole new level. The lack of a decisive refutation of Newtonian mechanics stood in the way of a full-fledged entry into the new physics. In addition to genius, the determination of a twenty-six-year-old was also required. There is a noticeable similarity here between the situations of Smoluchowski and Poincaré. Both made groundbreaking discoveries that changed fundamental paradigms in science, and both hesitated to take the decisive step.

According to Smoluchowski, Poincaré's works were of extraordinary value, his deliberations represent a counterbalance to pseudoscientific philosophies of nature, are characterised by brilliance and an abundance of pertinent remarks, and no other philosophy stimulates such critical reflection on the issues of the philosophies of nature and science[317]. As can be seen from the above discussion, becoming acquainted with Poincaré's philosophy prompted bold thoughts and influenced Smoluchowski's philosophy.

3.6. The monistic interpretation of nature

3.6.1. Richard Avenarius, Ernst Mach and Pierre Duhem

In the second half of the 19th century, two German philosophers, Richard Avenarius and Ernst Mach, developed the so-called monistic doctrine. The Greek word monos (μόνος) means 'one', 'only'. Avenarius and Mach arrived at their ideas independently, although their messages were similar. Mach—the creator of natural monism, set out his position in his most important work, *The Analysis of Sensations*, in which the term 'neutral monism' appears for the first time, as Jacek Jarocki writes in the article *Historyczne i systematyczne ujęcie monizmu neutralnego (Neutral monism—a historical and systematic approach)*. Natural monism arose on the basis of empirio-criticism and was associated

---

[316] H. Poincaré, *The Value of Science*, Authorised translation with introduction by George Bruce Halstead, New York, 1958, p. 111.

[317] See: M. Smoluchowski, *Two books from the field of 'philosophy of nature'*, op. cit., p. 291.

with the characteristic features of this philosophy. Mach believed that as a result of the application of sharp divisions, especially the distinction between material and mental substance, the division into a phenomenon and the thing itself, into substance and accidents, subject and object, which metaphysics brings with it, the so-called bridge problem arises. Mach solved this problem by denying the existence of substances, material objects and the spiritual 'self'. As a result, both the concept of matter and spirit became ontologically redundant. The element became the new 'metaphysical' unit. Elements are neutral, neither physical nor mental, although they constitute the occurrence of phenomena that are considered physical or mental. However, whether a given element is understood as physical or mental does not depend on its nature, but solely on the configuration of other elements[318].

Mach's concept assumed that there was nothing in the Universe except for elements between which functional relations exist. The task of science is to describe these elements and the relationships between them. The description should arrange the experiments economically so that as many functional relations as possible can be distinguished with a small amount of mental effort. The laws of science are merely an economic description of facts, and the basic rule of science should be the principle of economy of thought, determining the most concise possible description of phenomena to minimise the effort needed to present the facts.

Mach saw the difference between the physical and mental realms in the fact that all things that exist directly in space for everyone constitute the physical realm, while that which is given directly to only one person is merely a conclusion drawn by analogy and constitutes the mental realm. This distinction does not exist at the start of an experience, but arises only as a result of people interacting with each other.

One of Mach's key ideas was abolition of the distinction between physical and mental phenomena. In the book, *The Analysis of Sensations*, he describes his views, which he called "natural monism": a thing, a body, matter is nothing more than a relationship between elements. There are colours, sounds, smells and tactile properties that make up the "complex known as

---

[318] See: J. Jarocki, *Historyczne i systematyczne ujęcie monizmu neutralnego (Neutral monism—a historical and systematic approach)* 'Przegląd Filozoficzny' ('Philosophical Review'), 2015, No. 3, pp. 177–181.

our body"[319]. There are also colours, sounds, smells, tactile properties that make up the complex of will, memory images, etc. This complex includes images, impressions, thoughts, in a word: what is usually called the psyche. It is opposed, as 'I', to the corporeal world. On closer inspection, the elements of all complexes turn out to belong to the same type, but nevertheless the old idea of opposition between body and soul creeps into our consciousness[320].

An interesting fact is Mach's insistence that "metaphysical" concepts such as 'atom', 'force', or 'cause', which today are scientific concepts, be removed from science and metaphysics. At the same time, Mach's monism was a thoroughly metaphysical system.

Another scholar who made a significant contribution to the construction of monism in the 19th century was Wilhelm Ostwald. He built his monistic view of the world, called energetics, on the physical concept of energy. The emanation and attribute that causes the Universe to form a homogeneous whole is posited to be energy. The concept of energy, according to the monistic concept, was to cover the entirety of things; it referred to mental life and to all psychological facts in which Ostwald saw "mental energy".

Ostwald founded the Monists Association, but it was more of a cultural movement, opposing the ruling churches and promoting the ideas of a naturalistic approach to history, rather than acting with the intention of introducing a philosophical system. According to Smoluchowski, monistic attempts to solve the ontological problem, which was the problem of understanding the structure of the composition of matter, were misguided. The monistic movement left a positive mark on scientific thought in the second half of the 19th century, drawing attention to a fairly important postulate that the problem of the structure of matter should also be perceived in a broader perspective[321].

In addition to the aforementioned natural monism, in his last work, *Erkenntnis und Irrtum* (*Knowledge and Error*), Mach introduces a new theory,

---

[319] E. Mach, *The analysis of sensations, and the relation of the physical to the psychical*, Chicago, London, 1914, p. 9.

[320] E. Mach, *Analiza wrażeń i stosunek sfery fizycznej do psychicznej* (*The analysis of sensations, and the relation of the physical to the psychical*), trans. M. Miłkowski, Warsaw 2009, p. 9–14.

[321] See: M. Smoluchowski, *Ewolucja teorii atomistycznej* (*The evolution of atomic theory*), in: *The Writings of Marian Smoluchowski*, vol. 3, op. cit., p. 18.

which he calls the monism of cosmic history. Proponents of this view assumed that the reality that exists is only one entity or its manifestation; that there exists only one thing and that the multiplicity of objects that we see is mere appearance as behind the veil of these phenomena there exists that which is absolute, and this is all there is. Empirically, a given variety of phenomena is only an attribute or modification of what is absolute. All objects and phenomena are mere emanations, attributes, degrees of development of the one real entity. The entire Universe is a homogeneous whole.

Depending on the nature of a particular form of being (matter, spirit, idea, consciousness, self), there are different varieties of monism[322].

The oldest variety is the monism of substance. This is the Eleatic doctrine that originated in Greece in the 6th and 5th centuries BC. It was given ontological form by Parmenides of Elea in the famous statement that "being is one and eternal".

Ernst Mach's monism of cosmic history represented a specific version of this doctrine, alluding to the pioneer of another monistic notion, namely Heraclitus of Ephesus, who—in contrast to the Eleatics—recognised change as a fundamental principle of the world, claiming that "All things are passing, and nothing abides"[323].

Following on from Heraclitus, Mach saw the continuous historical process of happening as an immutable monistic oneness and called it the monism of cosmic history.

> According to Mach, the elements which constitute the universe are, primarily, sensations. These, being psychic phenomena, bear the impress of mere events, of processes pure and simple, without substrate, which is the distinctive characteristic of all psychic experiences. Mach is of the opinion that even physical phenomena, if thoroughly analyzed, are not represented as persisting substances, but merely as events between which uniform relations prevail, just like the psychical. He thus arrives at a real monistic theory which is free from materialism. The unifying principle at the center of all these reflections is the concept of Becoming[324].

---

[322] See: Z. Zdybicka, *Monizm* (*Monism*), in: *The Universal Encyclopedia of Philosophy*, vol. 9, op. cit., p. 354.

[323] See: W. Jerusalem, *Introduction to Philosophy*, trans. C.F. Sanders, New York, 1915, p. 179.

[324] Idem, p. 171–172.

Meanwhile, Wilhelm Ostwald, who considered energy to be the essential being, also referred to the monism of Heraclitus, but in ontological terms. In his book *Physics and Philosophy*, the outstanding German theoretical physicist Werner Heisenberg indirectly confirms this view:

> modern physics is in some way extremely near to the doctrines of Heraclitus. If we replace the word `fire' by the word `energy' we can almost repeat his statements word for word from our modern point of view. Energy is in fact the substance from which all elementary particles, all atoms and therefore all things are made. And energy is that which moves. Energy is a substance, since its total amount does not change, and the elementary particles can actually be made from this substance as is seen in many experiments on the creation of elementary particles. Energy can be changed into motion, into heat, into light and into tension. Energy may be called the fundamental cause for all change in the world.[325]

Modern science, perversely, is prepared to accept that both scholars are right. Both Mach's monistic thought and Ostwald's, who sees the energy as the primal being, were not entirely utopian in their essence, and their thinking about the monistic nature of reality turns out to be more penetrating than originally thought. Heisenberg's reference to the philosophy of Heraclitus is an indirect admission of a certain truth to both scientists' monistic concepts. Their assumptions about the essence of nature in philosophical terms do not contradict the hypotheses put forward in modern physics.

It was important that both Mach and Ostwald were at the same time opponents of the theory of the kinetic-atomic structure of matter. In his 1914 work *On thermodynamic fluctuations and Brownian motion*, Smoluchowski wrote that ambiguities or contradictions—inherent in the very indeterminism of the basic assumptions that the kinetic theory was accused of, especially by supporters of the energy school, which professed the absolute strictness of the laws of thermodynamics—contributed to the fact that at the end of the last century atomic theory was generally considered a relic doomed to obscurity[326].

The beliefs of both scientists had an impact on the development of physics at the end of the 19th century, all the more so because they were not alone

---

[325] W. Heisenberg, *Physics and Philosophy: The Revolution in Modern Science*, London, 2000, p. 30.
[326] See: M. Smoluchowski, *On thermodynamic fluctuations and Brownian motion*, op. cit., p. 272–273.

in their atomistic views. Another prominent opponent of atomic theory should be mentioned. It was Max Planck, which from today's perspective seems quite surprising.

As early as in high school, Planck was fascinated by the first law of thermodynamics, which became a model for him of how a physical law should be—universal and absolute. In 1879, after defending his doctoral thesis, he wrote a paper on the second law of thermodynamics, which he considered the fundamental law of thermodynamics. Entropy never decreases, and the entropy of the Universe constantly increases. Boltzmann argued that there always exists a negligible probability that a system will evolve to a less probable state, lowering its entropy.

According to Jeremy Bernstein, Poincaré postulated in 1890 that each initial state in a three-dimensional system eventually returns to its original state. Such a conjecture, Poincaré wrote, may pose a problem for the conventional view of entropy. It was these 'paradoxes' that caused Planck to join the anti-atomists. He changed his mind only at the turn of the century, while conducting research on the radiation of a perfectly black body[327].

The case of Planck is confirmed by the German physicist Arnold Sommerfeld (1868–1951) in the publication *Zum Andenken an Marian von Smoluchowski* (*In memory of Marian von Smoluchowski*) of 1917, comparing Planck's approach with Smoluchowski's. Planck, according to Sommerfeld, considered the second law of thermodynamics to be a strict law of nature, even despite interpretations that refuted it. He believed deeply in thermodynamics and although his formulation of quantum theory contributed immensely to the development of the statistical approach, he was not convinced of it.

Smoluchowski's statistical approach was resolute; he believed that the second law of thermodynamics is only a generalised formula which nature breaks at the microscopic level, but which remains valid in describing macroscopic phenomena[328].

---

[327] See: J. Bernstein, *Einstein and the Existence of Atoms*, 'American Journal of Physics' 2006, vol. 74. No. 10 p. 865
[328] See: A. Sommerfeld, *Zum Andenken an Marian von Smoluchowski* (*In memory of Marian von Smoluchowski*) 'Physikalische Zeitschrift' ('Physical Journal') 1917, No. 22, pp. 533–539.

Physicists, chemists and philosophers alike—Smoluchowski said during a lecture given at teaching seminars in 1913 in Lviv—differed fundamentally in their assessment of these two dominant concepts of the structure of matter. Wilhelm Ostwald, at the time the undisputed primary personality of physical chemistry, created the 'energetics' doctrine, which conceived of energy as the only reality, considered thermodynamics to be the science of energy, and rejected the mechanical view of the world along with the concept of atoms and molecules. Duhem played a similar role in France. These authorities were followed by a multitude of other scholars and students[329].

This contrary attitude to the structure of matter significantly influenced Smoluchowski's scientific position, as he recalled later—when a strong wave of opposition rose against atomism in around 1885, it came simultaneously from several sides. From a philosophical perspective, it was opposed by various critics of knowledge, sceptics, empiricists, positivists and empirio-critics, who accused atomism of not being an exact science because it was based on the hypothesis of the existence of atoms. They argued that we are unable to directly observe these supposedly smallest indivisible particles or their movements, so the hypothesis of their existence is speculative in nature, as if their monistic theories did not have speculative characteristics. Representatives of this view—Mach, Duhem, Ostwald, Karl Pearson (1857–1936) and others—were discouraged by the errors of the vulgarised atomism of the mechanistic materialists[330]. Proponents of these concepts argued that physical phenomena should be studied through their description, preferably using mathematical equations, while to search more deeply for the hidden mechanisms of phenomena is a pointless waste of time and effort. The slogan of "science free from hypotheses" was preached, extreme criticism towards assumptions inherent in kinematics and atomics was recommended, and calls were made for sobriety of thought and avoidance of all speculation, even the most heuristically efficient[331].

Today, concerns about the use of hypothesis in science seem naive. Hypothesis is one of the key tools in the methodology of science, and scientists

---

[329] See: M. Smoluchowski, *Today's state of atomic theory*, op. cit., p. 61.
[330] S. Loria, *Marian Smoluchowski and his work (1872–1917)*, op. cit., p. 36.
[331] See: M. Smoluchowski, *The current state of atomic theory*, op. cit., p. 62; idem., *The evolution of atomic theory…*, op. cit., vol. 3, p. 18.

have developed the ability to effectively apply this tool in practice. It was different at the end of the 19th century, when putting forward hypotheses was accompanied by some tension, especially in the case of explanatory hypotheses like atomic theory. This was due to the fact that, on the one hand, there were perceived benefits of using hypotheses, such as their effectiveness and efficiency in ordering the laws of observable phenomena, but on the other, suspicions arose about their metaphysical nature and they were perceived as epistemically uncertain.

Hypotheses were considered to be undefined experimentally, both horizontally as the evidence was not of a continuous nature, and vertically because new facts were constantly emerging that needed to be explained. These doubts were resolved in two ways: the first was to deprive hypotheses of cognitive values, treating them only as a way of organising knowledge and predicting it, and the second was to look for epistemic power in them, i.e. seeking a way to overcome the status of pure hypothesis. This second approach was developed by Perrin—to positive effect, as he treated it as a strategy to eliminate epistemic doubt, combining it with the 'epistemological openness' that characterised scientists with a philosophical bent[332].

> *A different approach was put forward by Mach, Duhem, and in part also William K. Clifford (1845–1879) and Pearson, who propounded a principle that was to some extent correct: that the ultimate and only goal of physics is to learn the laws of the phenomena available to us rather than to know the eternally hidden essence of all things. How passionate they were in their opposition to atomism is illustrated by the example by Ostwald. This later winner of the Nobel prize in chemistry was in such radical opposition to atomic science that he wrote a chemistry textbook in which he never used the word 'atom'. It was so spectacular that Smoluchowski noted this fact in* The evolution of atomic theory: *"He has indeed achieved no small feat in his chemistry textbook, outlining modern chemistry without using the concept of the atom or molecule. It is true that this book gives the impression of a very artificially conducted conversation, in which certain words considered* shocking *are deliberately avoided that would simplify and facilitate the entire discussion immensely if their use were permitted"*[333].

---

[332] See: S. Psillos, *Moving molecules above the scientific horizon: on Perrin's case for realism*, 'Journal for General Philosophy of Science' 2011, vol. 42 (2), p. 341.

[333] M. Smoluchowski, *The evolution of atomic theory*, op. cit., vol. 3, p. 18.

Duhem, a French physicist and philosopher of science and mathematics, believed that the purpose of science is not to provide explanations, but merely to systematise phenomena. He argued that metaphysical knowledge is knowledge beyond experience and he included hypotheses in this type of knowledge. He advocated for the scientific method to be limited to simple induction (refraining from hypotheses). Being an extreme opponent of the atomic hypothesis, he believed that it should be avoided or used only symbolically. In 1892, expressing his opposition to chemical atomism, he proposed developing the phenomenological theory of chemical composition, for example by introducing gram equivalents in place of atomic masses. Duhem never considered the atomic hypothesis to be an established scientific theory, despite knowing Perrin's arguments from his own experiments[334].

From the perspective of currently accepted theories, at the turn of the 19th and 20th centuries, we have quite a peculiar situation—on the one hand, the advocates of classical thermodynamics, supported mainly by adherents of monistic theories, exerted pressure in the scientific community, suppressed the intensifying discussion and thus contributing to the inhibition of the development of atomic theory. Smoluchowski, as a physicist, repeatedly criticised the positions of this group of scientists, writing that the energeticist school had outlived itself. On the other hand, however, for the 21st century natural philosopher, a certain philosophical farsightedness of these monistic concepts is not to be overlooked. The conflict arising in this situation is intriguing, as both eminent figures of 19th-century science (Mach and Ostwald), in defining both monisms, were ahead of their time in their hypotheses, at the same time being unable to accept either Boltzmann's hypothesis or Smoluchowski's arguments for the kinematic and atomic structure of matter, which were concepts constituting an intermediate link in the theories they constructed.

Mach's monism of cosmic history, referring to the continuous process of historical occurrence, is undoubtedly a theory that could be subscribed to by a contemporary philosopher of nature, while Ostwald's concept, assuming that the only homogeneous entity creating the universe is energy, has been accepted by many physicists.

---

[334] See: S. Psillos, *Moving molecules above the scientific horizon*, op. cit., pp. 242–243.

However, the conduct of monistic philosophers in the context of the breakthrough moment in physics that was the replacement of the theory of thermodynamics with the kinetic-atomic theory, not only inhibited the development of science, but was even—through their aggressive actions—somewhat destructive. Certain facts should be recalled here, such as Boltzmann's depression, which led to his suicide, or Smoluchowski's hesitation to publish the results of his research on Brownian motion.

Despite the unequivocal conviction of the supporters of classical thermodynamics as to the absolute rigour and conciseness of its theses, and their argument that to date not a single phenomenon had been found that violated the second law of thermodynamics, there were physicists and philosophers of nature who doubted the absolute validity of its claims. They believed that the evidence proffered related to phenomena occurring in the macroworld, perceived in everyday contact with nature, while in phenomena on the border of the macro- and microworlds, there are—admittedly, extremely rarely—deviations from the second law of thermodynamics.

3.6.2. Smoluchowski and monism

The prevailing philosophies of nature of the day, for which thermodynamics was the foundation of assumptions, were not able to explain the basis for deviations from the second law. In 1905, the long-standing dispute between atomic theory and thermodynamics regarding the absolute interpretation of Carnot's principle was finally settled. The perception of nature according to dogmatic thermodynamics turned out to be wrong, as Smoluchowski wrote in his essay *On thermodynamic fluctuations and Brownian motion*[335].

A review of Smoluchowski's publications from the period when he worked at the Warburg laboratory in Berlin shows that they lack references to atomic theory while the assumptions of phenomenological thermodynamics are cited. It is not known whether during this period Smoluchowski, under the influence of German scientists, supporters of Mach's monism and Ostwald's concept, succumbed to the influence of these then popular philosophies, or perhaps his atomistic beliefs were as yet not sufficiently crystallised to include thoughts on this subject in publications, or—perhaps—he did not want to

---

[335] See: M. Smoluchowski, *On thermodynamic fluctuations and Brownian motion*, op. cit., pp. 348, 350 and 351.

reveal too openly views unacceptable to the German academic community. Certainly, the threat of ostracism by the scientific community made him cautious in his statements. He wrote many times about his concerns over this problem, especially after 1906, mentioning the situation in science at that time.

Armin Teske confirms these circumstances: "Anyone who looks at his Berlin publication will see that Smoluchowski does not introduce atomic assumptions at all. It gives—as we say—a purely phenomenological presentation, independent of kinetic theory, mentioning it only here and there and giving the result in both 'languages'—one using a macroscopic formula and the second, formulated in terms of kinetic theory"[336].

This temporary influence of monistic thought left traces in Smoluchowski's philosophy of nature. It was noticeable even in the period when he had already become an advocate and a forerunner of the kinetic-atomic concept. His rationalist philosophy was underpinned by the views of an undeclared materialist monism, expressed more in the sphere of a proper understanding of atomic theory than perceived as a general philosophical concept. This monism was essentially something intermediate between materialistic monism, which considered the nature of being to be material, and attributive monism, assuming the existence of many substances characterised by a common nature. Not fitting into the concept of a narrowly understood definition of natural monism, Smoluchowski's atomism could function between ontological materialistic monism and attributive monism.

According to Klemensiewicz, Smoluchowski's first thoughts regarding kinetic theory may have appeared as early as during his sojourn with William Thomson in Glasgow, i.e. before his time in Berlin. There, the Polish physicist took part in scientific discussions with his colleagues, held every day during the so-called *teas* at which everyone met after lunch. During these meetings, kinetic theory was probably also discussed, but it seems that the very realistic treatment of electrons and other newly discovered particles by Thomson's school helped Smoluchowski in the future to overcome empirio-critical objections to atomic theory[337].

---

[336] A. Teske, *Marian Smoluchowski. Life and Works*, op. cit., p. 81.
[337] See: Z. Klemensiewicz, *Marian Smoluchowski, a memoir from forty years ago*, op. cit., p. 96.

Smoluchowski only leant unambiguous support to kinetic theory in 1897, after returning to Vienna, when he again entered the circle of scientists influenced by Boltzmann's thinking. This change influenced the direction of his research, which was determined by hypotheses assuming that the second law of thermodynamics, applied to the finite macroscopic system, is not absolute, but merely statistical.

Also, the hypothesis of a continuous increase in entropy in the context of the 'heat death of the universe' was an argument in favour of the kinetic-atomic structure of matter, because the apparent irreversibility of phenomena in thermodynamics determined a continuous increase in entropy, while kinematics permitted situations in which exceptional events could occur, a return to previous states, which clearly had to inhibit the increase in entropy.

Smoluchowski argued that although the system generally strives for a normal equilibrium position and usually deviates little beyond the range of mean fluctuations, if its movements are tracked for long enough, exceptional events must also be noticed connected with a loss of entropy; sometimes abnormal deviations from the equilibrium position must automatically occur, in which the potential energy increases at the expense of the thermal value. After all, the law (general probability of fluctuation) clearly states the probability of such an event during statistical equilibrium; the most abnormal state must also sometimes occur[338]. According to Smoluchowski, as the abnormality of the initial state increases, the average length of time that will elapse before the system returns to its initial state increases extremely quickly, but such a return is possible.

Despite Smoluchowski's critical attitude to monistic concepts as a physicist, his criticism was focused on the field of physics, while from the point of view of Smoluchowski the philosopher it was not so radical and the influence can be seen of monistic philosophy on the Polish physicist's worldview.

In the history of science, clashing concepts and hypotheses have usually constituted a forum for scientific discussions, becoming a driver of progress in science. In this case, the opposite was true. By exerting pressure on the scientific community, scientists of the monistic philosophy school suppressed the intensifying discussion, contributing to the inhibition of the

---

[338] See: M. Smoluchowski, *On thermodynamic fluctuations and Brownian motion*, op. cit., p. 344.

development of physics and preventing open discussion on atomic theory. Smoluchowski was critical of the actions of this group of scientists, writing that their philosophies were outdated, though to some extent he succumbed to this pressure and was aware of it. Years later, he wrote about it many times, with an unambiguously negative connotation, but nowhere is an open complaint to be found that his greatest, essentially historic, scientific success—the discovery of the cause of Brownian motion—was in some sense taken away from him, undoubtedly with his participation. In some of his works—as if between the lines—a regret can sometimes be detected that he failed to seize a good opportunity, but nowhere did he directly express his disappointment.

3.7. Attributes of atomic theory
3.7.1. Fluctuations

In his 1914 paper *On Thermodynamic fluctuations and Brownian motion*, Marian Smoluchowski once again recalls the revolution that had taken place over the previous decade, changing the attitude of science to the atomic-kinetic theory[339].

The changes confirmed that deviations from the normal course of phenomena actually occur as predicted by kinetic theory. *Macroscopic thermodynamic parameters are insufficient to determine the state of the material system, while random fluctuations are a recognisable manifestation of microscopic molecular phenomena sinceat the molecular level all matter is in constant motion.*

In this context, Smoluchowski writes:

studies on fluctuations have even greater general importance as they are directly related to the fundamental feature of kinetic theory, which, in contrast to the thermodynamic view, emphasises a certain indeterminism of macroscopic material phenomena, (we use the word "indeterminism" to indicate that the course of a phenomenon depends on circumstances, entailing the introduction of the concepts of chance and probability to the field of physics and expressed in the external form of this theory: using a statistical method of reasoning[340].

---

[339] Idem, p. 272.
[340] Idem, p. 270.

And further:

> We use the word indeterminism to signify that the course of a phenomenon is dependent on circumstances never accessible to direct experimental control, i.e. on the coordinates and velocities of all atoms; however, we assume that they depend on these quantities in the correct way. Therefore, it is not proper indeterminism in the philosophical sense of the word"[341].

The fact that matter is in equilibrium at the macroscopic level does not imply equilibrium at the molecular level. We are dealing here with the fluctuation of component particles that fluctuate around their positions of equilibrium. It was fluctuational phenomena that resolved the dispute between the kinetic-atomic theory and thermodynamics to the latter's detriment. The changes that occurred in scientific theories in the field of accepting the atomic structure of matter were preceded by changes in the philosophy of nature and philosophy of science.

In 1904, Smoluchowski started researching gas density fluctuations. First, he analysed the issue of fluctuations in an ideal gas in a state of thermodynamic equilibrium. It was the simplest, direct way of studying the phenomenon, which did not require reference to the models of statistical mechanics[342]. *Smoluchowski's research on fluctuations influenced the acceptance of indeterminism in microworld phenomena, and consequently—the introduction of the concept of chance and probability to physics and acceptance of the statistical method. They had a significant influence on the changes that occurred in the principles of scientific research methodology. In Smoluchowski's deliberations on the use of probability calculus in fluctuation research, the influence is discernible of Poincaré's views, which the Polish scholar appropriated for his own needs and developed.*

The statistical method used to calculate fluctuations and the related probability calculus have become important methods of theoretical research in physics. Smoluchowski emphasised many times that in the case of kinetic theory, indeterminism represented the greatest hurdle in overcoming the erstwhile understanding of the relationships occurring in nature. A consequence of the dominance of thermodynamics in physics was a deterministic perception of the course of phenomena. According to popular belief, the indeterminacy of elementary phenomena should result in indeterminacy in the resulting macroscopic phenomena, which did not correspond with everyday experience.

Commenting on the dilemmas of 19th-century scientific thought, Smoluchowski argued that our mental habits and established thought patterns limit the horizons of perception and of creative thinking in the formulation of new scientific theories. In one of his essays, he argued:

---

[341] Ibid.
[342] Idem, p. 274.

Today's astronomer accepts the Copernican system as an absolutely sure foundation. It seems impossible to the physicist to doubt the truth of the undulation theory of acoustic phenomena, although the philosopher will rightly consider each theory an 'image' of natural phenomena that does not concur with reality, but merely to a certain extent corresponds to our sensory impressions, brought about by these phenomena[343].

Theories that have gained credibility among physicists in more recent times, as Smoluchowski pointed out, include the atomic-kinetic theory of matter. It gained such glowing confirmation after a period of decline, and such unexpected evidence grew out of the accusations made against it, that its position as one of the surest and most important physical theories was firmly established. Evidence was provided mainly by studies of two types: on the phenomena of the properties of rarified gases, and on thermodynamic fluctuations344. Kinematic theory, and along with it atomic theory, came to be seen as much more universal, pushing thermodynamics into the realms of theory, albeit testable, though only within specific parameters.

Thermodynamics and all the laws of physics existing in this area before kinematics, insofar as they relate to material bodies, are verified as to the validity of the rules, albeit only in the case of the average course of phenomena345. However, the most important thing, Smoluchowski claims, is the experimental verification of kinetic theory's predictions, tested in the field of the physics of rarified gases and persuasive as to the correctness of accepted views on the structure of gases. In his research on gases, Smoluchowski employed a concept of the instrumentalist, which consisted of solving problems by first formulating a hypothesis, then verifying it logically, and finally falsifying it empirically.

Fluctuation in physics is defined as random deviations of a given physical quantity from its mean value. This applies in particular to random fluctuations that cannot be predicted. The fluctuation may manifest itself in the form of minor deviations in the density value in individual places, for example, in the gas system or in relation to the entire mass quotient of the gas and its volume. Thermodynamic fluctuations result from the presence of a finite number of particles in a given physical system, as a result of which the observed instantaneous parameters of the system state may differ from the values obtained from theoretical probabilities. Maxwell's famous 1860 law of molecular velocity distribution should be recalled here. He noticed that not only is the direction of each particle's velocity random, but so is the value of this velocity. The magnitude of the velocity oscillates around a certain mean value and only this mean value is precisely determined by the temperature of the gas.

---

[343] Idem, p. 268|
[344] Idem, pp. 268–269.
[345] Idem, pp. 273–274.

According to Smoluchowski, this is the first example of phenomena included today in the general concept of fluctuation346. Moving on to particular studies, specifically to the research of suspensions of particles visible under a microscope, we notice that new ways of observing them are emerging as it is possible to directly count the particles contained in equal volumes, notes Smoluchowski. In this way, it is possible to determine the number of particles contained in the same space at regular time intervals. It is therefore possible to empirically determine deviations from the mean, i.e. the number of particles that appear in a given space347.

An experiment of this kind was first conducted by Svedberg, who used a very simple experimental method: he observed colloidal suspensions using an ultramicroscope, successively: gold, mercury, selenium, gamboge, and even fats. By inserting suitable diaphragms into the ultramicroscope viewer, Svedberg left a small field of view, thus enabling the number of particles to be determined each time.

The scientific hypotheses considered by Smoluchowski, prompted by research, led him to new philosophical reflections not suited to positivist or empirio-critical thought but oriented more towards Poincaré's pragmatism and conventionalism. Recognising the enormous importance of thermodynamic methods in their proper field, i.e., within the scope of reversible thermal phenomena and physical chemistry, it must be admitted, as Smoluchowski did, that when their application is extended to the fields of electricity and irreversible phenomena, they have not proven as useful as expected, and that on certain disputed points they have been defeated by atomic-kinetic views[348].

According to the Polish physicist, some of the intentions of atomic theory's opponents were well grounded: "Mach, Duhem, in part also Clifford, Pearson and others, proclaimed the principle, which is partially correct, that the ultimate and only goal of physics is to know the laws of phenomena available to us, not to know the eternally hidden essence of everything[349].

Metaphysics, considered an important branch of philosophy, could not be an efficient tool for the study of nature, and in this respect Smoluchowski certainly agreed with the 'phenomenalists'. In his essay *Subject, task, method and the division of physics*, he wrote: "We must clearly realise that we are not interested in knowing the essence of things hidden behind appearances; that the task of physics is, as far as possible, a thorough and clear knowledge of the world of phenomena accessible to us. It is about examining these phenomena as thoroughly as possible and combining them into a whole that is comprehensible to our mind"[350].

---

[346] Idem, pp. 269–270.
[347] Idem, pp. 270–271.
[348] Idem., *Subject, task, method and the division of physics*, op. cit., p. 203.
[349] Idem, *The current state of atomic theory*, op. cit., p. 62.
[350] Idem, *Subject, task, method and the division of physics*, op. cit., p. 163.

3.7.2. Opalescence

It is natural that both fluctuation and physical phenomena based on random deviations, on unpredictable random fluctuations of molecules, were in Smoluchowski's field of interest. They confirmed the structure of matter foreseen in atomic-kinetic theory and provided further evidence proving the validity of the theory's assumptions. A special phenomenon of fluctuation that Smoluchowski studied was opalescence. This phenomenon, discovered by John Tyndall (1820–1893), was the subject of in-depth research by the Polish physicist, which resulted in the emergence of the hypothesis of the kinetic theory of opalescence. According to this theory, in gases in a critical state, opalescence arises as a result of light scattering in an inhomogeneous medium. This is the effect of the scattering of light in a turbid medium, resulting from the reflection of the light from the suspension particles.

According to Smoluchowski, the cause of opalescence is a sharp increase in density fluctuations near the critical temperature, which can be observed, for example, in gases near their condensation state. Opalescence can also be detected in liquid mixtures, near the so-called critical solubility point. This phenomenon is explained by the optical inhomogeneity of the medium, the particles of which are small in size compared to the wavelength of light absorbed and simultaneously partially scattered sideways. It is light energy propagating laterally, as linearly polarised light, that we call opalescence.

Thermodynamics was unable to convincingly explain the inhomogeneity of the liquid and gas phases, as this was inexplicable from its point of view. Only the hypothesis of Smoluchowski's kinetic theory of opalescence explained this problem in a clear and definitive way[351].

Opalescence was one of the main methods of determining the number of particles, along with the observation of Brownian motion and of the distribution of emulsion particles under the influence of gravity. As a result of density fluctuations occurring, a ray of light that had previously passed through the transparent medium unnoticed became visible[352]. This phenomenon was possible as a result of the occurrence of minor deviations in

---

[351] S. Loria, *Marian Smoluchowski and his work (1872–1917)*, op. cit., p. 16.
[352] See: M. Smoluchowski, *Liczba i wielkość cząstek i atomów*, (*The number and size of molecules and atoms*), 'Wiadomości Matematyczne' ('Mathematical News') 1913, vol. XVII, p. 322.

the density values at individual locations of the solution. Opalescence builds up near the critical state, taking the form of critical opalescence. Opalescence phenomena are an extremely sensitive indicator of critical parameters, so the quantities studied should be determined with extreme care. However, these phenomena constituted an important element of empirical research confirming the atomic structure of matter.

The issue of gas opalescence acquired fundamental importance, because together with disputes on the essence of suspensions and colloid and crystalloid suspensions and solutions, it gave a new impetus to research in the field of colloid solutions, which aroused the interest of chemists.

The kinetic-atomic theory of the structure of matter was confirmed by research on opalescence. In predicting a range of previously unknown phenomena (heat conduction in rarified gases, transpiration of diluted gases, radiometric forces), kinetic theory drew attention to the fact that Brownian motion constitutes a visible example of molecular movement.

In his paper *On thermodynamic fluctuations and Brownian motion*, Marian Smoluchowski states that the blue sky theory of John W. Strutt, Lord Rayleigh (1842–1919), is a special case of the theory of opalescence applied to an ideal gas. The difference in Smoluchowski's interpretation of the Tyndall phenomenon was that Rayleigh considered the cause of the turbidity to be the presence of extremely fine foreign bodies, for example dust, floating in the atmosphere, while Smoluchowski attributed the granularity to clusters of gas particles resulting from random fluctuations in air density. According to this hypothesis, the sky's blueness is due to the phenomenon of light scattering on the fluctuations of the molecules that make up the atmosphere, mainly O_2 and N_2[353].

> Einstein came to a similar conclusion. Both scientists made this discovery at the same time, independently of each other, however Smoluchowski notes that the calculations conducted in Einstein's method—although they highlight the necessary assumptions regarding the probability of the system of particles, which are also needed in Rayleigh's calculus—are not fully satisfactory as they are based on the semi-empirical, semi-theoretical Lorentz formula and on Maxwell's equations, while the mechanism of radiation phenomena consists of movements of electrons and a rigorous theory should take these into account[354].

---

[353] See idem., *On thermodynamic fluctuations and Brownian movements*, op. cit., pp. 288–290.
[354] Idem, p. 290.

The sky's blue colour is created as a result of the phenomenon of scattering by particles smaller than the wavelength of the scattered light. This phenomenon occurs when light passes through solids and liquids, however, it manifests itself most effectively in gases.

The probability of light scattering is inversely proportional to the fourth power of the wavelength. The intensity of the phenomenon, which is explained by an electric dipole being induced in the molecule (under the influence of incident light), is described numerically by the Rayleigh scattering ratio. The rising sun seems yellowish to us, and the setting sun red, because its light rays pass through a thicker layer of the atmosphere. Clouds are white because the tiny droplets of water that make them up are much larger than the aforementioned atmospheric molecules, and therefore the intensity of scattered light does not depend in this case on the wavelength.

The hypothesis of Smoluchowski's kinetic theory of opalescence was a consequence of the kinetic-atomic structure of matter, at the same time being an argument for the validity of this theory. Kinematics requires the formulation of hypotheses, wrote the Polish physicist in The current state of atomic theory, because they are the most powerful stimulus for progress in science, pointing to new paths of research[355].

3.7.3. Perpetual motion

Another issue considered in the context of kinetic-atomic theory was *perpetual motion*. Smoluchowski addressed this issue in 1912, and although it was a problem that lay mostly in the domain of physics, his considerations were also philosophical.

A consequence of the second law of thermodynamics was the negation of the possibility of building a *perpetual motion machine*; a machine whose principle of operation, contrary to the law of thermodynamics, would allow it to work for an infinite period of time. Hypothetical constructions were sought for two types of machine designs, called *perpetual motion machines* of the first and second types.

A perpetual motion machine of the first type is a machine that is to work without drawing energy from outside, i.e. it is to produce more energy than

---

[355] Idem, *The current state of atomic theory*, in: *The Writings of Marian Smoluchowski*, op. cit., p. 259 (lecture given at courses for teachers in Lviv on March 12, 1913).

it consumes. Therefore, it would be a machine operating contrary to the principle of energy conservation, in contradiction to the first law of thermodynamics, which states that energy is conserved in an isolated system.

A perpetual motion machine of the second type would be a machine that draws energy from the environment, converting it into mechanical work without causing an increase in entropy. Such a machine would be for example a heat engine that takes heat from the environment and then converts it entirely into work. Such an engine would not emit heat into the environment and would be 100% efficient.

Analysing the principles of kinematics, Smoluchowski posed several important questions, which he tried to answer in sequence. Is it possible to construct a *perpetual motion machine*? Is the second law of thermodynamics devoid of value? Is a so-called *perpetual motion machine* of the second kind, a device that constantly produces work at the expense of ambient heat, feasible?

> Smoluchowski cites the names of renowned physicists and chemists, such as Lippmann, Svedberg, Ostwald and Franz Richarz (1860–1920), who, in the context of kinetic-atomic theory, and especially in relation to Brownian motion, entertained the possibility of constructing a device producing work at the expense of the environment.

Smoluchowski described one of the designed machines as follows:

> The thing seems in principle very simple. It is enough if a partition equipped with a valve is placed in a vessel filled with emulsion, allowing particles to pass from one side to the other, but not the other way round. It would have to create, by itself, a difference in concentration and its dependent osmotic pressure, which could be used as a stimulus to power the machine[356].

> However, according to Smoluchowski, the possibility of building a perpetual motion machine *is illusory. In the case of the machine discussed above, the main obstacle is the impossibility of making the appropriate latches and valves.*[357].

The classic reference to the concept of *perpetual motion* is known as Maxwell's demon. Maxwell proposed a thought experiment:

---

[356] Idem, *On thermodynamic fluctuations and Brownian motion*, op. cit., p. 349.
[357] Idem., p. 349–350.

if we conceive of a being whose faculties are so sharpened that he can follow every molecule in its course, such a being, whose attributes are still as essentially finite as our own, would be able to do what is at present impossible to us. For we have seen that the molecules in a vessel full of air at uniform temperature are moving with velocities by no means uniform, though the mean velocity of any great number of them, arbitrarily selected, almost exactly uniform. Let us now suppose that such a vessel is divided into two portions, A and B, by a division in which there is a small hole, and that a being, who can see the individual molecules, opens and closes this hole, so as to allow only the swifter molecules to pass from A to B, and only the slower ones to pass from B to A. He will thus, without expenditure of work, raise the temperature of B and lower that of A, in contradiction to the second law of thermodynamics[358].

Smoluchowski was looking for a suitable valve that would enable the functioning of a Maxwellian demon and allow the use of random deviations from the norm generated by Brownian motion, thus enabling the conversion of stored energy into useful work. He was convinced that problems of a technical nature represented the chief obstacle to its construction. "I therefore admit," he wrote at one point, "that I personally do not consider it impossible to build a so-called *perpetual motion machine* (of the second type); today there are more such physicists, who until recently would have been considered utopians and heretics"[359]. At the Congress of Naturalists in Münster in 1912, he described the devices he had made in the paper *Experimentell nachweisbare, der üblichen Thermodynamik wiedersprechende Molekularphänomene* (*Experimentally verifiable molecular phenomena that contradict conventional thermodynamics*)[360] and in the book *Gültigkeitsgrenzen des zweiten Hauptsatzes der Wärmetheorie* (*Limits of applicability of the second law of thermodynamics*) published in Leipzig and Berlin in 1914[361].

---

[358] J.C. Maxwell, *Theory of heat*, London, 1871, pp. 308–309.
[359] See: M. Smoluchowski, *The evolution of atomic theory*, op. cit., vol. 3, p. 22.
[360] Idem, *Experimentell nachweisbare, der üblichen Thermodynamik wiedersprechende Molekularphänomene* (*Experimentally verifiable molecular phenomena that contradict conventional thermodynamics*), in: *The Writings of Marian Smoluchowski*, vol. II, op. cit., pp. 226–251.
[361] Idem, *Gültigkeitsgrenzen des zweiten Hauptsatzes der Wärmetheorie* (*Limits of applicability of the second law of thermodynamics*), in: *The Writings of Marian Smoluchowski*, vol. 2, op. cit., pp. 361–398.

An extract of Smoluchowski's deliberations was included in an article by Svedberg in 1907, when they had started corresponding intensively. In the article *The Svedberg and molecular reality* Milton Kerker notes that in 1907, Svedberg speculated on the possibility of using Brownian motion to build a *perpetual motion machine* of the second type. This is a further indication of Svedberg's view on the oscillatory nature of Brownian motion. Kerker also drew attention to a statement by Smoluchowski:

> This is in fact one of the many ways in which heat can be obtained from work (...), [it is – J.G.] not as completely unfeasible as it might seem, as an example of the capture of individual molecules by Maxwell's demon. Of course, it is a basic principle of statistical mechanics that the second law of thermodynamics determines the average direction of a large number of statistical events, although on a small scale a small number of events or individual events may contradict the average trend[362].

Over time, however, Smoluchowski came to the conclusion that the obstacle lay elsewhere, in two key issues; firstly, the device personifying the demon would have to be so small and so delicately constructed that it would itself be subject to Brownian motion, and secondly, if energy were to accumulate by chance, it would happen so extremely rarely that the efficiency of the engine constructed would be zero due to the time that would be necessary for these accidental situations to occur.

Smoluchowski arrived at the conclusion that in order for the second principle of thermodynamics, which had been raising doubts for some time, to prove useful, a significant change would have to be made to its content which would allow for the possibility of the second type of *perpetual motion* under certain conditions, but at the same time its permanent functioning would have to be ruled out. In this context, Robert Piotrowski writes: "There can be no automatic device that would constantly produce useful work at the expense of heat (coming from the body) at the lowest temperature[363]. Stanisław Loria proposed a definition to organise this idea: it is impossible to construct an automatically operating mechanism that would constantly

---

[362] See: M. Kerker, *The Svedberg and Molecular Reality*, 'Isis. A Journal of the Science Society' 1976, vol. 67 (21), pp. 201–202.
[363] R. Piotrowski, *Maxwell's Demon*, op. cit., p. 87

CHAPTER 3

provide useful work at the expense of heat drawn from the coldest body in the environment[364].

Smoluchowski drew attention to another aspect of Maxwell's paradox. In the world of microphysics, the demon cannot be replaced by any mechanical or electrical device, because it is not possible to predict fluctuations taking place in different parts of the system, which occur randomly and automatically.

Maxwell's demon is an intellectual surrogate which cannot actually appear. As a side note, a scientific solution to this paradox of thermodynamics was not proposed until 1929, independently by two physicists—the Hungarian Leó Szilárd (1898–1964) and the Frenchman Léon Brillouin (1889–1969).

At the end of the book, *Demon Maxwella. Dzieje i filozofia pewnego eksperymentu* (*Maxwell's Demon: The history and philosophy of a certain experiment*), Piotrowski notes that Brillouin's approach was the opposite of Smoluchowski's. The former argued that the inevitable source of interference was the demon itself. The Polish physicist, on the other hand, believed that the interference was caused by the movement of the observed particles[365]. Maxwell's demon would expend energy in measuring the velocity of molecules and, at the same time, being part of the thermodynamic system, would disturb the equilibrium by selecting molecules. The source of energy expended by the demon's measurements would undoubtedly also be part of this system. By decreasing the entropy in the air tank, the demon would increase the entropy of its energy source and the entropy balance of the entire system would still remain positive.

The construction of a *perpetual motion machine* remains to this day an illusion to which some devote many years of fruitless labour, although the nanotechnologies that have been developing in recent years, allowing for the construction of spectacular technological structures, for example two-dimensional structures at the level of single atoms, create new possibilities for the construction of materials and devices. At the molecular level, the probability calculus is distributed differently. The number of events occurring in a unit of time is incomparably greater than the number of events in the macro-world,

---

[364] See: S. Loria, *Marian Smoluchowski and his work (1872–1917)*, op. cit., pp. 35–36.
[365] See: R. Piotrowski, *Maxwell's Demon*, op. cit., p. 159.

and even the reversibility of some phenomena is noticeable. So can the debate over *perpetual motion* be concluded? One answer to the question is a discussion by Richard Feynment on the subject of Smoluchowski's ideas, which he presented in *The Feynman Lectures on Physics*. He compared the Maxwell demon to a ratchet and pawl, neither of which can systematically convert the internal energy from a reservoir into a specific action. Feynman wrote, "If we assume that the specific heat of the demon is not infinite, it must heat up. It has but a finite number of internal gears and wheels, so it cannot get rid of the extra heat that it gets from observing the molecules. Soon it is shaking from Brownian motion so much that it cannot tell whether it is coming or going, much less whether the molecules are coming or going, so it does not work"[366].

Modern computer simulations reveal fluctuation phenomena predicted by Smoluchowski and Feynman. The Polish physicist's observations suggest that Maxwell's demon should be buried and forgotten. But this did not happen, apparently because Smoluchowski's approach left open the possibility that it would somehow be possible to achieve a state in which eternal movement stimulates a machine operated by an 'intelligent' being.

---

[366] R.P. Feynman, R.B. Leighton, M. Sands, *The Feynman Lectures on Physics. VolumeI: Mainly Mechanics, Radiation, and Heat*, Reading MA 1963, chapter 46.

Transformation of thermodynamics into the kinetic theory of matter at the turn of the 19th and 20th centuries

4.1. Reflections on physics at the end of the 19th century

During a lecture given on December 1, 1924, at the University of Munich, Max Planck recalled the views prevailing in scientific circles at the end of the 19th century. It was widely believed that following the discoveries of the 19th century, we would enter the 20th century with the awareness that exact sciences like physics and chemistry had already achieved a very high level of development and were as mature as geometry, which had been developing for centuries. Planck recalled:

> As I began my university studies I asked my venerable teacher Philipp von Jolly for advice regarding the conditions and prospects of my chosen field of study. He described physics to me as a highly developed, nearly fully matured science, that through the crowning achievement of the discovery of the principle of conservation of energy it will arguably soon take its final stable form. It may yet keep going in one corner or another, scrutinizing or putting in order a jot here and a tittle there, but the system as a whole is secured, and theoretical physics is noticeably approaching its completion to the same degree as geometry did centuries ago[367].

Pieter Zeeman (1865–1943), who went on to win the Nobel Prize in 1902, made a similar statement, saying that in 1883 he was warned not to study

---

[367] See M. Planck, *Vom Reltivizm zum Absoluten* (*From the Relative to the Absolute*), lecture given in Munich on December 1, 1924 December 1, 1924, in: M. Planck, *Wege zur Physikalischen Erkenntnis: Reden und Vorträge* (*Where is Science Going? Speeches and Lectures*), trans. J. D. Wells, Leipzig, 1933, pp. 128–146.

physics as it was not a promising field—there was no room in it for anything new of significance as it was already a practically closed field of study[368]. Marcellin Pierre Berthelot (1827–1907), an outstanding French chemist, stated in 1885: *"Le monde est aujourd'hui sans mystère"* (The world is without mysteries today), while Albert A. Michelson (1852–1931), the American physicist of Polish-Jewish origin, winner of the Nobel Prize for physics in 1907, wrote in 1894: "The more important fundamental laws and facts of physical science have all been discovered, and these are now so firmly established that the possibility of their ever being supplanted in consequence of new discoveries is exceedingly remote… future discoveries must be looked for in the sixth place of decimals"[369].

Similar statements can be found in various publications devoted to the history of 19th-century physics. They send a certain message illustrating the way reality was perceived as the century drew to a close; an idea based on the illusion of complementary science, supported by the stable foundations of physics, as classical mechanics and thermodynamics were perceived. What prevented us from seeing in physics, if not its dormant potential, then at least the artificiality of the limitations imposed on it? The problem is complex, and one answer to it was presented by Marian Smoluchowski—people, he claimed, are by nature averse to constant uncertainty; they do not like doubt or the systematic mental effort that critical thinking requires, preferring to rely on an uncontested and safe faith in reality. Daily living prompts us to create certain cognitive frameworks within which our mind must function[370]. Smoluchowski's assertion, written over a hundred years ago, now seems to have changed. Today's physicists, most probably due to quantum mechanics, do not declare such certainty in the theories they posit. However, even today, there is a noticeable attitude of resistance to radical concepts, although of course not of the same magnitude as in the 19th century. Perhaps it is a natural need to maintain the stability of science, the principles of which are still

---

[368] See H. Casimir, *Haphazard Reality: Half a Century of Science*, New York 1984, p. 27.
[369] Speech by A. A. Michelson at University of Chicago, after H. de Swart, *Philosophical and Mathematical Logic*, Cham, Switzerland, 2018, p. 497.
[370] See M. Smoluchowski, *The Writings of Marian Smoluchowski*, vol. 3, op. cit., pp. 165–166.

used, though many scientists still prefer to build science safely on the basis of a created reality, rather than perceiving the limitations of this creation.

A certain conservatism of attitudes can be seen whenever changes occur in important paradigms, whether in knowledge, religion, science or even economics. This is a resistance to theories that change important conventions—it appeared with the Copernican and Darwinian theories and even during the industrial revolution in Britain. The mechanism of resistance also accompanied changes in science at the end of the 19th century, when the revolution in physics affected thermodynamics, which had previously seemed modern and complementary, as it was in its time. In his book *Brief Answers to the Big Questions*, Stephen Hawking (1942–2018) stated: "At the beginning of the twentieth century, we understood the workings of nature on the scales of classical physics that are good down to about a hundredth of a millimeter"[371]. In kinematics, this was not enough; a change had to occur that most physicists at the end of the 19th century did not understand.

In *A Brief History of Time*, Hawking writes:

> in 1928, physicist and Nobel Prize winner Max Born told a group of visitors to Göttingen University, "Physics, as we know it, will be over in six months." His confidence was based on the recent discovery by Dirac of the equation that governed the electron. It was thought that a similar equation would govern the proton, which was the only other particle known at the time… However, the discovery of the neutron and of nuclear forces knocked that one on the head too[372].

Not even thirty years had passed and again physicists were predicting the end of science, Hawking says ironically. However, in the same paragraph, he states: "I still believe there are grounds for cautious optimism that we may now be near the end of the search for the ultimate laws of nature"[373]. As you can see, the optimism of scientists is resistant to any reflections arising from history.

At the end of the 19th century, there was a change in the scientifically widely accepted philosophy of positivism, which was a good fit in the framework of classical physics and thermodynamics. Id quod positum est ("that which is defined")—this formulation of Comte's reduced scientific thinking

---

[371] S. Hawking, *Brief Answers to the Big Questions*, New York 2018, p. 104
[372] Idem, *A Brief History of Time. From the Big Bang to Black Holes*, New York 1998, p. 172.
[373] Ibid.

to what was presented to us, what is empirically and mathematically available. Positivism was a 'positive' philosophy that eliminated metaphysical reasoning from scientific thought, operating in a different order than scientific thinking and often constituting an obstacle to building correct scientific methodologies. This trend postulated abandoning the study of the fullness of being in favour of what is available empirically. Ultimately, a narrowed perception of reality, limited to what is available to the eye, became in the 20th century a hurdle to the creation of new scientific and philosophical concepts. An example of other, similarly narrowing narratives was Mach's phenomenological theory or Marx's materialistic concept. Philosophy went from one extreme—metaphysical orthodoxy, to the other—principled positivism.

Leszek Kołakowski identified four basic features as characteristic of positivism and defining the concept of knowledge and science.

The first—the rule of phenomenalism—assumed that there was no real difference between the essence and the phenomenon; it excluded from science explanations of phenomena that refer to invisible, supernatural things, the existence of which cannot be confirmed experimentally. We have the right to record what is actually revealed by experience, while any opinions about secret entities, of whose existence experiences are supposed to be manifestations, are not credible, and disputes between positions on issues beyond the realm of experience are purely verbal in nature. Bound in such a strict framework, the rules of phenomenalism raise doubts as to whether the questions we pose are among those through which we are looking for a 'mechanism' beyond the manifestation of occult beings, or whether they are of an unacceptable metaphysical nature.

The second rule—the rule of nominalism—implies, in turn, the exclusion of the existence of objects other than those that are specific and available to experience. The objects denoted by general concepts are only tangible objects. From the point of view of nominalism, all abstract knowledge is a means of organising, a shorthand recording of the multiplicity of experiences, and has no independent cognitive function in the sense that, as an abstract, it would give us access to any realm of reality hidden from empiricism.

The third is the rule of denying any cognitive value in all forms of judgment or normative statements, which shows that we have no right to suppose that our value judgements about the human world have a scientific rationale or, in general, any rationale other than our free choice. Therefore, no experience can force us, by any logical means, to accept statements containing orders or prohibitions indicating that something should or should not be done.

The fourth rule—of methodological unity—assumes that regardless of the scientific discipline, the method of acquiring knowledge is always the same, based on experience and the general principles of processing the data derived from this experience. In an extreme situation—it was believed—a reduction should be expected of all knowledge to the physical sciences, as well as the translation of all scientific theorems into theorems concerning physical dependencies in nature and the translatability of all terms into physical terms[374].

Taking the above indications in a broader perspective, it can be said that positivism was a minimalist philosophy, sticking to facts, avoiding hypotheses and scientific speculation. In a narrower sense, it was a philosophy accepting only external facts, which—as Comte argued—were considered the only object of reliable knowledge. The consequences of this new philosophy were very interesting—starting from Galileo and Newton, a mechanistic concept of perceiving physical reality developed in the form of classical mechanics. Thermodynamics and electrodynamics also developed, which worked perfectly in the macro-world along with elements of the philosophy of positivism.

Positivism introduced a natural tendency to systematise and organise reality according to already accepted or predictable theories, based on experience and proven methodology. It was a an unsurprising tendency as this is the nature and value of science. This inclination, which also occurred in earlier periods of scientific development, constitutes a certain regularity of researchers' self-limitation in inquiries aimed at changing existing paradigms and is usually supported by the elders of the academic community. This way of thinking has been broken by outstanding individuals such as Copernicus, Georg Friedrich Bernhard Riemann (1826–1866), Maxwell, Boltzmann and Einstein, who created—as Popper put it in *Objective Knowledge*—a new theory that absorbs the old one as its approximation, at the same time correcting it, contradicting it and explaining it from a different angle[375].

---

[374] See L. Kołakowski, *Filozofia pozytywistyczna. Od Hume'a do Koła Wiedeńskiego* (*The Philosophy of Positivism. From Hume to the Vienna Circle*), Warsaw 1966, pp. 11–16.

[375] See K.R. Popper, *Objective knowledge*, op. cit., p. 29.

CHAPTER 4

A change took place in physics at the turn of the 19th and 20th centuries, as a result of which classical thermodynamics, called phenomenological, was supplanted by the concept of statistical physics, which attempted to explain laws and phenomena at the molecular level. Thermodynamics, which focuses on the study of energy changes resulting from physical and chemical actions, addressed thermodynamic phenomena at the macroscopic level, on the basis of accepted axioms and the experiments conducted. It studied the relationships between macroscopic quantities that characterised the system as a whole. It assumed that system parameters such as temperature, volume, pressure, internal energy, and mass would be sufficient to describe the phenomena, and that a change in one of these parameters would affect other values.

Classical mechanics, which describes the kinematics dealing with the movement of bodies; dynamics, which analyses the influence of interactions on the movement of bodies; and statics, which is concerned with the study of the equilibrium of material bodies based on the principles of Newtonian dynamics, constituted the foundations of classical mechanics until the end of the 19th century, explaining in practical terms the correct behaviour of most bodies, and were considered sufficiently accurate theories in principle to explain the reality around us.

Despite full concordance with other 'classical' theories, such as electrodynamics and thermodynamics, at the end of the 19th century certain contradictions came to light, for example regarding entropy, i.e., the thermodynamic function of state, or the speed of light, which was believed to be constant for all observers in a vacuum, while the research of some physicists and emerging theories indicated a greater complexity of reality than had hitherto been assumed. Further discoveries were made that successively changed science. However, for most scientists, these discoveries continued to function within the realms of classical mechanics and thermodynamics, which led them to believe they were slowly approaching the end of knowledge. It was not until the beginning of the 20th century that the scientific community understood that this was not the demise of physics, but merely the closure of a certain era in physics, which ended in a rapid acceleration of the development of this field of research and it being led in a completely new direction.

The statistical physics that was entering science analysed energy changes by examining microscopic parameters of the system. It sought their mean values and addressed the relationship of these values to macroscopic parameters describing the system as a whole. Basic quantities describing the behaviour of molecules, such as mean kinetic energy, velocity, particle mass, magnitude of momentum, and the number of molecules striking a surface or fitting within a unit of volume, make up the kinetic-molecular description. The starting point of this description is the movement of molecules and their interaction. This represents a tool for explaining the laws and phenomena studied by classical thermodynamics at the level of molecular research.

Physical reality transpired to be more complex and positivist philosophy was an obstacle to the further development of science, especially physics. The rules of phenomenalism, nominalism and methodological unity, as well as the principle of *Id quod positum est* had an inhibiting effect on physics at the end of the 19th century. The legacy left by Newton, notes Roger Penrose (b. 1931), brought the perception of reality through a mechanistic philosophy, i.e. the so-called mechanistic vision of the world. This reduced, for example, the perception of the interaction of molecules creating reality to the action of central forces. According to this assumption, any two molecules are subjected to forces of attraction or repulsion acting centrally along the lines connecting these molecules. It was assumed that all physical phenomena could be explained using Newton's three laws[376].

It is characteristic of the 1880s that many physicists claimed their science had come to an end, that it had been used up "like an old mine", but in 1900 only traditionally thinking physicists persisted in this belief and by 1914 there was a new physics that posed so many questions that it could not answer them all. It was exciting for the group of physicists who became involved in it and saw its unlimited possibilities, but probably only a few scientists noticed the change.

In order to fully understand the situation in which scientists like Boltzmann, Poincaré, Smoluchowski or Einstein found themselves, who struggled with the matter of thermodynamics in order to finally make a transformation in physics, replacing thermodynamics with kinematics and proving the atomic structure of matter, it is desirable to trace the development of thermodynamics. The timeline presented below of the discoveries made in thermodynamics is intended to show the scope and dynamics of

---

[376] A. Einstein, *5 prac, które zmieniły oblicze fizyki* (*5 works that changed the face of physics*), trans. P. Amsterdamski, Warsaw 2005, pp. 21–22.

the changes that took place in this field of physics in the 19th century, as well as to inspire questions about how science was created in the 20th century, how the understanding of the concept of science itself has changed over the last hundred years, and how the understanding of credibility of scientific knowledge has evolved and changed, as well as questions about the evolution of the criterion of truth in science.

A peculiar consequence of the turmoil accompanying the discoveries was the emergence of a singular paradox—most scientists of the late 19th century were convinced that the apex of scientific development was approaching, when in fact it was merely the end of the era of thermodynamics. For most, these same arguments testified to the end of science, while for a few they were a factor in bringing about a revolution that extended to the basic paradigms of knowledge: time, space and matter.

A comparison of two great scientists shows how perverse the assessments of the method of creating science can be. Ernst Mach, considered one of the fathers of positivism with a strongly anti-metaphysical attitude[377], argued that the role of theoretical physics comes down to the economic description of observed facts. He was opposed to the building of hypotheses and the inductive method. According to Bernstein, his stubbornness had a negative impact on the development of physics at the end of the 19th century[378]. It was similar in the case of Aristotle, creator of the common-sense method used in the natural sciences, which Smoluchowski called "an insanity inhibiting the development of physics", advocates of which were inclined towards baseless speculation devoid of experimentation[379]. As Kuhn writes, it was burdened with many limitations, causing a tendency to explain nature in terms of strivings and desires[380]. According to Smoluchowski, the indisputable position of Aristotle's philosophy held back the development of science for centuries. About 2,200 years separate the two scientists, both were powerhouses of knowledge, their concepts being supported by many other thinkers and scientists, and yet—despite their opposing views—both were accused of inhibiting scientific development. It is hard to find a reasonable justification

---

[377] See W.P. Grygiel, *Stephen Hawking and Roger Penrose's dispute over reality*, op. cit., p. 32.
[378] See J. Bernstein, *Einstein and the existence of atoms*, op. cit., p. 864.
[379] See M. Smoluchowski, *Object, task, method and the division of physics*, op. cit., pp. 177–178.
[380] See T. Kuhn, *The Copernican Revolution*, op. cit., p. 151.

for these seemingly irrational beliefs and yet it appears that the criterion of utility proposed by Smoluchowski, which will be discussed in more detail in chapter six, explains this supposed inconsistency persuasively.

The glimpse into the era of discoveries presented below is an attempt to seek answers to the questions and doubts raised, and to investigate the cause of this paradox. At the same time, it is an exciting journey through more than a hundred years of experience in acquiring subsequent resources/levels of knowledge about reality. Certain examples of changes and discoveries go beyond the scope of thermodynamics even in quite a broad sense, but have been cited nonetheless due to their importance in the history of 19th-century physics, as they had a key role in changing views on classical physics. It is a record of both the discoveries that built the position of thermodynamics in physics and those that reduced thermodynamics/this science to the macro-scale physics.

4.2. Selective calendar of discoveries in 19th century physics

To capture the beginning of the process of discovery in physics, it is necessary to go back to the 18th century, to 1783, when Antoine Lavoisier (1743–1794) discovered oxygen and explained the process of combustion in his work Réflexions sur lephlogistique (Reflections on phlogiston), criticising the theory of phlogiston and proposing the caloric theory. Lavoisier's experiments were among the first to use a quantitative method, which provided the basis for the formulation of the law of conservation of mass. In 1787, he published Méthode de nomenclature chimique (Method of Chemical Nomenclature), which became a basic textbook for chemists. Lavoisier was the first to burn a diamond under a magnifying glass, using sunlight focused through the lens to prove that the precious stone was pure carbon. He substantiated the thesis that combustion is a chemical reaction involving the combination of various substances with oxygen. In the 18th century, chemists thought that heated water turned into air, which they believed to be the only real gas. Lavoisier showed that there are three states of matter: solid, liquid, and gas, and that any element can exist in each of them.

In 1791, Pierre Prévost (1751–1839), a Swiss professor of physics, proved that all bodies radiate heat—regardless of whether they are cold or hot. In 1792, he published Sur l'équilibre du feu (On the equilibrium of heat), contributing to the explanation of the nature of heat.

In 1824, Nicolas Léonard Sadi Carnot (1796–1832) analysed the efficiency of steam engines using caloric theory, thereby giving rise to the science of thermodynamics and laying the foundations for the second law of thermodynamics. He formulated a principle later called the Carnot-Clausius principle, which enables determination of the maximum efficiency of a thermal machinebased on the temperatures of its hot source and of its cooling source. He also introduced the concept of reversibility, whereby the driving power can be used to create a temperature difference in the engine.

An important date is the year 1827, in which Robert Brown (1773–1858) discovered so-called Brownian motion (in an experiment using pollen and dye in water). This was a significant discovery that had serious consequences for physics, because 78 years later it became the basis for the formulation of the fundamental proof of kinetic-atomic theory. Brown studied microscopically small particles of material floating in liquid and performing tiny irregular movements, visible only through a powerful microscope, which he called molecular motion[381]. He was not the first to make such observations; in 1784 Jan Ingenhousz (1730–1799), a Dutch doctor, chemist and plant physiologist, was the first to describe the movements of carbon molecules on the surface of water. However, it was Brown who conducted systematic research[382] on tiny starch grains suspended in liquid. It was the Scottish botanist who showed that irregular vibrating movements are made by fine particles of organic or inorganic substances, and these particle vibrations came to be called Brownian motion after him.

In 1843, James Prescott Joule (1818–1889) experimentally determined the mechanical correlate of heat by defining the amount of work required to produce a unit of heat, known as the mechanical equivalent of heat. Using different materials, he proved that heat is a form of energy regardless of the substance heated.

In 1847, Hermann von Helmholtz (1821–1894) proved that work cannot be continuously produced from nothing and that the energy that appears lost is in reality transformed into heat energy. Continuing this idea, he formulated

---

[381] See M. Smoluchowski, *The evolution of atomic theory*, op. cit., p. 21.
[382] See J. Bernstein, *Einstein and the existence of atoms*, op. cit., p. 864.

the principle of the conservation of kinetic energy. Helmholtz was an example of a 19th-century positivist-minded scientist who believed that all knowledge was derived from the senses. Moreover, he thought that science should be reduced to laws of classical mechanics, which in his view encompassed matter, force and energy as the whole of reality. In 1854, Helmholtz introduced the concept of the heat death of the universe, which was later developed by Kelvin.

Michael Faraday (1791–1867) was a scientist widely considered the greatest experimenter of all time. In his work *On the Various Forces of Nature and Their Relations to Each Other*[383] he wrote:

> I was formerly a bookseller and binder, but am now turned philosopher, which happened thus: Whilst an apprentice, I, for amusement, learnt a little of chemistry and other parts of philosophy, and felt an eager desire to proceed in that way further. After being a journeyman for six months, under a disagreeable master, I gave up my business and, by the interest of Sir H. Davy, filled the situation of chemical assistant to the Royal Institution of Great Britain, in which office I now remain; and where I am constantly employed in observing the works on nature, and tracing the manner in which she directs the order and arrangement of the world.[384]

Fascinated by the natural world, he asked:

> "For what study is there more fitted to the mind of man than that of physical sciences? (…). And what is there more capable of giving him an insight into the actions of those laws, a knowledge of which gives interest to the most trifling

---

[383] The work *On the Various Forces of Nature and Their Relations to Each Other* was created as a record of lectures given by Faraday at the Royal Institution in the 1860s aimed at a young audience. The publisher, physicist William Crookes, declared: "The lectures were published as they were given, *verbatim et literatim*. A careful and competent rapporteur noted their content, and the manuscript that was produced after deciphering the notes was then developed by the publisher with the utmost care". Quote after: M. Litwinowicz-Droździel, *Indukcje I przepływy. Michael Faraday – mikrostudium o romantycznej nauce* (*Inductions and Flows. Michael Faraday – a Microstudy of Romantic Science*), 'Wiek XIX. Rocznik Towarzystwa Literackiego im. Adama Mickiewicza' ('The Nineteenth Century. Annual of the Adam Mickiewicz Literary Society'), 2015, vol. VIII (L), p. 99.

[384] M. Faraday, *The Philosopher's Tree. Michael Faraday's Life and Work in His Own Words*, ed. by P. Day, London 1999, p. 255, quoted after: M. Litwinowicz-Droździel, Inductions and flows, op. cit., pp. 93–94.

phenomenon of nature, and make the observing student find—tongues in trees, books in the running brooks, sermons in stones and good—in everything"[385].

In 1831, Faraday discovered the phenomenon of electromagnetic induction, followed by the phenomenon of self-induction, diamagnetism, and paramagnetism. In 1834, he discovered electrolysis and introduced the nomenclature to describe it. "I need new names," he wrote, "to express my discoveries in the science of electricity, however without involving an excessive theory which I won't know how to master"[386]. The scale of the problem is illustrated by the fact that Faraday created basic concepts such as electrolysis, the electrolyte, the cation, anion, cathode and anode.

Faraday discovered that during the electrolysis process, the mass of the substance deposited when an electric current runs through an electrolyte solution is related to the charge flowing. This relationship—called Faraday's first law—states that the mass of a substance deposited on an electrodes is directly proportional to the product of the current flow intensity, the flow time and the electrochemical coefficient. The discovery of this law contributed to an interesting remark made many years later in London in 1881 by Helmholtz, who stated: "The most startling result of Faraday's Law is perhaps this: If we accept the hypothesis that the elementary substances are composed of atoms, we cannot avoid concluding that electricity also, positive as well as negative, is divided into definite elementary portions, which behave like atoms of electricity"[387]. (Faraday Lecture (1881). In 'On the Modern Development of Faraday's Conception of Electricity', Journal of the Chemical Society 1881, 39, 290. - ej) Helmholtz's fame meant that the content of his lecture was widely known and often cited, and in German literature the value of the elementary charge became known as the Helmholtz elementary quantum[388].

---

[385] Ibid., p. 1710, quoted after: M. Litwinowicz-Droździel, *Inductions and flows*, op. cit., p. 89.

[386] Ibid., p. 1049, quoted after: M. Litwinowicz-Droździel, *Inductions and flows*, op. cit., p. 93.

[387] Faraday Lecture (1881). In *'On the Modern Development of Faraday's Conception of Electricity'*, 'Journal of the Chemical Society' 1881, 39, 290.

[388] A.K. Wróblewski, *Długie narodziny elektronu (The long birth of the electron)*, 'Wiedza i Życie' ('Knowledge and Life') 1998, No. 5, http://archiwum.wiz.pl/1998/98052300.asp (accessed: 30/04/2020).

Faraday's second law concerns the electrochemical coefficient, and states that it equals the ratio of the mass of a given substance deposited on the electrodes to the product of the ion charge and Faraday's constant.

In 1845, by making the discovery of paramagnetism and proving that diamagnetism is a universal property of matter, Faraday took a major step towards the emergence of electrodynamics. He then introduced the concept of field lines of force and advanced the theorem that electric charges act on each other through such a field and discovered the magneto-optical phenomenon. The scientist made discoveries of the laws of physics directly, through observation and experimentation, without being guided by the theories and beliefs of others: "All this is a dream. Still examine it by a few experiments. Nothing is too wonderful to be true, if it be consistent with the laws of nature; and in such things as these, experiment is the best test of such consistency"[389], as he pointed out. He was a genius of the experiment—not having a full mathematical education, he illustrated observed phenomena such as the lines of force of a magnetic field. His lack of deep mathematical knowledge did not become an obstacle for him, but more of a supporting element, because—as Maxwell claimed—Faraday had more time to do his work and express ideas in natural language.

Faraday's discoveries in the field of electrodynamics were of great importance in two ways—theoretical and practical. Firstly, Faraday's laws remain to this day of fundamental importance to the theory of electromagnetism. Secondly, electromagnetic induction was used to generate electric current, which Faraday himself demonstrated by building the first generator and the first model of an electric motor. Modern electric generators are, of course, more complex, but they are based on the same principle discovered by Faraday.

In 1846, Faraday organised a series of lectures to popularise science. An invited guest who was to give one of the evening lectures at The Royal Institution of Great Britain pulled out at the last minute, leaving Faraday with a full hall. On the spur of the moment, Faraday offered to fill in and spontaneously gave the talk *Thoughts on Ray Vibrations*. Invoking the concept of point atoms and their infinite fields of force, he suggested that the lines of electric and magnetic force associated with these atoms might, in fact, serve as the

---

[389] Laboratory journal entry #10,040 (19 March 1849); published in *The Life and Letters of Faraday Vol. II*, ed. Henry Bence Jones, London, 1870, p. 248.

medium by which light waves were propagated. Many years later, on the basis of this same concept, Maxwell developed the theory of the electromagnetic field. Faraday believed in what he called the unity of the forces of nature. He saw this unity as a manifestation of a single, universal force and believed that all the forces of nature should be mutually interchangeable.

Throughout his life, Faraday was an active member of a small Protestant church of the benign Sandemanian Christian sect. He was very devout and saw science and the exploration of nature as an extension of faith. Although he devoted much of his time to research, he always managed to find time to meet his brothers in faith to pray together. John Tyndall, who knew Faraday personally, said the scientist's Sunday meetings with other Christians gave him the strength to work hard during the week.

A characteristic feature of Faraday's straightforward personality is his attitude towards lectures, for which he prepared with extraordinary care. In 1813, in a letter to the preacher Benjamin Abbott (1793–1870), he described his thoughts on this subject. He believed that the items needed to conduct a lecture and its accompanying demonstrations (and only they) should be on a table; all had to be in order and visible (unless their invisibility and subsequent appearance were somehow justified). The speaker should be clearly visible, cannot be obscured, must be present in every sense of the word. Opacity raises suspicions—the scientist appears like a suspicious schemer and a cheat. The language of the lecture also had to be suitable: 'smooth, harmonious, simple and easy'. The experiments he conducted for public display, Faraday did not call 'experiments' but 'illustrations'. They should be characterised by minimalism— everyday items should be sufficient to conduct them (paper, a candle, water, a cup, a pencil, a piece of string; they could also be glass, a child's dummy or raisins); the scientist should approach them freely, in a natural way. Failure is acceptable as only conjurers and jugglers are always successful in their arts. 'Illustrations' are by no means intended to portray the horror or majesty of nature, nor the infallibility of the science that claims to represent it. They are intended help show that the learning process starts with being mindful of the world and observing the natural phenomena that reveal themselves constantly and spontaneously, and we need many experiments only because we are too clumsy to notice the phenomena when they manifest themselves. Faraday writes to Abbott about the most important purpose of his lectures—kindling the flame of knowledge; it should appear in the audience when they participate in a spectacle of science and they should leave the lecture hall in a state of mental exaltation[390].

In 1845, Kirchhoff put forward the so-called Kirchhoff's laws, which enable the calculation of the currents, voltages and resistances of electrical networks. In collaboration with Robert Bunsen (1811–1899), he developed chemical spectroscopy, which divides light into different wavelengths, enabling the determination of the chemical composition of objects such as the Sun and other stars. They showed that each element, when heated to a state of incandescence, emits characteristic coloured light. When refracted by a prism, this light has an individual wavelength pattern specific to each element. Using this new research apparatus, they discovered two new elements, in 1860 and 1861. Their discoveries marked the dawn of a new era, introducing a different way of searching for undiscovered elements. Kirchhoff established that when light passes through a gas, it absorbs those wavelengths that it would emit after heating. He used this principle to explain numerous dark lines (Fraunhofer lines) in the Sun's spectrum. In 1859, he proved that the energy emission from a perfectly black body is merely a function of temperature and frequency.

At the turn of 1855 and 1856, James C. Maxwell published *On Faraday's Lines of Force*, referring to Michael Faraday's experimental research.[391] Maxwell's main goal in researching electricity and magnetism was to create a mathematical framework underpinning Faraday's experimental results and his ideas on field theory. The four Maxwell equations are recognised along with Newton's laws of motion and Einstein's theory of relativity as the most fundamental contribution to physics. Of all the discoveries of the 19th century, Maxwell's work had the most profound effect on the physics of the next century. His electromagnetic theory and the field equations related to it paved the way for Einstein's theory of special relativity. It was Maxwell who initiated the idea that in the 20th century took the form of quantum theory.

In the mid-19th century, electrical and magnetic phenomena were already fully known; only a satisfactory theory encompassing the whole issue was missing. In the years 1861–1865, Maxwell formulated a theory of electromagnetism, unifying the interactions of electricity and magnetism

---

[390] See M. Litwinowicz-Droździel, *Inductions and flows*, op. cit., pp. 99–100.
[391] J.C. Maxwell, *On Faraday's Lines of Force*, 'Transactions of the Cambridge Philosophical Society' 1864, vol. X, part I, no. III, pp. 27–83.

and proving that electricity and magnetism are two aspects of the same phenomenon—electromagnetism.

The Maxwell equation introduced in 1861 showed that electrical and magnetic fields propagate in a vacuum in the form of a wave at the speed of light. In *A Dynamical Theory of the Electromagnetic Field* he wrote: "This velocity is so nearly that of light, that it seems we have strong reason to conclude that light itself (including radiant heat, and other radiations if any) is an electromagnetic disturbance in the form of waves propagated through the electromagnetic field according to electromagnetic laws"[392].

Although Maxwell relied on mechanical analogies, his work contributed to the invalidation of mechanicism. The purely mechanical image of the world was destroyed by the 'great revolution', which will always be associated with Faraday, Maxwell and Hertz. However, an essential contribution to the revolution must be attributed to Maxwell. Since his time, we have believed that physical reality is described by continuous fields. It can be concluded that this altered conception of reality is the most profound and fruitful change in physics since the days of Newton.

It was Maxwell's research on electromagnetism that assured him a place among the greatest scientists in history. In the preface to *A treatise on electricity and magnetism* (1873), *incidentally* the best exposition of this theory, Maxwell stated that his main task was to transform Faraday's physical ideas into mathematical form. In an attempt to illustrate Faraday's law of induction (a changing magnetic field gives rise to an induced electromagnetic field), Maxwell constructed a mechanical model. He claimed that the proposed model led to the emergence of the appropriate 'displacement current' in a dielectric medium that could be the seat of transverse waves. By calculating the velocity of these waves, he discovered that they are very close to the speed of light and argued that it was difficult to avoid the conclusion that light is made up of transverse waves of the same medium that gives rise to electrical and magnetic phenomena[393].

Maxwell also made great contributions to other fields of physics, for example, in 1859, with his essay *On the stability of the motion of Saturn's Rings*, he won the University of Cambridge's Adams Prize to the tune of 130 pounds.

---

[392] Idem, *A Dynamical Theory of the Electromagnetic Field*, The Royal Society 1865, p. 466.
[393] Idem, p. 497.

In it, he argued that the rings must be made up of many small particles, which he called "brick-bats", orbiting independently of each other[394]. This was a conclusion that was confirmed more than a century later by the first Voyager spacecraft.

According to Roger Penrose, Maxwell and Faraday are responsible for the third revolution in our views of physical reality. The first occurred in ancient Greece, when the geometry of Euclid and the basics of statics of rigid bodies emerged. The second was in the 17th century, and was pioneered by Galileo and Newton, who explained the movement of heavy bodies, by referring to the forces acting between the particles of which they are composed. The third, 19th-century revolution showed that the concept of particles is insufficient to describe nature; the existence of continuous fields permeating space also needs to be taken into account[395].

William Thomson, later Lord Kelvin, published works on theories of electrical and magnetic phenomena from 1842. In a dissertation published in 1845, as in several subsequent papers, Kelvin made reference to Faraday's method of analysing electrical phenomena. In 1857, he observed a change in electrical resistance occurring under the influence of a magnetic field (the magneto resistance phenomenon). While conducting research on the propagation of electrical impulses along cables, he developed a theory of electrical oscillations. He conducted preliminary studies on the relationship between magnetism and electricity, which were later developed by Maxwell. He presented the theory of oscillating currents to explain the phenomenon of electricity.

Originally, Thomson's beliefs were based on the assumption that the phenomena that as a result of action generated force, i.e. electricity, magnetism and heat, are the result of the action of an invisible material in motion, but over time he changed his mind and presented a mathematical framework that underpinned experimental results in various fields such as heat, mechanical motion, the motion of fluids (gas or liquid), electricity and magnetism. He was the first to suggest that there are mathematical analogies between the types of energy, thus playing an important role in the syntheses made at

---

[394] See https://pl.qwe.wiki/wiki/James_Clerk_Maxwell (access: 2.05.2020).
[395] See R. Penrose, *Foreword*, in: A. Einstein, *5 works that changed the face of physics*, op. cit., pp. 7–8.

that time. His success—as a scientist bringing together different forms of energy—puts him in the same position in 19th-century physics as Newton holds in the physics of the 17thcentury, and Einstein of the 20th. Thomson sought to unify the theories, though he doubted it would be possible in his lifetime or even in the future.

He was the first physicist to try to treat mathematically Faraday's concept of lines of force, which led Maxwell to the problems of the electromagnetic field. In the 1860s, Thomson used his knowledge of electricity in practical inventions, being involved in work on laying the first telegraph cable on floor of the Atlantic Ocean.

In 1848, Thomson developed an absolute temperature scale. The proposed scale was to be independent of the physical properties of any particular substance. Absolute zero was to be the lowest possible temperature, which according to him was -273.15°C. This was later dubbed in his honour 0 K (zero on the Kelvin scale).

In 1851, he formulated one of the rules of the second law of thermodynamics, stating that heat cannot be totally converted into work. He recognised that a key issue in the interpretation of this law is the explanation of irreversible processes. He introduced the concept of the internal energy of a thermodynamic system, and formulated the principle of energy dissipation. In 1851, in his mathematical treatise On the Dynamical Theory of Heat, he wrote that if entropy always increases, the universe will eventually reach a state of uniform temperature and maximum entropy, from which it would be impossible to extract any work. He described his theory as the Heat Death of the Universe, which contributed to lively discussion between physicists on the subject.

In 1847, Thomson analysed Joule's argument on the mutual exchange ability of heat and mechanical work and their mechanical equivalence and established that different forms of mechanical, electrical and heat energy are essentially the same and can be exchanged.

In 1856, he discovered the so-called Thomson phenomenon, consisting in the release of heat during the flow of electric current in a homogeneous conductor and determined the amount of heat released, which is proportional to the difference in temperature, current intensity and time of its flow, as well as to the type of conductor.

Lord Kelvin had a huge influence on the shape of science in today's world, hence it is intriguing how wrong he was in some of his predictions. In 1895 he stated that "heavier than air flying machines are impossible"[396], and in 1897 that "radio has no future"[397]. He is also credited with a statement made in a speech to the British Association for the Advancement of Science in 1900: "There is nothing new to be discovered in physics now. All that remains is more and more precise measurement"[398]. In 1901, he added that "X-rays will prove to be a hoax"[399]. Kelvin's statements are shocking; made 20 years earlier they would have been no surprise, but coming at the turn of the century they were a manifestation of intellectual conservatism. However, in the interests of complete honesty, it should be added that according to Davies and Brown, the aforementioned quote about the end of physics was wrongly attributed to Kelvin. There is no evidence that he uttered such a sentence, and the quote is a paraphrase of the words of the already cited Albert A. Michelson, who in 1894 concluded that fundamental laws and facts in physics had already been discovered and it was unlikely they would be supplemented as a result of new discoveries. Perhaps this speech by Kelvin was confused with the lecture *Two Clouds*, given to the British Association for the Advancement of Science in 1900, where he highlighted areas that later transpired to be revolutionary[400]. These were exciting and turbulent times for physics. Many scientists went down in history, while others—such as Smoluchowski—were forgotten.

In 1862, scientists discovered the so-called Joule-Thomson effect, which explains the fact that the temperature of a real gas decreases when it expands from an area of higher pressure to an area of lower pressure. The Joule-Thomson effect was used to build the refrigeration industry in the 19th century.

In 1865, Rudolf Clausius (1822–1888) introduced the modern macroscopic concept of entropy. His most famous article, entitled Über die bewegende Kraft der Wärme (On the Moving Force of Heat), was published in 1850. It concerned the laws governing the relationship between heat and mechanical work. Clausius noted that there is a clear contradiction between the concept

---

[396] G. Dryden, J. Vos, *The Learning Revolution*, Stafford, 2005, p. 232.
[397] S.J. Marshall, *Shaping the University of the Future*, Wellington 2018, p. 156.
[398] W. Isaacson, *Einstein. His life and universe*, Warsaw 2014, p. 214.
[399] See https://pl.wikiquote.org/wiki/Lord_Kelvin (access: 2.05.2020).
[400] P. Davies, J. Brown, *Superstrings: A Theory of Everything?*, Cambridge 1988, p. 575.

of energy conservation advocated by Joule, Helmholtz and Thomson, and the principle presented by Carnot in an article published in 1824. Clausius restated the two laws of thermodynamics to overcome this contradiction. The first states that there is a constant relationship between the work performed and the heat thus generated, or vice versa—by absorbing heat, we can produce work. The second is connected with the observation that whenever heat is converted into work, it is always accompanied by a certain amount of heat flowing from a warmer to a cooler body. This article caught the attention of the scientific community and is often credited with placing the field of thermodynamics on a firm footing. In 1864, Clausius published a collection of his articles, and a year later introduced the word 'entropy' into the lexicon of physicists. Entropy is purely mathematical in nature and is typically found to increase in any process involving heat exchange. Clausius chose the word 'entropy', derived from the Greek en and tropein, meaning 'content transformative' or 'transformation content'.

In 1879, Slovenian physicist Josef Stefan discovered a law according to which the total radiation from a black body is proportional to the fourth power of its temperature, and in 1884 a student of his—the Austrian physicist *Ludwig* Boltzmann—also theoretically deduced this fact from thermodynamics, taking into account the second law of thermodynamics and the kinetic theory of gases. This theory was called the Stefan-Boltzmann law. Stefan was one of the first physicists to fully understand Maxwell's electromagnetic field theory and one of the few people outside of England to propagate it. According to Boltzmann, apart from Stefan, only Helmholtz realised of the importance of Maxwell's theory. Other physicists were distrustful of it and even rejected it, mainly due to the emergence of incomplete and underdeveloped vector analysis. Boltzmann published a series of articles in the 1870s in which he showed that the second law of thermodynamics, which concerns the exchange of energy, can be explained by the application of the laws of mechanics and probability theory to the movements of atoms. In 1872, he derived a theorem for the temporal development of the distribution function in phase space.

Boltzmann developed a general law of energy distribution in different parts of a system at a certain temperature and the theorem on equipartition of energy; the so-called Maxwell-Boltzmann distribution law. This law states that the average amount of energy involved in each direction of an atom's motion is the same. Boltzmann also derived an equation for the change in energy distribution between atoms resulting from atomic collisions and laid the foundations for statistical mechanics. In doing so, he explained that the second law is essentially statistical and that the system approaches a state of thermodynamic equilibrium (uniform distribution of energy throughout the system), since equilibrium is the most probable state of a material system.

Boltzmann's greatest achievement was the development of statistical mechanics, which explains and predicts how the properties of atoms (such as mass, charge, and structure) determine the visible properties of matter (such as viscosity, thermal conductivity, and diffusion). His work on statistical mechanics was severely attacked and long misunderstood, but his conclusions were ultimately supported by discoveries in atomic physics. It was recognised that fluctuation phenomena, such as for instance Brownian motion (the random movement of microscopic particles suspended in a solvent), can only be explained by statistical mechanics. The randomness of the behaviour of atomic research objects, changing positions in an uncontrolled way, can only be described by probability. Boltzmann's entropy is proportional to the logarithm of the (average) number of states available to the system under given external conditions, denoted by W. The inverse of W can be thought of as the probability of occupying each available state. Using the example of an ideal gas, Boltzmann showed that for an isolated system, entropy defined in this way increases, reaching the maximum value asymptotically.

He thus challenged the fundamentals of thermodynamics, proposing acceptance of the fact that not everything can be defined and calculated with hundred-percent certainty, and that some results can only be given approximately, with a certain probability.

The idea that matter and all complex things—water, fire, life—are subject to entropy and probability caused major changes in the world of physics, but not without great resistance among scientists, including Ostwald and Mach, Boltzmann's two greatest opponents.

The equation $S = k \times log\, W$ became a symbol of the end of the era of thermodynamics and the culmination of Boltzmann's scientific achievements. It was inscribed on the tombstone of the Austrian physicist, a pioneer of statistical mechanics, and—as Hans Thirring put it during the unveiling of the monument—it will remain valid after all the tombstones have crumbled, turning to dust with the passage of centuries. This formula shows the dependence of entropy on the probability of possible thermodynamic states of matter and contains the κ constant, later called the Boltzmann constant.

In 1889, Boltzmann used the second law of thermodynamics to calculate the temperature dependence of an ideal substance that emits and absorbs all frequencies. An object absorbing light of all colours is black and became known as a black body. The Stefan-Boltzmann law, developed in 1879, describes the total heat power emitted by a perfectly black body at a given temperature.

Quantum theory in its current form was introduced in 1900[401] by Max Planck, who is known as its creator, but Boltzmann and Planck are referred to as the father and mother of the quantum. According to Arnold Sommerfeld, "for the atomistically oriented mind of Ludwig Boltzmann, quantum theory would have been the true playground"[402]. Planck proposed quantum theory using Boltzmann's statistical theory, but it was Boltzmann who first used the concept of energy quantisation in his work as early as 1872, twenty-eight years before Planck's publications, dividing the system's energy into very small discrete packets. Boltzmann invented this quantisation as a kind of mathematical trick to allow the use of combinatorial equations in probability calculations. Energy quanta did not appear again in the final equations, but there is no doubt that it was Boltzmann, through his approach, who helped pave the way for quantum theory[403].

Non-equilibrium thermodynamics or the thermodynamics of irreversible processes, which is also of great importance in philosophy, concerns a thermodynamic process causing an increase in the sum of entropy of the system and its surroundings. The name suggests that it is impossible to reverse

---

[401] In the years 1897–1899, Max Planck presented his theses relating to quantum theory several times. In 1899, in *Sitzungsberichte der Königlich Preussischen Akademie der Wissenschaftenzu Berlin. Mitteilung* (*Proceedings of the Royal Prussian Academy of Sciences in Berlin. Communication*) (No. 5, p. 440) he gave the size of the Plank constant. Historically, it has been assumed that the date of introduction of quantum theory is December 14, 1900, i.e. the date of publication of the paper *Über irreversibleStrahlungsvorgänge* (*On Irreversible Radiation Processes*), 'Annalen der Physik' ('Annals of Physics'), 1900, vol. 306, No I. 1, pp. 69–122.

[402] *Debates of the Czechoslovak Academy of Sciences: Social Sciences Series*, vol. 80, Prague 1970, p. 29.

[403] See A. Eftekhari, *Ludwig Boltzmann (1844–1906)*, https://pdfs.semanticscholar.org/5c96/924ab515da7ebb6cb7601ec916099b03aed0.pdf (access: 28/03/2020), p. 21.

an irreversible process, however, due to the statistical nature of thermodynamic phenomena, the reverse process is possible, though the probability of its occurrence is close to zero. The logical consequence of adopting the kinetic theory of matter was acceptance of Boltzmann's hypothesis about the "reversibility of phenomena". As a consequence, ice in a glass of water, hypothetically, according to the Boltzmann hypothesis and kinetic theory, can cool by itself and at the same time heat the water in the glass, but this phenomenon is so unlikely that it is overlooked in practice[404].

Boltzmann had the greatest influence on the study of nonequilibrium states, irreversibility and irreversible processes, by combining the kinetic theory of gases with thermodynamics. He proposed an equation known worldwide as the Boltzmann equation. It is characteristic that he built models based on the behaviour of atoms, even though the existence of atoms had not been proven at the time. The credibility of Boltzmann's statistical mechanics was confirmed by the discoveries of atomic physics. It was concluded that the phenomenon of Brownian motion can be explained only by statistical mechanics. In 1905–1906, Einstein and Smoluchowski published papers confirming the validity of the thesis of the atomic structure of matter.

After the publication of Boltzmann's work *Vorlesungenüber Gastheorie* (*Lectures on Gas Theory*) in 1898, the following report appeared in a German scientific journal: "Kinetic theory, as we know, is as wrong as various mechanical theories of gravity and especially misunderstands the principle of energy conservation; however, if someone really wants to get acquainted with it, let him pick up Boltzmann's work"[405]. Participating in discussions with Boltzmann at the University of Vienna, Mach cut off the talks with the statement: "I don't believe that atoms exist!" In subsequent academic debates, Mach repeatedly took the floor, asking Boltzmann: *Eineshaben Sie gesehen?* ("Have you seen one?")[406]. The extremely critical opinions of the physics community exacerbated Boltzmann's already less than ideal state of mind[407], which showed depressive tendencies, and ultimately, most probably, were a contributing factor in his suicide.

---

[404] Idem, p. 20.
[405] M. Smoluchowski, *The current state of atomic theory*, op. cit., p. 61.
[406] See J. Bernstein, *Einstein and the existence of atoms*, op. cit., p. 864.
[407] Boltzmann died by suicide. His death was most likely caused by depression resulting from undiagnosed bipolar disorder.

CHAPTER 4

In 1893, Wilhelm Wien (1864–1928) formulated a displacement law for the perfect black body radiation spectrum. In 1894, he published a paper on temperature and the entropy of radiation, in which the terms temperature and entropy were extended to radiation in empty space. He went on to define an ideal body, called the black body, which completely absorbs all radiations. In 1896, he published a formula to determine the composition of black body radiation. A little later, it was proven that this formula is only valid for short waves, but Wien's work enabled Max Planck to solve the problem of radiation in thermal equilibrium by means of quantum physics. For this work, Wien was awarded the Nobel Prize in Physics in 1911.

Among the researchers of cathode rays, English engineer Cromwell Fleetwood Varley (1828–1883) was probably the first to formulate the hypothesis in 1871 that these are small charged particles of matter ejected from the cathode. The direction of their deviation in a magnetic field indicated their negative charge. Among German physicists, there were no supporters of this concept, with the exception of Emil Johann Wiechert (1861–1928), who considered cathode rays to be charged particles. Wiechert argued that cathode rays are particles with a mass 2,000–4,000 times smaller than a hydrogen atom and cannot be the atoms known from chemistry. Believing them to be something like electrical bodies independent of atoms, ae was the first to clearly formulate a hypothesis on the existence of subatomic particles and gave the limits of their mass[408].

The basic conclusion resulting from the equations formulated in 1864 by Maxwell (determining the relationship between the quantities characterising the electromagnetic field in any medium and the sources of this field, i.e. electric charges and currents) was the existence of electromagnetic waves. In 1886, Heinrich Hertz discovered them and in 1887–1888 proved that electricity travels in waves at a frequency that can be calculated. Maxwell's theory, which had aroused much scepticism in scientific circles, was confirmed by this research. In appropriately designed experiments, conducted in the lecture hall of the Technical University of Karlsruhe in 1886, the German scientist produced the first waves (called Hertz waves) using an electrical oscillator (called a Hertz oscillator). He was convinced that these waves would have no practical application, believing they were irreversibly dispersed into space.

---

[408] A.K. Wróblewski, *The long birth of the electron*, op. cit.

The experiments he conducted were the starting point that led to the invention of wireless telegraphy. In 1894, Guglielmo Marconi (1874–1937) built an apparatus for the wireless transmission and reception of electromagnetic waves. The transmitter included an improved Hertz oscillator, a Morse telegraph key and a receiver. In 1896, Marconi sent a radio telegraphic signal over a distance of 3 km and patented a radio transceiver apparatus, the so-called wireless telegraph. Hertz used a rapidly oscillating electric spark to create ultra-high frequency waves. He proved that the waves generated in this way caused similar electrical oscillations in a distant wire loop. He also showed that light waves and electromagnetic waves are identical, thus paving the way for the development of radio, television and radar.

In 1892, Dutch physicist Hendrik Lorentz published *Die elektromagnetische Theorie von Maxwell und ihre Anwendung auf bewegte Körper*[409] (*Maxwell's electromagnetic theory and its application to bodies in motion*), in which he put forward the idea of introducing a discrete structure of electricity into Maxwell's equations. He assumed the existence of ether as an unchanging dielectric, devoid of internal motion and not subject to mechanical forces, and a substance composed solely of elementary particles of negative or positive electricity. According to Lorentz, a charged body has excess charges of some kind. The electromagnetic field observed by macroscopic instruments is the result of the statistical superposition of elementary fields generated by separate charged particles. Lorentz's microscopic field laws, when averaged, give a macroscopic field described by Maxwell's laws. When, in the autumn of 1896, a student of Lorentz, Dutch physicist Pieter Zeeman, discovered the phenomenon of the splitting of spectral lines in a magnetic field, Lorentz could immediately give a theoretical explanation for this phenomenon within the framework of his 'electron' theory. In 1892 Lorentz used the phrase 'charged particles', and in 1895 he changed this name to 'ions', and only from 1899 did he begin to refer to them as 'electrons'.

However, Lorentz's electron theory was not effective in explaining the negative results of the Michelson-Morley experiment, an effort to measure

---

[409] H.A. Lorentz, *Die elektromagnetische Theorie von Maxwell und ihre Anwendung auf bewegte Körper*, (*Maxwell's electromagnetic theory and its application to bodies in motion*) 'Archives néerlandaises des sciences exactes et naturelles' ('Dutch Archives of the Exact and Natural Sciences') 1982, no. 25, pp. 363–551.

Earth's velocity relative to a hypothetical ether by comparing the speed of light from different directions. In an attempt to overcome this hurdle, in 1895 Lorentz introduced the concept of local time (different times in different locations) and concluded that moving bodies approaching the speed of lightcontract in the direction of motion. He found that if, instead of Galileo's transformations describing the transition to a moving frame of reference, others are used (dubbed 'Lorentz transformations' by Einstein in his honour), Maxwell's equations regarding the propagation of light remain unchanged. Lorentz transformations cause the equations of mechanics to change, which seemed absurd at the time, but Einstein proposed such a modification of the equations of mechanics that the Lorentz transformations did not change their form, so it can be concluded that to some extent Lorentz was a forerunner of this theory.

On November 8, 1895, Wilhelm Conrad Röntgen (1845–1923) discovered a new type of radiation: X-rays, which were named Röntgen rays or Roentgen rays after their discoverer. After confirming his discoveries, in 1896, Röntgen published an article On a New Kind of Rays. The mysterious radiation had the ability to pass through many materials that absorbed visible light. He wrote:

> A kind of relationship between the new rays and light rays appears to exit; at least the formation of shadows, fluorescence, and the production of chemical action point in this direction. Now it has been known for a long time that, besides the transverse vibrations which account for the phenomena of light, it is possible that longitudinal vibrations should exist in the ether, and according to the view of some physicists must exist. It is granted that their existence has not yet been made clear, and their properties are not experimentally demonstrated. Should not the new rays be ascribed to longitudinal waves in the ether? I must confess that I have in the course of this research made myself more and more familiar with this thought, and venture to put the opinion forward, while I am quite conscious that the hypothesis advanced still requires a more solid foundation[410].

This discovery and its almost immediate application to all kinds of medical imaging earned Röntgen an honorary medical degree. In 1896, he won the Rumford Medal, awarded by the Royal Society in London. In 1901, when the

---

[410] W.C. Röntgen, *On a New Kind of Rays*, Science 3, no. 59 (1896): 227–31. p. 231

first Nobel Prize in physics was awarded, it was Röntgen who received it. The scientist refused to take out any patents, so that the world could freely use his work, and he never earned a penny from it. He wrote: "I believe that—in line with a good scientific tradition—inventions and discoveries should benefit society at large and are not to be reserved for individuals through patents, licences and such"[411]. At the time of his death, after the inflation that followed World War I, he was almost bankrupt.

Joseph John Thomson (1856–1940) presented the results of research on cathode rays during a speech at the Royal Institution in London on April 30, 1897. This was the next step in the discovery of the electron, but the scientist stubbornly stuck to the name 'corpuscle' for many years[412]. He assumed that an atom is a ball of matter with a positive charge, in which electrons are suspended. Until the components of atoms were discovered in 1911, they were thought to consist of subatomic particles, called protons and electrons. However, it was unclear how these protons and electrons were arranged inside the atom. Thomson chose the 'plum pudding' model, in which electrons and protons are evenly distributed. He wrote in the book *Recollections and Reflections*:

> At first there were very few who believed in the existence of these bodies smaller than atoms. I was even told long afterwards by a distinguished physicist who had been present at my lecture at the Royal Institution that he thought I had been 'pulling their legs.' I was not surprised at this, as I had myself come to this explanation of my experiments with great reluctance, and it was only after I was convinced that the experiment left no escape from it that I published my belief in the existence of bodies smaller than atoms[413].

The most important success of Thomson's work was the conclusion in 1897 that all matter, regardless of its source, contains particles of the same kind, which are much less massive than atoms, and constitute their parts.

---

[411] See https://www.uu.nl/en/organisation/wilhelm-conrad-rontgen (accessed: 8/04/2020).

[412] Irish physicist and astronomer George Johnstone Stoney (1826–1911), who introduced the term although not the word as early as 1874, initially using the term electrine, is considered the first researcher to introduce the term 'electron' to physics as the https://pl.wikipedia.org/wiki/1826basic unit of the amount of electrical energy.

[413] J.J. Thomson, *Recollections and Reflections*, London 1936, p. 341.

These were electrons, which he initially called corpuscles. This discovery was the result of an attempt to resolve long-standing controversies about the nature of cathode rays, which arise when an electric current passes through a vessel from which most of the air or other gas has been pumped out.

Between February and May 1896, Antoine Henri Becquerel (1852–1908) discovered radioactivity. Soon after, the scientist demonstrated that the source of the effect is the uranium contained in a sulphate crystal. The phenomenon called radioactivity once again shook the scientific world. Researchers set themselves the goal of discovering what this mysterious radioactivity was and what its features were. Polish physicist Wiktor Biernacki (1869–1918) wrote about it as follows:

> Becquerel discovered that uranium salts emit invisible rays that pass through aluminium and black paper opaque to light, as well as dissipating electrical charges. Just as phosphorescent bodies glow for some time (more than 12 hours) after being briefly illuminated, so the uranium salts also emit these invisible rays, but for much longer—for months. Some physicists call this phenomenon hyperphosphorescence. Pure uranium acts similarly, even more powerfully. Many other bodies have been found to emit similar invisible rays, including zinc sulphide, calcium sulphide, etc. All these bodies phosphoresce visibly; but although they cease to glow visibly, for a very long time they continue to emit these invisible rays, which have properties similar to those of Röntgen rays[414].

Joseph J. Thomson, an English physicist associated with the Cavendish Laboratory at the University of Cambridge, was awarded the Nobel Prize in Physics in 1906 for his theoretical and experimental research on the electrical conductivity of gases, which led to the discovery of the electron. During the Nobel lecture he said:

> In this lecture I wish to give an account of some investigations which have led to the conclusion that the carriers of negative electricity are bodies, which I have called corpuscles, having a mass very much smaller than that of the atom of any known element... (...) The arguments in favour of the rays being negatively charged particles are primarily that they are deflected by a magnet in just the same way as moving, negatively electrified particles. We know that such particles, when a magnet is placed near them, are acted upon by a force whose direction as at right

---

[414] W. Biernacki, *Nowe dziedziny widma* (*New spectra domains*) Warszawa 1898, p. 78.

angles to the magnetic force, and also at right angles to the direction in which the particles are moving.[415].

A significant contribution of Thomson to science was his teaching. Seven of his assistants won the Nobel Prize in Physics, as did his own son. One of Thomson's best-known students was Ernest Rutherford, who replaced him at the Cavendish Laboratory.

Between June and December 1896, Pierre Curie (1859–1906) and Maria Skłodowska-Curie discovered two radioactive elements: polonium and radium. The substances identified two years after Becquerel's discovery turned out to be a much stronger source of radiation than uranium. At a meeting of the French Academy of Sciences on April 12, 1898, in Paris, Skłodowska-Curie presented the results of her work:

> I have studied the conductance of air under the influence of the uranium rays discovered by M. Becquerel, and I examined whether substances other than compounds of uranium were able to make the air a conductor of electricity. In this research I employed a parallel-plate condenser; one of the plates was covered with a uniform layer of uranium or of another finely pulverized sample. One establishes a potential difference of 100 volts between the plates. The absolute value of the current which traversed the condenser was measured by means of an electrometer and a piezoelectric quartz. I examined a large number of metals, salts, oxides, and minerals (…).
>
> All the uranium compounds studied are active, and are, in general, more active to the extent that they contain more uranium. The compounds of thorium are very active. Thorium oxide surpasses even metallic uranium in activity. It is remarkable that the two most active elements, uranium and thorium, are the ones which possess the greatest atomic weight… Two minerals of uranium, pitchblende (a uranium oxide) and chalcolite (uranyl copper phosphate) are much more active than uranium itself. This fact is most remarkable, and suggests that these minerals may contain an element much more active than uranium (…). To interpret the spontaneous radiation of uranium and thorium, one could imagine that all space is constantly traversed by rays analogous to Röntgen rays but much more penetrating and unable to be absorbed except by certain elements with high atomic weight such as uranium and thorium[416].

---

[415] Thompson's Nobel Lecture of December 11, 1906, in: *Nobel Lectures: Physics, 1901–1921*, Amsterdam 1967, pp. 145–153.

[416] Rays Emitted by Compounds of Uranium and of Thorium, Marie Sklodowska Curie, presented by M. Lippmann, Comptes Rendus 126, 1101-3 (1898), trans. Carmen Giunta. https://www.ias.ac.in/article/fulltext/reso/006/03/0094-0096.

Physicists studying uranium, radium and polonium observed that the radiation is not of a uniform nature. They distinguished three types of radiation: alpha, beta and gamma. In making this discovery, they created the concept of radioactivity. In her Nobel Lecture, Skłodowska-Curie said:

> now, only 15 years after Becquerel's discovery, we are face to face with a whole world of new phenomena belonging to a field which, despite its close connexion with the fields of physics and chemistry, is particularly well-defined. In this field the importance of radium from the viewpoint of general theories has been decisive. The history of the discovery and the isolation of this substance has furnished proof of my hypothesis that radioactivity is an atomic property of matter and can provide a means of seeking new elements. This hypothesis has led to present-day theories of radioactivity, according to which we can predict with certainty the existence of about 30 new elements which we cannot generally either isolate or characterise by chemical methods. We also assume that these elements undergo atomic transformations, and the most direct proof in favour of this theory is provided by the experimental fact of the formation of the chemically defined element helium starting from the chemically-defined element radium[417].

During the Nobel Prize award ceremony in 1911, the chairman addressed Maria Skłodowska-Curie:

> Madam. In 1903, the Swedish Academy of Sciences had the honour of conferring upon you the Nobel Prize for Physics for the part which you, together with your late husband, took in the momentous discovery of spontaneous radioactivity.
> This year, the Academy has decided to award you the prize for Chemistry in recognition of the eminent services you have rendered to this science by your discovery of radium and polonium, by your description of the characteristics of radium and its isolation in the metallic state, and by your research into the compounds of this remarkable element.
> During the eleven years in which Nobel Prizes have been awarded, this is the first time that the distinction has been conferred upon a previous prizewinner. I beg you, Madam, to see in this circumstance a proof of the importance which our Academy attaches to your most recent discoveries, and I invite you, Madam, to receive the prize from His Majesty the King, who has graciously consented to present it to you.[418]

---

[417] Skłodowska-Curie Nobel Lecture of 11 December 1911, https://www.nobelprize.org/prizes/chemistry/1911/marie-curie/lecture/ (access: 9.04.2020).

[418] Speech by Erik Wilhelm Dahlgren, Chief Librarian of the National Library and President of the Royal Swedish Academy of Sciences, 10 December 1911, https://www.nobelprize.org/prizes/chemistry/1911/ceremony-speech/ (accessed: 9/04/2020).

On December 14, 1900, Max Planck published a paper, *Über irreversible Strahlungsvorgänge* (*On Irreversible Radiation Processes*)[419], which changed the face of physics. He put forward the revolutionary idea that the energy emitted by a resonator can only take discrete values, or quanta. The energy for a resonator of frequency v is hv, where h is a universal constant, now called Planck's constant. Planck's new thesis saw the light of day following afternoon tea at the Planck household on October 7, 1900, during which Heinrich Rubens told Planck that the results of measurements showed a deviation from the predictions of Wien's second law, defining the radiation distribution of a perfectly black body. This led Planck to the thought of improving Wien's formula and consequently led to the formulation of a new theory of black body radiation. He presented the results of his deliberations on October 19, 1900, at a meeting of the German Physical Society in Berlin. This was a turning point in the history of physics. The improvements introduced by Planck formed the foundations of quantum mechanics. Planck's work on quantum theory was published in ‚*Annalen der Physik*', ('Annals of Physics') and the scientist summarised the achieved results in two books: *Thermodynamik*[420] (*Thermodynamics*) and *Theorie der Wärmestrahlung*[421] (*The Theory of heat Radiation*). This was not only Planck's most important work, but also a turning point in the history of physics. The significance of this discovery, with its far-reaching impact on classical physics, was not at first appreciated. However, proof of its validity gradually became overwhelming as the application of this theory explained many discrepancies between observed phenomena and classical theory.

Surprising in this context is the issue of Planck's attitude towards atomic theory, of which he was an opponent (as mentioned earlier). He changed his mind only at the turn of the century while conducting research into black body radiation[422].

---

[419] M. Planck, *On Irreversible Radiation Processes*, op. cit., pp. 69–122.
[420] Idem, *Thermodynamics*, Leipzig 1897.
[421] Idem, *The theory of heat radiation*, Leipzig 1906.
[422] See J. Bernstein, *Einstein and the existence of atoms*, op. cit., p. 864.

Planck had to give up one of his most cherished convictions, that the second law of thermodynamics was an absolute law of nature. In place of that, he was forced to accept Boltzmann's interpretation, that it is only a statistical law. This is surely why Helge Kragh, a contemporary historian of the philosophy of science, quite aptly called Planck a "reluctant revolutionary"[423].

Planck decided to become a theoretical physicist at a time when theoretical physics was not yet recognised as a scientific discipline. He concluded that the existence of physical laws presupposes that "the outside world is something independent from man, something absolute, and the quest for the laws which apply to this absolute appeared (…) as the most sublime scientific pursuit in life"[424].

Between March 18 and December 19, 1905, Albert Einstein wrote six papers that altered the nature of physics. They concerned the theory of relativity, the photoelectric effect, and the theory of Brownian motion. According to Roger Penrose, by publishing the first five works, Einstein started the fourth revolution in physics, manifesting in the way of perceiving nature (there was discussion earlier of the three previous revolutions). John J. Stachel (b. 1928) a physicist and director of the Boston University Center for Einstein Studies, believes that these five papers caused the year 1905 to be known in the general discourse of physics as the 'miraculous year' analogous to 1666, known as the *annus mirabilis*. It is widely held that in 1666 Newton wrote a series of works that created the foundations of physical and mathematical theories that revolutionised 17th-century science. More thorough research has proven that this concerned the period from 1665 to 1667 as well as the year 1668, in which he achieved basic results in mathematics and physics. Twenty-one years later, Newton wrote, "I was in the prime of my age for invention and minded Mathematics and Philosophy more then at any time since"[425]. Similarly, Einstein's works published in 1905 laid the foundations for a revolution in 20th-century science[426].

---

[423] H. Kragh, *Max Planck: The Reluctant Revolutionary*, 'Physics World' 2000, vol. 13, NO. 12, pp. 31–36.

[424] Max Planck *A Scientific Autobiography* (1948), in 'Scientific Autobiography and Other Papers', trans. Frank Gaynor (1950), London, p. 13.

[425] See I.B. Cohen, *Introduction to Newton's 'Principia'*, Cambridge 1971, p. 291.

[426] J. Stachel, *Introduction*, in: A. Einstein, *5 works that changed the face of physics*, op. cit., p. 15.

On April 30, 1905, Einstein completed his doctoral dissertation *Eine neue Bestimmung der Moleküldimensionen* (On a new determination of molecular dimensions). As he himself recalled, one of the reviewers of his work, the mathematician Heinrich Burkhardt (1861–1914), sent his dissertation back to him with the comment that it was too short. Einstein was unconcerned by this remark, added one sentence and submitted it again on July 20 at the University of Zurich. Einstein's doctoral thesis was accepted without any further comments. It went on to become one of his most cited works, although it does not concern the theory of relativity.

May 11, 1905, saw the publication of an article by Einstein, *Über die von der molekularkinetischen Theorie der Wärmegeforderte Bewegung von in ruhenden Flussigkeitensuspendierten Teilchen*[427] (*On the Motion of Suspended Particles Postulated by the Molecular-Kinetic Theory of Heat*), which explained Brownian motion. It contained novel physical concepts arguing for the atomic structure of matter. Many years later, John Stachel, the publisher of a multi-volume collection of Einstein's writings, claimed that the research he conducted showed the visible influence on the above-mentioned Einstein paper of another paper: *On irregularities in the distribution of gas molecules and their effect on entropy and the equation of state*, written in 1903 by Marian Smoluchowski. Einstein had reviewed a book in which that paper was included. Smoluchowski's work contained suggestions related to research and conclusions on Brownian motion which the Polish physicist had already fully developed in 1903 and with which Einstein undoubtedly became familiar. Unfortunately, Smoluchowski did not publish his work until 1906, as a result of which all the glory for the discovery of the phenomenon went to Einstein (this issue will be discussed in chapter seven).

---

[427] A. Einstein, *On the Motion of Suspended Particles Postulated by the Molecular-Kinetic Theory of Heat*, op. cit., pp. 549–560.

On June 30, 1905, Einstein's best-known work was published: *Zur Elektrodynamik bewegter Körper*[428] (*On the Electrodynamics of Moving Bodies*), simultaneously giving great importance to a physical constant—the speed of light, which initiated the era of relativistic physics. In it, Einstein formulated the theory of special relativity.

September 27, 1905, saw the publication of the fourth and shortest paper *Ist die Trägheit eines Körpers von seinem Energieinhalt abhängig? (Does the Inertia of a Body Depend Upon its Energy Content?)*[429]. It contained the derivation of the most famous formula in the world: $E=mc^2$, although it did not appear in the article in this form as Einstein only performed the proof leading to the equation.

On March 18, 1905, Einsten published *Über die heuristische Sichtweise bezüglich Emission und Transformation von Licht*[430] (*On a Heuristic Point of View Concerning the Production and Transformation of Light*). He believed that only this publication had truly revolutionary value. He showed in it that light can be treated not only as a wave but also as a collection of particles—quanta of energy. This paper is currently cited as the work that contained the explanation of measurement results of the photoelectric phenomenon, but its content is much broader and concerns in large measure dark body radiation, including analyses of radiation entropy. Einstein thus became one of the founders of quantum physics.

On December 19, 1905, Einstein finished and sent to the editorial office of *Annalen der Physik* a second paper as a complement to his earlier deliberations on Brownian motion—*Zur Theorie der Brownschen Bewegung*[431] (*On the theory of Brownian motion*).

On July 20, 1906, Marian Smoluchowski published a paper on Brownian motion in *Annalen der Physik* under the title *Zur kinetischenTheorie der*

---

[428] Idem, *On the Electrodynamics of Moving Bodies*, 'Annalen der Physik' ('Annals of Physics') 1905, No. 17, pp. 891–921.

[429] Idem, *Does the Inertia of a Body Depend Upon its Energy Content?*, 'Annalen der Physik' ('Annals of Physics') 1905, No. 18, pp. 639–641.

[430] Idem, *On a Heuristic Point of View Concerning the Production and Transformation of Light*, 'Annalen der Physik' ('Annals of Physics') 1905, No. 17, pp. 132–148.

[431] Idem, *On the theory of Brownian motion*, op. cit., pp. 371–381.

Brownschen, Molekularbewegung und der Suspensionen[432] (*An outline of the kinetic theory of Brownian motion and suspensions*). This paper, together with Einstein's earlier publications, constitutes a breakthrough in the perception of the atomic hypothesis proclaimed by some physicists and a resolution to the dispute ongoing at the time on the structure of matter. Previously, before Einstein, Smoluchowski had published the work Über Unregelmäßigkeiten in der Verteilung von Gasmolekülen und deren Einfluß auf Entropie und Zustandsgleichung[433] (On irregularities in the distribution of gas molecules and their influence on entropy and the equation of state), which he included in a commemorative book published on the occasion of Ludwig Boltzmann's 60th birthday[434]. In it he wrote: "While in the theory of gases, deviations of individual molecular velocities from mean values are taken into account through the law of velocity distribution, in terms of the distribution of molecules in the molecular space, homogeneity is generally assumed and—it seems to me—the influence of irregularities in the spatial distribution is underestimated. In order to examine this situation more closely, I would like to make certain suggestions"[435].

---

[432] M. Smoluchowski, *Zur kinetischenTheoric der Brownschen, Molekularbewegung und der Suspensionen* (*An outline of the kinetic theory of Brownian motion and suspensions*), 'Annalen der Physik' ('Annals of Physics') 1906, No. 21, pp. 756-780; I.B. Cohen, *Introduction to Newton's 'Principia'*, op. cit., p. 291.

J. Stachel, *Introduction*, in: A. Einstein, *5 works that changed the face of physics*, op. cit., p. 15.

A. Einstein, *Über die von der molekularkinetischen Theorie der Wärmegeforderte Bewegung von in ruhenden Flüssigkeitsuspendierten Teilchen* (*On the Motion of Suspended Particles Postulated by the Molecular-Kinetic Theory of Heat*), op. cit., pp. 549–560. M. Planck, *Über irreversible Strahlungsvorgänge*, op. cit., pp. 69–122.

M. Planck, *Thermodynamics*, op. cit.

Idem, *The Theory of heat Radiation*, op. cit.

See J. Bernstein, *Einstein and the existence of atoms*, op. cit., p. 865; W. Biernacki, *New spectral domains*, op. cit., p. 78.

J.J. Thomson, *Recollections and reflections*, op. cit., p. 341; C. Röntgen, *Übereineneue Art von Strahlen*, op. cit.

[433] See M. Smoluchowski, *On irregularities in the distribution of gas molecules and their influence on entropy and the equation of state*, op. cit., pp. 626–641.

[434] *Festschrift Ludwig Boltzmann. Gewidmet zum sechzigsten geburtstage* (*Ludwig Boltzmann Festschrift (commemorative book). Dedicated to his sixtieth birthday*), Leipzig 1904.

[435] See M. Smoluchowski, *On irregularities in the distribution of gas molecules and their influence on entropy and the equation of state*, op. cit., pp. 626–641.

Einstein and Smoluchowski discovered that the nature of Brownian motion is purely physical. Atoms and molecules, in constant chaotic motion, surround the suspension particle and collide with it. As a result of the impact of molecules, the particle performs irregular movements at a velocity that changes rapidly in terms of magnitude and direction. The number of molecules striking on one side is usually different from the number striking on the other. As a result, the particle receives continuous impulses causing it to move slightly, noticeable at the mesoscopic level. Due to the high frequency of impacts (in the order of 10^{20} per second), an appropriately averaged particle displacement is created. A mathematical description of Brownian motion was one of the elements that contributed to the emergence of the theory of stochastic processes. Jean Perrin quantitatively tested the Einstein-Smoluchowski formula, showing that the mean squared shift in a given direction is proportional to time. On the basis of these observations, he determined the value of the Avogadro number.

In 1906, Walter Hermann Nernst (1864–1941) posited the third law of thermodynamics, assuming it is impossible to achieve a temperature of absolute zero (zero Kelvin) in a finite number of steps if a non-zero absolute temperature is taken as a starting point.

In 1909, Greek mathematician Constantin Carathéodory (1873–1950) developed an axiomatic system of thermodynamics. He presented in it all previously formulated laws of thermodynamics using a mathematical tool.

In April 1911, Ernest Rutherford performed the so-called Rutherford experiment, considered by physicists among the ten most beautiful experiments. He passed positively charged alpha particles through a very thin gold foil. Most of these particles passed through it unhindered, which indicated the presence of empty space inside the atoms. Some alpha particles were deflected slightly, which, according to Rutherford, indicated interactions with other positively charged particles within the atom. Other alpha particles were scattered at large angles, and few were even reflected back towards the source. Only a positively charged and relatively heavy particle of the disc forming the "core of the atom", as the nucleus was assumed to be, can explain this outcome of the experiment. Negative electrons, compensating for the positive charge of the nucleus, regarded as travelling in circular orbits around the nucleus, and the force of attraction between the electrons and nucleus Rutherford compared to the force of gravity between revolving planets and the Sun. Most of the atom was a void that did not create any resistance to the passing alpha particles. In this way, Rutherford experimentally confirmed the existence of the atomic nucleus.

On April 8, 1911, Heike Kamerlingh Onnes (1853–1926) discovered that pure metals such as mercury, tin, and lead become superconducting at extremely low temperatures. He published his findings in November of that year under the title Über die plötzliche Veränderung der Rate, mit welcher der Widerstand von Quecksilber verschwindet (On the Sudden Change in the Rate at which the Resistance of Mercury Disappears).

In July 1913, *The Philosophical Magazine and Journal of Science* published the first part of Niels Bohr's work entitled *On the constitution of atoms and molecules*[436], and the next two parts were published in September and November. Bohr referred to the Rutherford atom model, which consisted of a heavy, positively charged nucleus with much lighter, negatively charged electrons orbiting it at a considerable distance. Despite the fact that according to the principles of classical physics such a system should be unstable, Bohr made important assumptions about electrons orbiting in atoms in specific circular orbits. According to the first assumption, electrons cannot emit any radiation, and their energy can be stable. The second assumption envisaged the possibility of the electron changing from one orbit to another, with this leap being accompanied by the emission or absorption of radiation of a certain frequency. Bohr's papers were received with disbelief by supporters of classical physics, and two future Nobel Prize winners—Otto Stern (1888–1969) and Max Theodor Felix von Laue (1879–1960)—vowed that "if this nonsense of Bohr should prove to be right in the end, we will quit physics"[437].

4.3. Conclusions after the transformation

The development of physics in the 20th century tested the importance of the achievements in physicists at the turn of the century. Reflections on a given period in science are valuable from a time perspective as the perception of

---

[436] N. Bohr, *On the Constitution of Atoms and Molecules*, 'Philosophical Magazine and Journal of Science' 1913, vol. 26, No. 1, pp. 1–25.

[437] J.P. Toennies, H. Schmidt-Böcking, B. Friedrich, and J. C. A. Lower, *Otto Stern (1888–1969): The founding father of experimental atomic physics* in: 'Annalen der Physik' ('Annals of Physics') (Berlin) 523, No. 12, 962–964 (2011), p. 1048.

changes taking place in science alters from the viewpoint of a new, subsequent era. This idea is illustrated by a comparison of data from the history of physics from two studies classifying discoveries from the turn of the 19th and 20th centuries. In 1910, German physicist Felix Auerbach (1856–1933) published the book *Geschichtstafeln der Physik* (*Historical Tables of Physics*)[438], while in 1983 Jurij Chramow (born 1933), a science historian, published *Биографияфизики* (*A Biography of Physics*)[439]. The two works are separated by 73 years of development in physics and from a comparison of the authors' assessments we can see the changes that have occurred in the evaluation of discoveries made at the turn of the century.

According to Auerbach in 1910, 44 discoveries made in 1899 and 69 made in 1900 were considered important to physics. In 1983, Jurij Chramow repeated Auerbach's study, and from his perspective, 17 important discoveries were made in physics in 1899, and in—only 16.

From a list of 113 discoveries considered important in 1910, after 73 years only seven were recognised while as many as 26 discoveries made in 1899–1900 on Chramow's 1983 list did not appear on Auerbach's. The seven recognised entries were: 1) Planck's formula for black body radiation (the theory itself was not mentioned by Auerbach!); 2) experimental confirmation of this law by Heinrich Rubens and Ferdinand Kurlbaum (1857–1927); 3) the discovery of gamma rays by Paul Ulrich Villard (1860–1934); 4) the discovery of beta ray deviations in an electric field (Friedrich Ernst Dorn, 1848–1916; Becquerel); 5) the discovery that beta rays are negatively charged particles (Pierre Curie and Maria Skłodowska-Curie); 6) measurement of the e/m ratio for beta rays, yielding a result very similar to the that for cathode rays (Becquerel); 7) the discovery by Pyotr Lebedev (1866–1912) of light pressure predicted by the Maxwell theory.

Looking back at the run of successes in physics in the second half of the 19th century, one may wonder at the apparent scepticism about its future. Certainly, 1900—due to the work of Planck, as well as 1905—with the work of Einstein—are breakthrough dates in the history of physics, but did not the works of Maxwell, Kelvin and Boltzmann signal to the scientists of the

---

[438] F. Auerbach, *Geschichtstafeln der Physik* (*Historical Tables of Physics*), Leipzig 1910.
[439] Ю.А. Храмов, *Биографияфизики* (*A Biography of Physics*) Kyiv 1983.

time the existence of the complexity of space, time and matter? The reason cannot be as straightforward and trivial as an error of prediction since too many eminent scientists drew similar conclusions at the time.

The paradox of the late 19th century—when for many physics had reached its peak, and for a few it heralded a new era—is intriguing. Breakthroughs and progress in science, especially in physics and mathematics, are owed largely to young minds. Both *annus mirabilis* and the 'miraculous year' share a common attribute; in 1666 Newton was 24 years old and was a student, and in 1905, Einstein was 26 years old and worked at a patent office. Both stood at the beginning of their scientific path, though the works they published were at the same time the culmination of that path.

It is not difficult to see by carefully reading the above that many important discoveries have been made by scientists when they were only in their twenties. Their success is often due to the fact that—as the saying goes—they did not know that what they were doing was impossible. Kirchhoff posited the so-called Kirchhoff's Law at the age of 21. James Clerk Maxwell published the work *On Faraday's Lines of Force* when he was 24. Maria and Pierre Curie discovered two radioactive elements at the age of 29, Niels Bohr was 28 when he published *On the Constitution of Atoms and Molecules*. Many other examples could be cited.

"The owl of Minerva begins its flight only with the onset of the dusk," not at dawn, claims Georg Hegel in the preface to *Elements of the Philosophy of Right*[440]. *Might he have been wrong?* After all, it is not only the two non-incidental cases of Newton and Einstein that testify to this. These events constitute the antinomy of the aforementioned thought of Hegel as it was "at dawn" that the genius of both young physicists revealed itself. Hegel adds:

> philosophy, at any rate, always comes too late (...). As the thought of the world, it appears only at a time when actuality has gone through its formative process and attained its completed state... When philosophy paints its grey in grey, a shape of life has grown old, and it cannot be rejuvenated, but only recognized by the grey in grey of philosophy; the owl of Minerva begins its flight only with the onset of dusk[441].

---

[440] G.W.F. Hegel, *Elements of the Philosophy of Right*, Ed. Allen W. Wood, Cambridge, 2003, p. 23
[441] Ibid.

We can look at Hegel's metaphor from many angles and each time understand it differently. The Owl of Minerva is philosophy itself which, in contrast to the exact sciences, always comes, as Hegel writes, too late. That philosophical thought "takes flight at dusk" is the culmination of the philosopher's deliberations.

However, this statement could also refer to an understanding of nature's complexity and progress in science. Wisdom comes at dusk, at a time when the hitherto achievements of science are already in their closing stage, when the process of forming the paradigms of a given era has ended. Reflection summarising a period in science is possible when its dusk has already fallen. Only the darkness ending the 'day of the era' allows its achievements to really be known.

Such an understanding of Hegel's thought, supported by the research of Auerbach and Chramow, prompts and attempt to understand a paradox—the predicted end of physics in the 19th century. The Owl of Minerva flies at the twilight of an era ending, heralding a new age in science. Foretelling the end of physics was not such a great mistake, as we know from today's perspective, as it was not about the whole of physics but about the era of mechanistic physics and thermodynamics. Michelson was not wrong in saying that "[t]here is nothing new to be discovered in physics now. All that remains is more and more precise measurement"[442]. He lived at a time when Newtonian physics and thermodynamics reigned, an era that Planck and Einstein ended with their publications. A similar situation occurred at the end of the Middle Ages, when few people realised they were dealing with the decline of old paradigms. In the 19th century, the changes were heralded by the development of statistical mechanics by Boltzmann, while the beginning of the revolution in the worldview of the 16th century was *De revolutionibus orbium coelestium*.

Hegel's observation applies more to philosophical thought, which gains depth through the perspective of time, while the exact sciences, such as physics and mathematics, require reflection at dawn, when the mind is not yet overloaded with conventions.

---

[442] Speech of A.A. Michelson at the University of Chicago, after: A.K. Wróblewski, *Historia fizyki: od czasów najdawniejszych do współczesności* (*The history of physics: from antiquity to modernity*), Warsaw 2006, pp. 395–396.

The change in the perception of knowledge in the 19th century also related to the practice of it as a subject. Referring to his life path, Faraday stated: "I was formerly a bookseller and binder, but am now turned philosopher"[443]. In writing this, he meant his research in the field of physics. No significant physicist since Faraday has claimed that philosophy was the principal field of their research, even though it enjoyed considerable success in the second half of the 19th century and in the 20th. However, it was already a time when physics was building its position as the dominant science. This fact highlights a change in the order of science problems. Throughout the 20th century, according to Tadeusz Gadacz, there are about a hundred outstanding philosophers who created new concepts of perceiving reality, leaving their mark on knowledge. During this period, the focus of scientists' interests shifted towards the exact sciences, especially physics, astrophysics, cosmology, computer science and biology, in which hypotheses, theories and constructs were created that changed the perception of reality.

To sum up the paradox problem outlined above, it should be noted that in each era there is an exaggerated, unjustified belief in the exceptional value of current achievements of science, which does not concur with assessments made from the perspective of the subsequent period, and only philosophical ex post reflection allows a more realistic evaluation, as in the case of Hegel's Owl of Minerva. The question that arises is: which discoveries of the end of the second decade of the 21st century are really important? It would be good to draw up such a list, which would most likely guarantee that in 80 years the name of its author will be cited in analyses of the scientific achievements of the previous century.

---

[443] M. Faraday, *The Philosopher's Tree*, op. cit., p. 1235.

Smoluchowski's approach to the philosophical aspects of causality and chance

5.1. Causality

Causality is the link that connects intuition with the world of pure concepts; hence its dual role, hence the contradictions it bristles with. It is the crack through which the light of reason penetrates the chaos of impressions, imbuing it with order and harmony. Therefore, the assumption will always remain unshakable: to understand means to find a causal relationship. The critical requirements for establishing this relationship in everyday life and in non-methodical thought differ from those concerning knowledge[444].

Through this philosophical reflection, Władysław M. Kozłowski[445] illustrates the difficulties faced by generations of philosophers, starting with Aristotle, in trying to capture the essence of the causal relationship. It underscores the distinction that exists between the everyday and methodological approaches to an event occurring in nature.

---

[444] W.M. Kozłowski, *Przyczynowość jako základní pojęcie przyrodoznawstwa* (*Causality as a fundamental concept of natural science*), Warsaw 1906, p. 41.

[445] Władysław Mieczysław Kozłowski (1858–1935) – philosopher, naturalist, and author of publications in philosophy and other fields, the number of which exceeds 850. Pragmatist – in 1911 he translated *Pragmatism* and the *The Dilemma of Determinism* by William James. He devoted numerous articles and lectures to pragmatism. In 1901, he obtained his habilitation at the University of Lviv, after which he was to take over the second chair of philosophy, as the first was held by K. Twardowski. Unfortunately, the *venia legendi* were not approved by the Minister of Education in Vienna. In 1913, at the invitation of the Kraków Philosophical Society, he gave a series of lectures on pragmatism. His interest in pragmatism and causality, as well as the convergence of places and dates, suggest that Smoluchowski was familiar with his lectures.

CHAPTER 5

This chapter attempts to interpret Marian Smoluchowski's approach to the principle of causality. Its key aim is to show the scholar's contribution to bringing deliberations on the principle of causality from the philosophical and familiar spheres into the methodological realm, as well as to examine the consequences arising from the principle's introduction into science.

The two different conceptions of the relationship between cause and effect presented by the German philosopher Edmund Wilhelm Hermann König (1858–1939) are the starting point for understanding Smoluchowski's approach to the issue of causality. In his paper *The development of the causal problem in contemporary philosophy*, published in 1888–1890, which analysed the historical development of the concept of causality, König argues that attempts to define the essence of causality were conducted in philosophy in two ways—in the context of its metaphysical and epistemological perception. Pre-Kantian philosophy is mainly concerned with the metaphysical understanding of causality, while post-Kant—almost exclusively epistemological. König writes that the metaphysical conception is associated with understanding causality as inherent in things themselves and being reduced to the study of how things can interact with each other. The epistemological understanding is based on the assumption that the source of the causal link lies in the person perceiving it, which consequently involves a search for the foundation on which the concept of causality is based. König's proposed division of the periods of perceiving causality differs somewhat from that commonly accepted in philosophy, whereby Hume's concept is seen as a turning point in the understanding of causality[446].

However, König's periodisation is valuable in that it constitutes an *a priori* argument for the division of causality proposed by Smoluchowski—into epistemological and ontological. The difference is that—according to the Polish physicist—the understanding of causality inherent in things themselves relates to formal ontology, to formalised ontological systems created using mathematical and logical theories, rather than to ontology understood metaphysically, defined by König as pre-Kantian. Smoluchowski's epistemological

---

[446] In *An Enquiry Concerning Human Understanding*, in the third chapter *The Association of Ideas*, Hume assumed that between events, treated as cause and effect, at most a temporal sequence can be stated.

understanding of causality does not exclude ontic aspects, thus enabling the use of mathematical tools in the study of cause-effect relationships.

To illustrate an important element of Smoluchowski's causality, the thoughts of two scholars should be mentioned. The first is the physicist Mario Augusto Bunge (1919–2020), for whom causality, along with other types of determination, is not only a certain property of our knowledge of reality, but can also be an objective property occurring in nature. Bunge argues that causal conditioning is one of many forms of determination[447]. The philosopher Stefan Amsterdamski (1929–2005) places within the epistemological understanding of causality the ongoing dispute between determinism and indeterminism as to whether for every type of natural event known to us it is possible to indicate such parameters of the state of the appropriate physical system that at the same time determine the occurrence or non-occurrence of the appropriate type of event[448]. The outcome of a possible event can be determined by science, but the answer cannot be positive in the ultimate sense as it can never be ruled out that unambiguous prediction will prove impossible[449].

Bunge believes that the multiplicity of causal determinants leads to statistical determination, which may result in quantitative self-determination[450]. He argues that determinism in a broad sense is an ontological theory consisting

---

[447] See M. Bunge, *O przyczynowości. Miejsce zasady przyczynowej we współczesnej nauce* (*Causality: The place of the causal principle in modern science*), Warsaw 1968, p. 31; W.M. Kozłowski, *Causality as a fundamental concept of natural science*, op. cit., p. 41; G.W.F. Hegel, *Elements of the Philosophy of Right*, op. cit., p. 21.
M. Bunge, *Causality*, op. cit.
M. Faraday, *The Philosopher's Tree*, op. cit., p. 1235.

[448] See S. Amsterdamski, Z. Augustynek, W. Mejbaum, *Prawo, konieczność, prawdopodobieństwo* (*Law, necessity, probability*), Warsaw 1964, pp. 89–90.

[449] *Nota bene*, Amsterdamski refers to Smoluchowski at this point, proving that the probabilistic answers proposed by the Polish physicist do not contradict the objectivity of the statements. Based on this idea of Smoluchowski's, it is necessary to distinguish, on the one hand, the issue of the objective nature of probabilistic statements from the issue of the objective and relative nature of any separation of the subject of scientific research and, on the other, the issue of the objectivity of probabilistic statements from issues related to the dispute between determinism and indeterminism. This is a condition for understanding the meaning of probabilistic statements. See S. Amsterdamski, Z. Augustynek, W. Mejbaum, *Law, necessity, probability*, op. cit., p. 90.

[450] See M. Bunge, *Causality*, op. cit., pp. 45–46.

of a genetic principle and a regularity principle. In both the above cases—in Bunge's, where the arguments support an ontological interpretation of causality, and Amsterdamski's, in which the arguments support an epistemological interpretation of causality—the arguments are valid.

According to Bunge, the principle of causality is a special case of the principle of ontic determinism. It includes those cases that involve unambiguous determination by external conditions[451]. Amsterdamski's causality principle bases the occurrence or non-occurrence of an event on its comprehension by the perceiver[452]. From the point of view of Smoluchowski's deliberations, what is interesting is the principle of regularity, which states that nothing happens in an unconditioned and completely irregular way, i.e. in a way that is arbitrary, that is not subject to specific laws. Which principle of regularity we are dealing with—whether in an ontic or epistemic understanding of an event—is largely determined by the establishment of the law we want to use to describe the event.

Anna Lemańska concurs with Smoluchowski's theory and answers the question more succinctly—nothing happens in nature for no reason; everything must be conditioned by something. So the causality principle can be formulated as follows: there exist no events in nature that are not subject to any regularities, and hence, for example, absolutely random events. The principle of determinism and the principle of causality state that it is possible to predict future events and to reconstruct the past, and that there is order in nature. Two aspects of these principles can also be distinguished: epistemological—related to the possibility of forecasting, and ontological—concerning the structure of natural reality[453]. This shows why Smoluchowski's reflections on the essence of causality are focused on the above two aspects of the issue—epistemic and ontic. This twin perception of causality is rooted in the dual character of the perception of nature, i.e. both in learning the effects of its action in relation to the supposed causes, and in understanding the existential side of the event, open to exploratory study.

---

[451] Idem, p. 40.
[452] Referring to König – the epistemological understanding is based on the assumption that the source of the causal bond lies in the perceiver.
[453] See A. Lemańska, *Determinizm* (*Determinism*) www.kul.pl/files/57/encyklopedia/lemanska_determinizm.pdf (access: 29/03/2020), pp. 3–4.

Since its inception, philosophy has made successive attempts to elucidate the essence of events occurring in nature, doing so mainly in the ontological aspect—originally understood metaphysically—as well as epistemologically. Although both aspects were previously philosophical, in entering into considerations on causality, Smoluchowski sees an opportunity to make the problem scientific, enabling the use of mathematical instruments. This becomes possible through the assumption that epistemic determinism speaks about the possibilities of predicting a course of events based on knowledge of the conditions in which they occur. It examines cause and effect in terms of their mutual relationships. Ontic determinism deals with the construction of a system of reality, referring to existential relationships. In methodological terms, by assuming a specified certainty of the occurrence of physical phenomena and a certain constancy of causal relationships, it allows the search for a method on the basis of which specific predictions could be made.

In studying the epistemic form, Smoluchowski analyses the cognitive features that constitute the common understanding of causality. This study involves the application of methodologies enabling a de-anthropomorphised understanding of the essence of causality, determination of the scope of operation of formulated theories and their verification in terms of the constructed theorems of physics.

In the epistemic view of causality, an event is considered to have been explained, writes Smoluchowski, if it is reduced to causes whose mode of action is sufficiently familiar to us and therefore seems understandable. Such an explanation is not merely a common way of clarifying the causes of events occurring in everyday life, but also a typical explanation of this problem in the natural and historical-philosophical sciences. It is based on the law of causality, which is the result of an unconscious belief acquired in the process of building life experience. The law, which is almost instinctive, holds that every event has a cause, and the same causes have the same effects. However, if the concept of cause thus-defined is considered, it must be noted that it contains (as in the concept of purposefulness) certain anthropomorphic elements, transferred from the realm of the human psyche to the external inanimate world[454].

---

[454] See M. Smoluchowski, *Object, task, method and the division of physics*, op. cit., pp. 167–168.

Starting from Smoluchowski's reference to Aristotle's principle of causality, which had consistently dominated philosophy, and which claims that this approach is subjectivist and does not contribute much to the sciences, we see a change in his views on the properties and nature of causality. The epistemic understanding of the principle of causality, in which every event has its cause, and the same causes produce the same effects, based on anthropomorphic associations, did not allow for an objective scientific study of the problem. This dysfunction of the previous philosophical analysis of causality was intensively discussed in scientific circles, with Smoluchowski also participating.

Władysław M. Kozłowski was the other philosopher whose views influenced Smoluchowski in various ways. In 1906 he wrote:

> It should be remembered above all that the primary effort of knowledge is the elimination of qualities as absolutely distinct and irreducible to one another. This is possible in two ways: 1. The ontological one involves accepting as possible a qualitatively-free substratum, the states of which serve to explain quality. 2. The mathematical approach consists in reducing qualitative differences to quantitative ones[455].

Smoluchowski also called for a differentiated quantitative and qualitative approach in the study of phenomena[456].

Kozłowski was a pragmatist and although his understanding of causality was more philosophical in nature, his remarks may have inspired the philosophical physicist. The statement that "Causality in the world of phenomena, due to the ontological formation of our basic concepts about it, naturally assumes the nature of action, whatever critical-philosophical concept we have about it"[457] is the very start of thinking about the phenomenon of causality in terms of analysing the understanding of chance. Indeed, the most significant change Smoluchowski made in his perception of causality is his understanding of the role of chance in cause and effect. According to Smoluchowski, chance is a specific type of causal reference[458]. However, this short definition does not explain its essence.

---

[455] W.M. Kozłowski, *Causality as a fundamental concept of natural science*, op. cit., p. 29.
[456] See M. Smoluchowski, *The Self-Study Handbook*, vol. I, op. cit., p. 38.
[457] W.M. Kozłowski, *Causality as a fundamental concept of natural science*, op. cit., p. 30.
[458] M. Smoluchowski, *On the concept of chance and the origin of laws of physics based on probability*, op. cit., p. 37.

So what is chance? Smoluchowski formulated the question more precisely: what characterises the essence of chance? When treating chance as a negation of regularity, certain contradictions appeared that were a source of dilemmas and, moreover, such an understanding of chance could not be reconciled with the determinism generally prevailing at the time. There was an attempt to get out of this situation by assuming that despite the correct causal relationship between cause and effect, the type of this relationship is unknowable as the phenomenon is too complex, hence the impression of an apparent break from regularity[459]. Determinism requires cause and effect to be treated as events related to one another through internal relations of necessity, therefore the seeming lack of necessity is an apparent phenomenon resulting from the fact that some causes are unknowable[460]. Chance is then defined as a hidden causal relationship existing between cause and effect.

In his work *Zagadnienie przypadku (The Problem of Chance)*, Joachim Metallmann (1889–1942) defines chance as that which is unnecessary, or that of which we do not know the causes, often that which is not in line with our intentions and plans. Sometimes it is the same as something insignificant, sometimes something unforeseen, even if *ex post* it is explicable, it is still something that cannot be fully explained. Where does this come from? From the fact that chance is comprehensible, that it makes sense only against the background of a certain regularity; it is the result of an attempt to rationalise experience, an attempt that encounters resistance, becomes conscious in connection with the establishment of regularity. In a word, chance is correlative to determinism, it's just that the correlation is overtaken by a simpler and psychologically closer negation[461].

In excessively complex phenomena, chance manifests itself as an apparent break from regularity, but in science such a definition was impossible to accept. As Smoluchowski writes in his essay *On the concept of chance and the origin of laws of physics based on probability*, the probability of a certain event can depend only on the conditions affecting that event's occurrence, but

---

[459] Idem, p. 29.
[460] Idem, p. 30.
[461] See J. Metallmann, *Zagadnienie przypadku (The problem of chance)*, 'Przegląd Współczesny' ('Modern Review'), 1933, year XII, vol. XLIV, p. 90.

cannot depend on the extent of our knowledge[462]. Therefore, Smoluchowski posits the removal of the subject from the concept of chance, which in turn was to result in the objectification of the concept itself. Strict natural science is interested not in subjective statements and presumptions, but in objective or mathematical probability, i.e. the relative frequency of the occurrence of given random events. In this narrower sense, the concept of probability becomes available in strictly mathematical terms[463].

> A similar problem was presented by the Polish physicist in *Notes on the concept of chance in physical phenomena*. Smoluchowski puts forward the thesis that the subject of physics is the laws of nature and the assumption that causality and determinism constitute the antithesis of randomness is erroneous. How, then, can chance arise within phenomena proceeding in accordance with the immutable laws of nature? How can physics describe chance using deterministic laws? Is it possible for chance to occur in a nature governed by deterministic laws? If essentially unpredictable chance, being a negation of causal regularity, plays a certain role in the phenomena of physics, how can the correct course of these phenomena be predicted?[464]
>
> Smoluchowski saw that finding answers to such questions lies in a correct understanding of chance, which may be the key to introducing probability calculus into the exact sciences. Indeterminism is a fundamental concept for deliberations in the field of probability. He tries to define chance, departing from the common understanding, giving it the status of a scientific concept. He therefore further analyses the concept of chance, this time in the context of broadly accepted determinism. This issue is addressed in his works, such as in the article On the concept of chance and the origin of laws of physics based on probability, in which he raises questions on the relationship of chance with regard to the stable laws of physics. Asked what events fall within the scope of the applicability of probability calculus, Smoluchowski answers that they are usually said to be those events whose occurrence depends on chance[465]. Science studies objective regularities, and scientific laws must assume the existence of a causal relationship, which does not mean that the concept of cause can be replaced by the concept of scientific law. Scientific laws answer not only the questions of when, how and where, but also try to find an answer to the question of why, constituting a tool for explaining events occurring

---

[462] M. Smoluchowski, *On the concept of chance and the origin of laws of physics based on probability*, op. cit., p. 31.
[463] Idem, p. 30.
[464] See M. Smoluchowski, *Notes on the concept of chance in physical phenomena*, op. cit., p. 74.
[465] Idem., p. 29.

in nature. However, probabilistic statistical forecasts can become an additional instrument for explaining events.

Bunge states that the probabilistic nature of statistical forecasts does not at all indicate their lesser cognitive value in relation to scientific laws. Moreover, statistical forecasts have the advantage over other types of forecast that they allow the formulation of assumptions about collectives, because they make it possible to answer questions against which causal laws regarding individual events are helpless. He points out that statistical laws are incomplete in the sense that they do not allow the prediction of the individual behaviour of, for example, particles. They should therefore be complemented by other types of laws. He further adds that statistical laws are no less complete than other types of scientific law, and their incompleteness is merely a different kind of completeness[466].

Smoluchowski had been convinced of the molecular and kinetic nature of Brownian motion since 1900[467]. In this case, the search for evidence to confirm the kinetic-molecular theory overlapped with the search for the cause of Brownian motion. The evidence for the existence of a causal relationship between the energetic state of the particles, leading to fluctuating collisions, and the noticeable movements of the gamboge was convincing and obvious to Smoluchowski. This was a typical situation in which the epistemic recognition of cause and effect provided an explanation of the event occurring. According to Bunge's principle of causality, the effect is constantly and necessarily correlated with the cause, because it is generated by the cause. Smoluchowski applied the principle of causality to the phenomenon of Brownian motion in the same way[468].

---

[466] See Z. Hajduk, *O przyczynowości: miejsce zasady przyczynowej we współczesnej nauce. Mario Bunge* (*Causality: the place of the causal principle in modern science. Mario Bunge*), trans. S. Amsterdamski, 'Studia Philosophiae Christianae' ('Studies in Christian Philosophy') 1969, No. 5/2, pp. 217–225.

[467] M. Smoluchowski, *On thermodynamic fluctuations and Brownian motion*, Warsaw 1914, p. 299.

[468] In diffusion theory, he associates the macroscopic viscosity phenomenon with the microscopic notion of the mean free path of a molecule. In his works devoted to the issue of Brownian motion, he wrote explicitly that he was trying to explain the internal mechanism of diffusion and link it to the phenomena of molecular movements. In his view, the macroscopic phenomenon of diffusion is a manifestation of molecular movement or density fluctuations.

According to Bunge, "Almost every philosopher and scientist uses his own definition of cause, even if he has not succeeded in formulating it clearly"[469]. Zenon Roskal claims that Smoluchowski built his own concept of causality: "[O]n the one hand, Smoluchowski himself nominally accepts the positivist criticism of the causality principle, but at the same time, on the other hand, in his scientific activities, especially in attempts to explain Brownian motion, he practically overrules this criticism"[470].

Smoluchowski treated the proof of the causality of Brownian motion as an argument for epistemic causalism. He proved the causes of these movements mathematically. He built a mathematical apparatus enabling scientific investigation of the causes using statistics. As Stanisław Loria writes, he created a research method consisting in drawing up statistics of the displacements achieved by particles at specific times[471]. The mathematical proof explaining the origins of Browian motion was an important step leading to considerations on causality, used in scientific methodology, not merely a philosophical dissertation. Smoluchowski treated this solution as an incomplete example, not including, for example, Mach's functionalism, the causally independent aspect of causality, research on which he had had initiated earlier, considering the ontic interpretation of chance.

The Polish scholar initially identified three conditions a phenomenon must meet in order to be called randon. These are: "small cause – big effect"[472], "different causes – same effects"[473] and "the probability of a certain value of

---

[469] M. Bunge, *Causality and Modern Science*, New York, 2009., p. 31.

[470] Z.E. Roskal, *Marian Smoluchowski's approach to the causality principle in research on Brownian motion*, op. cit., p. 6.

[471] Smoluchowski understood that if the molecular-kinetic theory, using statistical methods, provides for the possibility of processes consisting in departing from the normal course of phenomena in the macro-world, it is the researcher's job to theoretically analyse the conditions under which they can be expected to be implemented and to predict and reproduce a quantitative-hypothetical picture of their course. One of the phenomena particularly suitable for this purpose, namely Brownian motion, had already been highlighted by Helmholtz. But physics owes to Smoluchowski the development of a mathematical method allowing this motion to be described and studied in detail. See S. Loria, *Marian Smoluchowski and hiw work (1872–1917)*, op. cit., pp. 5–38.

[472] M. Smoluchowski, *Notes on the concept of chance in physical phenomena*, op. cit., p. 448.

[473] Idem, *On the concept of chance and the origin of laws of physics based on probability*, op. cit., p. 41.

the initial state is determined by the $\Phi(x)$ regular function, which does not have many maxima or minima"[474]. The first condition states that a small change in the input parameters causes major changes to the effect. A small change is understood in a relative sense—it is small compared to the possible range of changes. The second condition describes that fact that in the range of initial states there exist many configurations that lead to the same final state. The third condition imposes constraints on the function describing the probability distribution of the occurrence of a certain initial value. Subsequent analysis by Smoluchowski showed that the adoption of this third assumption proved unnecessary[475].

As indicated above, despite his critical attitude to epistemic causality, Smoluchowski does not give up its application to scientific research. Accepting positivistic criticism, he adopts for research purposes the principle of causality "cleansed of obscure human and metaphysical impurities, as the quintessence of all experiences and observations, all of which confirm the unchanging regularity of nature"[476]. He emphasises its value in the form he outlines, arguing: "We also acknowledge as absolutely right those who hold the assumption of causality in this form to be a cardinal condition of thinking about nature"[477]. This idea was more clearly presented by Max Kistler, in whose concept of the causal relationship the production of an effect by a cause takes place in accordance with the physicalistic concept, in which causality is reduced to the transfer of a certain specific physical quantity, i.e. energy or momentum[478].

Smoluchowski's study of Brownian motion is a search for an epistemic cause— according to Bolesław Gawecki (1889–1984), the so-called causal causality[479]—of the vibrations of molecular particles resulting in the transfer over time, through collisions, of the energy of solvent particles to particles

---

[474] Idem., *Notes on the concept of a case in physical phenomena*, op. cit., p. 449.
[475] See P. Polak, *The concept of chance in the writings of Marian Smoluchowski*, in: *Krakowska filozofia przyrody w okresie międzywojennym* (*Kraków philosophy of nature in the interwar period*), op. cit., p. 454.
[476] M. Smoluchowski, *Self-Study Handbook*, vol. I, op. cit., pp. 24–25.
[477] Ibid.
[478] See Z.E. Roskal, *Zasada przyczynowości w kontekście eksperymentalnych badań ruchów Browna* (*The principle of causality in the context of experimental research on Brownian motion*), 'Filo-Sofija' 2017, No. 37, p. 71.
[479] See B.J. Gawecki, *Zagadnienie przyczynowości w fyzce* (*The issue of causality in physics*) Warsaw 1969, p. 106.

in suspension, thus leading to vibrations visible under the microscope. In his article *Marian Smoluchowski's Contributions to the Philosophy of Causation*[480], Zenon Roskal argues that in his search for the epistemic cause of Brownian motion, Smoluchowski used the ontological interpretation of the causality principle, thereby becoming a pioneer of the transferential concept of the causal relation, known as *The Transference Theory of Causation*. The essence of this theory is described by Douglas Ehring in an article of the same title, citing, among others, the writings of Jerrold Aronson. Aronson assumed that in the theory of causal transference, the most important concept is "the cause object possesses a quantity (e.g., momentum, energy, heat, etc.) which is transferred to the effect object". This theory applies to "mechanical" cases of causation and comes down to the following points of relation: (1) If "A causes B", then B denotes a change in the object, an unnatural change (a change that cannot be taken into account without reference to the behaviour of other bodies); (2) If "A causes B", then at the moment of B's occurrence, the object that causes B is in contact with the object that undergoes a change; (3) Before B's occurrence, the body in contact with the effect has a certain quantity (of, for example, velocity, momentum, kinetic energy, heat), which is transferred to the object of the effect (when contact is established) and manifested as B. The epistemic[481] cause—being the source of causal transfer—refers to the ontological formula of the existence of the cause-object and effect-object.

Ehring's considerations translate into the way in which Smoluchowski, proceeding to present a solution to the problem of Brownian motion, uses the physical concept of causation. Smoluchowski does not write about it directly, but the velocity achieved by microscopic bodies is associated with the transfer of momentum, which is consistent with the concept of cause as a reaction to the energy supplied.

Kistler creates a principle that he calls the Causal Criterion of Reality. It states that the capacity to interact causally—or to contribute to determining causal interactions—is not only the ultimate metaphysical ground for the existence of an entity, but also provides a criterion for determining the nature

---

[480] See idem, Marian Smoluchowski's contributions to the philosophy of causation, 'Organon' 2018, no. 50, p. 12

[481] D. Ehring, *The Transference Theory of Causation*, 'Synthese' 1986, vol. 67. No 2, p. 249.

of that entity[482]. In searching for the epistemic cause of the movements of gamboge particles vibrating in reaction to the energy transferred to them, Smoluchowski arrived at the discovery of the actual cause of Brownian motion. The cause, which is the thermal fluctuation vibrations of solution particles that have energy, initiates the transfer of energy, causing Brownian motion, and constitutes—according to Kistler's concept—the criterion determining the nature of this entity, and therefore the ontic cause of this entity.

More than a hundred years after Smoluchowski, the intricate complexity of the problem of causality was put quite simply by the mathematician Fernando Corbalán, who claimed that we live in a sea of ambiguity, from which only a few islands of certainty protrude, and understanding the world forces us to find a way to master chance, one of the last lands we have left to conquer. Chance, which brings so much chaos into a life based on order and security[483].

5.2. Probability calculus

The concepts of chance and probability operate daily in colloquial language and we intuitively sense their meanings, but the scientific approach requires that these meanings be clarified. The problem repeatedly emphasised by Smoluchowski was the lack of an in-depth analysis of the functioning of the concept of chance in nature that would enable its clarification and its optimal meaning to be grasped, as well as the lack of development of a basis for using probability calculus in physics. "Despite this enormous extension of probability calculus's range of applicability, a strict analysis of the basic concepts of this calculus has made little progress; today it is surely still correct to say that no other branch of mathematics is built on such unclear and shaky foundations,"[484] wrote Smoluchowski a hundred years ago, adding that "within the framework of axiomatised calculus, the concept of probability

---

[482] M. Kistler, *The causal criterion of reality and necessity of the laws of nature*, 'Metaphysica' 2002, No. 3 (1), pp. 57–86.

[483] See F. Corbalán, G. Sanz, *Poskromienie przypadku. Teoria prawdopodobieństwa* (*The taming of chance. Probability theory*), trans. K. Rejmer, Warsaw 2012, p. 10.

[484] M. Smoluchowski, *On the concept of chance and the origin of laws of physics based on probability*, op. cit., p. 28.

itself is not *explicitly* defined. It is treated as an undefined primary concept, on which conditions are imposed in the form of axiomatic systems"[485].

> Chance plays a significant role in physical phenomena; for it to be treated scientifically required the use of probability calculus methods, even though the state of this branch of mathematics was unsatisfactory. However, this was indispensable as the study of chance cannot be conducted according to the common understanding of the term, since that would exclude it from the scope of calculus methods[486].

At the beginning of the article On the concept of chance and the origin of laws of physics based on probability, Smoluchowski introduces the reader to the history of probability calculus, highlighting that at the beginning it was used mainly in the study of social and biological phenomena, but that more recently a very important field of application of calculus in physics had been found, not only in the theory of levelling out errors in physical measurements, but also in the very core of physics, in the system of theoretical physics.[487] In his essay Notes on the concept of chance in physical phenomena, he stated that probability calculus was increasingly coming to the fore as the most appropriate mathematical method for the purposes of this science: "Not only the kinetic theory of matter, but also electronics, the theory of radiation, [and] the science of radioactivity use this very method in their fundamental research"[488].

In 1857–1860, Clausius and Maxwell were the first to introduce probability calculus into the kinetic theory of gases[489]. This was natural as calculus—as Smoluchowski writes—is used to study multitudinous phenomena[490]. The physicist, the philosopher, the mathematician —each looks for something different in probability. The physicist assesses that which he considers random

---

[485] S. Amsterdamski, Z. Augustynek, W. Mejbaum, *Law, necessity, probability*, op. cit., pp. 8–9.

[486] See M. Smoluchowski, *Uwagi o roli przypadku we fizyce* (*Notes on the role of chance in physics*), Kraków 1917, 'Zagadnienia Filozoficzne w Nauce' ('Philosophical Problems in Science') 2017, p. 295.

[487] See M. Smoluchowski, *On the concept of chance and the origin of laws of physics based on probability*, op. cit., p. 27.

[488] Idem, *Notes on the concept of chance in physical phenomena*, op. cit., p. 74.

[489] M. Smoluchowski, *On the concept of chance and the origin of laws of physics based on probability*, op. cit., p. 27.

[490] Idem, *Notes on the role of chance in physics*, op. cit., p. 280.

as equally probable, comparing the consequences of calculation with experience[491]. The philosopher turns his attention primarily to the subjective, psychological side of the concept of probability, analyses its theoretical and cognitive significance, studies how probable sentences can be included in the system of formal logic alongside true and false sentences, but is not in the habit of considering more closely the question of to what genre objective facts forming the basis of probable sentences belong[492]. A mathematician deals for example with the theory of various games of chance, i.e. the formal side, calculating the probability of a complex phenomenon on the basis of component, elementary phenomena[493].

Historically, the roots of probability calculus stretch back to ancient times, but the Renaissance saw a breakthrough with the calculus of probability entering mathematicians' sphere of interest. The source of this interest was gambling, which was widely practised at that time. This need of the hour meant that at the turn of the 15th and 16th centuries, there was significant interest in creating the elements of probability theory and many of the enlightened minds of that era, such as the eminent mathematicians Giovanni Francesco Peverone (1509–1559), Girolamo Cardano (1501–1576) and Galileo himself, considered the mathematical dilemmas of gambling[494].

Blaise Pascal (1623–1662) and Pierre de Fermat (1601–1665) are considered the founders of modern probabilistics and were also inspired by games of chance. Pascal's interest in stochastic processes was prompted by a chance encounter with the gambler Antoine Gombaud, known as 'Chevalier de Méré' (1607–1684). He sought solutions to several dilemmas encountered in card and dice games which were de facto mathematical problems.

The classic problem of a game of chance, resolved by Pascal for Chevalier de Méré and entering the history of probability theory, was a situation in which during a game of dice, betting on the appearance of a six once within four throws of a die, the gambler who bet against the occurrence—that a six would not appear—won more often than he lost. Wanting to make the game more complicated so that players would not be aware of the system

---

[491] Idem, p. 282.
[492] M. Smoluchowski, *On the concept of chance and the origin of laws of physics based on probability*, op. cit., p. 30.
[493] See idem., *Notes on the role of chance in physics*, op. cit., p. 282.
[494] See F. Corbalán, G. Sanz, *The taming of chance*, op. cit., p. 40.

he was using, *de Méré* added another die and bet on a double six on four rolls of two dice. He assumed that the extra die represented only six times more possibilities, so two dice would have to be rolled six more times. This reasoning proved false as after adding another die, he lost more often than he won. *In his reasoning, Chevalier de Méré had used the arithmetic rule of three*, which enables the calculation of a fourth valueon the basis of the remaining three known values. When betting on at least one six on four rolls, he won, but when, according to the rule of three, he rolled two dice twenty-four times and bet on throwing at least two sixes, he lost. Pascal found the solution. He wrote to Fermat, who lived in Toulouse, asking what he thought of his conception. Fermat concluded his positive response with the famous statement: "I plainly see that the truth is the same at Toulouse and at Paris[495]."

Another important figure that made a huge contribution to the development of probability calculus was Pierre Simon de Laplace, a French astronomer and mathematician. He expressed his views in the work Théorie analytique desprobabilités (The analytic theory of probabilities), published in 1812.

Being an extreme determinist, Laplace published in 1814 *Essai philosophique sur les probabilités* (*A Philosophical Essay on Probabilities*), in which he included elements of the theory of scientific determinism. In the so-called confession of faith, he argued that the current state of the Universe should be considered the result of its previous state as a cause, constituting the result of the past and a cause of the future. For the needs of his deliberations, he created a figure of omniscient intellect, which later became known in science as Laplace's demon. It was a mind that at a given moment knew all the forces of nature and the current position of all the bodies making up the Universe and which would be sufficiently powerful to analyse the data and describe the movements of all the bodies in the Cosmos—from the heaviest to the lightest atoms. To such an intellect, nothing would be uncertain and it would see both the past and the future.

Laplace arrived at this mental conception by assuming that "all events are as necessary as the revolutions of celestial bodies; only ignorance of the laws connecting them is the cause of the introduction into nature of such concepts as purpose or chance, which are being removed ever more from science with

---

[495] Idem, p. 49.

the development of our knowledge of nature"[496]. According to this concept, "The curve described by a simple molecule of air or vapor is regulated in a manner just as certain as[497] the planetary orbits". Nothing would be uncertain to it and it would see both the past and the future.

According to Fernando Corbalán, Laplace was not talking about some omniscient mind. The author of *The Taming of Chance* believes that the French mathematician meant the demon metaphorically. It was intended to be a scientific method of calculating the probability of an event's occurrence with the aid of probability calculus, enabling the prediction of nature's behaviours and the learning of its laws[498]. The essence of Laplace's idea is a situation in which the demon knows the future. In probability calculus, this is an extreme chance, a state in which we foresee the occurrence of a given event with 100-percent certainty. In such a situation, the essence of the problem is missed. We cannot talk about calculating probability when we have a hundred percent certainty of a certain movement of particles occurring. We talk about probability when the possibility of a given phenomenon is probable. In writing about the demon, then, was Laplace thinking about calculus?

It is possible, although according to Smoluchowski,

> chance of that type is removed from all *a priori* calculation and as such can never be the basis for applying probability calculus. Because as long as we do not know the causes with sufficient precision (…), nothing at all can be foreseen regarding the outcome. But when we know, the result can be predicted with certainty so that there remains no place for probability[499].

However, the demon's powerful mind, which could determine all the consequences of the laws of nature, can be reduced to a calculus of probability developed to know these laws of nature as well as possible[500]. All the more so since even assuming the 100-percent efficiency of the Laplace demon, i.e. assuming a situation in which we managed to know in physics with absolute

---

[496] W. Krajewski, *Marian Smoluchowski's worldview*, op. cit., pp. 104–105.
[497] See *Historia matematyki (The history of mathematics)*, vol. 3, *Matematyka osiemnastego stulecia (Themathematics of the 18th century)*, ed. A.P. Juszkiewicz, Warsaw 1977, p. 164.
[498] See F. Corbalán, G. Sanz, *The taming of chance*, op. cit., p. 54.
[499] M. Smoluchowski, *On the concept of chance and the origin of laws of physics based on probability*, op. cit., p. 33.
[500] See F. Corbalán, G. Sanz, *The taming of chance*, op. cit., p. 54.

accuracy the momentary state of the system of all atoms and calculate its changes, it must be concluded that probability calculus retains its value[501].

According to Smoluchowski, even absolute knowledge of the initial state of all gas particles would not eliminate the usefulness of probability calculus to describethat system and therefore the utility of the concept of probability itself. It can therefore be tempting to state that for Maxwell's demon[502], the concept of probability calculus would be useful as a convenient tool for describing complex systems[503]. Despite the acceptance of the existence of a demon whose predictive abilities surpass those of a person, it must be assumed that he would not be able to determine all intermediate states without the help of the calculus of probability. Today, in the age of quantum physics, we know Smoluchowski's final suggestion to be true. It is not possible to predict the movement of particles with mathematical certainty; it will always be hampered by the uncertainty of velocity (or position). Hawking notes that even predicting aspecificset of positions and velocities ceases to be possible once the existence of black holes is taken into account[504].

For Smoluchowski, Laplace's concept was too philosophical. He did not agree that chance should be understood as a sign of incomplete knowledge

---

[501] See M. Smoluchowski, *Notes on the role of chance in physics*, op. cit., pp. 290–291.

[502] Smoluchowski recalls another concept of the demon—proposed by Maxwell. This definition was provided in 1867: "if we conceive a being whose faculties are so sharpened that he can follow every molecule in its course, such a being, whose attributes are still as essentially finite as our own, would be able to do what is at present impossible to us. For we have seen that the molecules in a vessel full of air at uniform temperature are moving with velocities by no means uniform, though the mean velocity of any great number of them, arbitrarily selected, is almost exactly uniform. Now let us suppose that such a vessel is divided into two portions, A and B, by a division in which there is a small hole, and that a being, who can see the individual molecules, opens and closes this hole, so as to allow only the swifter molecules to pass from A to B, and only the swifter molecules to pass from A to B, and only the slower ones to pass from B to A. He will thus, without expenditure of work, raise the temperature of B and lower that of A, in contradiction to the second law of thermodynamics". Maxwell's demon was created for construction of a perpetual motion machine. The essence of its action comes down to interference with nature at the level of cause, so it is in possession of knowledge, just like Laplace's demon.

[503] See P. Polak, *The concept of chance in the writings of Marian Smoluchowski*, op. cit., p. 459.

[504] See S.W. Hawking, *Brief Answers to the Big Questions*, op. cit. p. 127.

about phenomena, which would mean that chance is mainly subjective. In physics, such an understanding of chance would not be useful as it would not be subject to empirical verification and, in addition, it would limit the pursuit of new facts and finding relationships between them; it would be more philosophical than scientific in nature. It would not be suitable for use in the exact sciences as they describe something that is independent of the researcher[505]. The argument suggests Smoluchowski's search for an ontological way of understanding chance, one that could be researched empirically and, consequently, on which a mathematical description of the event could be performed.

The echo of Laplace's views still resonated clearly in both physics and philosophy at the beginning of the 20th century. Smoluchowski emphasises that at that time a common belief prevailed that "a given phenomenon was caused by a properly functioning but unknown cause or chain of causes, the links between which are not more closely known or cannot be strictly controlled"[506], and that "chance is only the name given to our lack of awareness; probability would not exist for an omniscient being"[507].

Such an understanding of chance made it impossible to apply scientific methodology to the study of causality. Smoluchowski was aware of this and, wanting to introduce mathematical relations to his considerations, began analysing the ontic aspect of the problem of chance, i.e. the existential aspect of an effect's occurrence. He specified how chance should be understood in order for it to provide a basis for the application of scientific methodology. His attempts went in the direction of defining ontic causality using probability calculus. He presented the obstacles encountered through the example of an artilleryman firing a cannon.

> The gunner fires a projectile from a certain point at a speed of c, at an angle of elevation a, to hit a point located at a distance x. Taking the problem theoretically, we exclude air resistance, accidental inaccuracies in positioning, etc. We can therefore easily calculate the quantity x as a function of the initial data: $x = f(c, a)$, or vice versa: $a = F(r, c)$; we are then convinced that the point x will be hit, thanks

---

[505] See M. Smoluchowski, *Notes on the concept of chance in physical phenomena*, op. cit., pp. 75–76.
[506] Idem., *Notes on the role of chance in physics*, manuscript, p. 4.
[507] Ibid.

to the elevation's calculated angle of inclination; we are also sure that using the elevation angle contained between the boundaries a1 and a2, some point of the target, stretching from $x1$ to $x2$, will be hit. However, if the artilleryman lacks sufficient data to predict the result (e.g. he does not know the initial speed c and-cannot calculate the correct elevation angle a, then he must fire 'blind'. We can then say in coloquial language that chance determines whether the target (ranging from $x1$ to $x2$) will be hit; however, we cannot say anything about the probability of this occurrence unless we have some further indications as to how the gunner set up the gun. It is perhaps a field suitable for psychological speculation, but not for physical calculations. Chance, in this sense, is a negation of rational regularity and must be excluded from science[508].

Elsewhere, Smoluchowski again formulates the previously asked questions, presenting them in a more mathematical form. He questions under what conditions the phenomenon y, caused by a properly functioning cause x, can be considered accidental, in other words—he considers in what situations causes x, acting randomly, can yield the correct and specific result y. In connection with the above, a twin-problem situation has arisen in physics which needs to be solved. In the essay On the concept of chance and the origin of laws of physics based on probability, Smoluchowski answers these questions by analysing the problem of chance through a specific example. This property occurs in a particularly characteristic way in all these situations involving a state of unstable equilibrium.

Let us imagine, as he writes, the 'perfect' gaming die, placed on one of its corners—the slightest deviation of the centre of gravity from the vertical determines on which of the three planes converging at the bottom the die will rest. Which number shows on top depends on chance. Mathematically, this is expressed as follows: the effect y (the number appearing at the top) depends on the cause x (the location of the centre of gravity) in such a way that the function $y = f(x)$ shows discontinuity for the appropriate equilibrium value x_0. The cause in this coincidence consists of two variables. First, by projecting onto the horizontal plane the centre of gravity O and the three edges converging at the lower corner E, we can see that in the projection obtained in this way, the distance $r = OE$ determines the speed at which the

---

[508] See M. Smoluchowski, *Notes on the concept of chance in physical phenomena*, op. cit., pp. 76–77.

die will topple. Secondly, the direction of the vector *OE* with respect to the three edges (which can be defined with a certain angle *O*) will determine what number will appear on the top. Chance of this kind, according to Smoluchowski, also falls outside any *a priori* calculation, so can never be the basis for applying probability calculus. Until we know with sufficient accuracy the causes determining the direction and numerical value of the vector *OE*, nothing can be predicted concerning the effect. But when we know, the result can be predicted with absolute certainty so that no room remains for probability[509].

In order to demonstrate specific features of an event in which the values enabling the use of probability calculus are known, Smoluchowski cites an example illustrating such a situation. A shooter fires a fixed shotgun at a spinning circular disc divided into sectors and painted alternately black and white. Whether he hits a black or white sector depends on the moment he pulls the trigger. The disc can be spun so fast that certainty of firing a shot on target can be eliminated. At whatever moment the shooter pulls the trigger, the time that has elapsed from taking the decision to fire will vary within certain limits such that the probability of the shot occurring at time t is expressed by the function $\varphi(t)$ (perceptibly different from zero in the range of t to $t + \tau$), however for this function it must be assumed that there are no unique features. If there are many revolutions of the disc in the time range τ, the influence of the individual form that the function $\varphi(t)$ could have disappears and the probability of hitting a white or black sector depends on the relative size of the fields[510].

Chance for which probability can be calculated, the so-called normalised chance, differs significantly from chance in the wider sense in that the effect shows a certain regularity through frequent repetition of the phenomenon regardless of the type of cause[511]. In complex situations, the task of probability calculus is not to explain the probability of a given event, but to detect it on the basis of another probability—assumed as the known probability

---

[509] Idem, p. 33.
[510] See M. Smoluchowski, *On the concept of chance and the origin of laws of physics based on probability*, op. cit., p. 35–36.
[511] Idem, 34.

of a simpler phenomenon that causes it[512]. We are dealing here with chance suited to the calculation of probability. It is distinguished from chance in the broader sense by the important feature that the effect is characterised by a certain regularity with frequent recurrence of the phenomenon, regardless of any particular type of cause[513].

Amsterdamski makes the same statement differently, noting that

> the objectivity of probabilistic statements does not consist in the fact that they refer to the only appropriate, fullest or best description of an experimental situation in which we reflect on the occurrence or non-occurrence of a given event: in other words—that they are based on 'one' adequate reading of the reference class. Probabilistic statements are objective because they describe the relations occurring between events grouped into classes on the basis of one or another set of their properties in such a way as to allow a prediction, at least probabilistic, of events occurring in this system[514].

This makes it possible to express the predicted event in a mathematical, empirically verifiable, formula. According to Amsterdamski, when considering the issue of the relationship between deterministic and probabilistic laws, the indeterminist argues that probabilistic laws are not reducible to deterministic ones. This means that, if not for all, then at least for certain events to which the relevant probabilistic laws apply, there are no deterministic laws, there are no situational factors that uniquely determine them[515]. The obviousness of these considerations is emphasised by the opposite case, in which an event that occurred, even if it were not probable, would nonetheless be clearly defined.

5.3. Epistemic causality

Grasping the ontological aspect of causality is an important step towards the scientific study of certain types of phenomena in nature. We obtain

---

[512] Idem., p. 37.
[513] Ibid.
[514] S. Amsterdamski, Z. Augustynek, W. Mejbaum, *Law, necessity, probability*, op. cit., p. 89.
[515] Idem, page 114.

a concept of the notion of causality that enables the application of new mathematical methods to the description of an event: statistical calculations and probability theory. Adopting a deterministic position, we treat cause and effect as a constant necessity linked by internal relations connecting partial events. An ontic approach to causality makes it possible to describe physical events using probability calculus. This is due to the possibility of determining mathematical relations connecting the occurrence of partial events with the effect.

In the understanding of Alexius Meinong (1853–1920) "there exists a causal relationship between a given cause and effect, but it is unknowable to us because the phenomenon is too complex. Therefore, we are dealing with an apparent break from regularity, and chance is referred to as 'a partial cause unknown to us'"[516]. Smoluchowski perceived the issue described by Meinong differently. He believed that ignorance of the partial cause was not a barrier to calculating probability. It is possible to calculate the effect caused by a "partial cause unknown to us", referring to the law of large numbers, a rule that cannot be proven, but which has also turned out to be impossible to refute empirically[517].

The law of large numbers causes irregularities brought into the world by random events to disappear in the overall result. Our mind probably cannot reconcile itself to adopting such a rule just because its validity has been confirmed here and there[518]. In his 1906 book *Principles of Probability Calculus*, Władysław Gosiewski (1844–1911) defines the law of large numbers as follows:

> One can always think of such a large number of trials (μ) that each of the exclusive events (ai), the probabilities of which are generally variable from trial to trial (ps(ai)), would be repeated so many times (ai) that the ratio of this number to the number of all trials (ai/μ)) differs as little as one likes from the average value of all

---

[516] See M. Smoluchowski, *On the concept of chance and the origin of laws of physics based on probability*, op. cit., p. 29.
[517] Idem, p. 31.
[518] Idem, pp. 30–31.

probability trials of this event (1/μΣsps (*ai*)), with a probability as close to certainty as one likes (*P*)[519].

Gosiewski writes that the probability of exclusive events, i.e. those where the probability of the expected effect of the next trial is independent of the effect of the previous one, is obvious and simple. However, the case in which the probability of the expected effect of each subsequent trial depends on the effect of the preceding one requires further explanation. In this case, there is no longer a simple dependence of probability on the order of execution of the trial, but each trial to which the probability belongs prepares the probability for the one following it. Thus, as a complement to the law of large numbers, we obtain the following theorem: if the probability of the expected effect of each subsequent trial depends on the effect of the previous trial, then the expected result of trials, as well as the probability of this result, are the same as if the probability of the expected effect of each subsequent trial did not depend on the effect of the previous trial, but acquired current values in all trials, depending only on the order of the trial[520].

Gosiewski refers to the research of Andrey Andreyevich Markov (1856–1922), citing the so-called Markov property. This is a property of stochastic processes consisting in the conditional probability distribution of future states of the process, which are determined solely by its current state, with no stochastic process memory. The probability distribution of future process states depends solely on the current state. Future states of the process are conditionally independent of past states. Examples of Markov processes are the description of Brownian motion and the process of a radioactive substance emitting particles. A Markov process is "a stochastic model describing a sequence of possible events in which the probability of each event depends only on the state attained in the previous event"[521]. This statement by Gosiewski is of great importance for understanding

---

[519] W. Gosiewski, *Rules of calculus of probability*, Warsaw 1906, p. 95.
[520] Idem, p. 95–97.
[521] V. Givindaraju, V.V. Raghavan, D. Rao, *Big Data Driven Natural Language Processing Research and Applications*, in: *Handbook of Statistics*, Elsevier 2015, vol. 33, pp. 203–238.

Smoluchowski's concept in the context of the possibility of mathematically determining the occurrence of chance using probability calculus. Chance eligible for calculation of its effect through probability theory depends only on a sufficiently frequent occurrence of a situation and its repetition. The effect does not depend on the probability of the occurrence of a previous event, but only on the order of the next one. A safe situation is when we are dealing with relative frequencies falling within the scope of the law of large numbers, and the essence of chance comes down to the fact that the cause of the event is consistent with the law of probability.

Smoluchowski writes that we can talk about mathematical probability when the function $y = f(x)$, representing the causal relationship between random cause x and effect y, has the property that a certain set of values for x, in an arbitrary distribution, always corresponds to approximately one and the same distribution of the corresponding values for y. The term "approximately" is intended to express the fact that a strict identity of y distributions can be expected only in infinitely numerous special cases, i.e. in numerous sets[522].

Probability is a mathematical concept, hence every probabilistic theory is associated with the issue of its physical and operational interpretation. The frequency interpretation of probability is generally accepted. It states that in a set of randomly selected, identical systems to which a given theory applies, individual dynamic possible behaviours should occur with relative frequencies, proportional to theoretical probabilities, and this consistency is the better the more individual cases are taken into account. This is not so much an operational as simply a physical meaning of the abstract, mathematical concept of probability[523], writes physicist Jan J. Sławianowski. The scholar thereby confirms that we build probabilistic consistency based on the law of large numbers, which enables physical and operational interpretation with the greatest possible use of individual cases.

---

[522] See M. Smoluchowski, *On the concept of chance and the origin of laws of physics based on probability*, op. cit., p. 44.
[523] See J.J. Sławianowski, *Przyczynowość w mechanice kwantowej*, Warszawa 1969, pp. 37–38.

In discussing the nature of chance, Ernst Mach's functionalism cannot be ignored[524], as the initiation of Smoluchowski's idea of creating the theory of an ontic understanding of causality can be seen in a particular approach to Mach's concept. He presented the idea of replacing an epistemic understanding of causality with a relationship called functionalism[525], in which the so-called causal relationship is a sequence of successive phenomena that are functionally dependent on one another. Mach understood function in the ordinary mathematical sense, which resulted in removing the concepts of cause and effect from science as vague and ambiguous, and replacing them with the mathematical concept of a function describing the dependence of the characteristics of phenomena on each other.

Positivists and neopositivists regarded David Hume as their spiritual patron, especially in terms of understanding causality. Hume's undeniable merit is forming the basis of a causal law for physics based on experience. A retreat to dogmatic positions has since become an impossibility[526], but the very thesis of causality proved to be more controversial. The legitimacy of Hume's concept has been questioned many times, and his idea has often been challenged[527].

---

[524] This concept is interesting for at least four reasons. Firstly, Mach had a negative attitude towards the classical understanding of causality and, being a celebrated figure in the world of science, especially physics and philosophy, he had an influence on the perception and definition of the problem through the scientific circles of his time. Secondly, Smoluchowski certainly knew Mach's philosophical concepts, including those concerning causality; he was interested in his philosophy in terms of the discourse ongoing at the time on the essence of matter. Thirdly, in 1895, Ernst Mach took the chair of philosophy at the University of Vienna and often entered into disputes over kinetics and atomics with Ludwig Boltzmann, who headed the physics faculty. At this time, 1895, Smoluchowski was defending his doctorate in physics and the natural order of things and being at the university undoubtedly took direct or indirect part in the disputes between the two scientists, the subjects of which were of fundamental importance to him. Fourthly, in his scientific work, Smoluchowski paid a lot of attention to the issues of chance and causality, building the basis of probabilistic methodology in physics. It seems inconceivable in this situation for the Polish physicist to not have had an interest in Mach's views on causality and hence it must be asked what influence Mach's philosophy had on Smoluchowski's views.

[525] See B.J. Gawecki, *The issue of causality in physics*, Warsaw 1969, p. 179.

[526] Ibid.

[527] See Z.E. Roskal, *The principle of causality in the context of experimental research on Brownian motion*, op. cit., p. 59.

Mach resumed his work in the field of causality, following the direction set by the Scottish philosopher, and came to the conclusion that the concept of causality as a pre-scientific and anthropomorphic term should be eliminated entirely from physics. The causal explanation of phenomena should be replaced by a mathematical description written in the form of a mathematical function[528].

Mach's concept constituted a particular continuation of the idea of Hume's critique of causality, which assumed that between events treated as cause and effect, at most a temporal sequence could be established. Mach went further in his deliberations than Hume and abandoned the temporal consequence in causal relationships. He believed the functional dependence of phenomena to be mutual and reversible, and that subsequent relationships occurring in nature could be defined as bilateral and simultaneous dependencies—hence they could be expressed as mathematical functions. Time, according to Mach, can be omitted in causal relationships, since all relationships—both spatial and temporal—irrevocably come down to the functional dependence of phenomena.

In *An Enquiry Concerning Human Understanding*, Hume wrote:

> In a word, then, every effect is a distinct event from its cause. It could not, therefore, be discovered in the cause, and the first invention or conception of it, *a priori*, must be entirely arbitrary. And even after it is suggested, the conjunction of it with the cause must appear equally arbitrary; since there are always many other effects, which, to reason, must seem fully as consistent and natural. In vain, therefore, should we pretend to determine any single event, or infer any cause or effect, without the assistance of observation and experience[529].

Mach's assertion is taken straight from Hume, who argued that the fact that 'B followed A' does not entitle us to construct a cause-effect relationship, that 'A caused B', because there are many other effects in this relationship that may seem consistent and natural. On the basis of experience, we learn the constancy of the something following on from something else, but, contrary to popular belief, we do not know the inevitability of this relationship.

---

[528] See B.J. Gawecki, *The issue of causality in physics*, op. cit., p. 84.
[529] D. Hume, *An Enquiry Concerning Human Understanding*, Ed. Eric Steinberg, Second Edition, Indianapolis/Cambridge, 1993, p. 19.

It transpires that this is merely our addition, which has its roots in the fact that we become accustomed to a certain course of phenomena[530].

Michał Heller and Tadeusz Pabjan do not share Mach's view on the critical perception of the causal relationship, arguing: "Against (...) criticism of the principle of causality stand scientific practice and the effectiveness of the scientific method, based on building mathematical models of physical reality, in which formal symbols are assigned (using appropriate rules) to the physical world"[531]. However, it should be emphasised that Mach's intuition concerning time was correct according to some modern scientists since, as Heller and Pabjan claim:

> The logical implication between the model's mathematical symbols indicates that between the physical events (or properties of the world) that correspond to the symbols, such a result also occurs rather than merely a temporal sequence (in the mathematical model, the time parameter can be dispensed with entirely). It is true that the logical implication is not in itself causation, but it effectively models causation[532].

Disagreeing with Hume and Mach, in *Time and Causality* Heller emphasises the importance of the causal relationship, writing that:

> The concept of causality—contrary to the assertion of various positivist and post-positivist philosophers—is well established in physics. The theory of relativity—both special and general—talks about the so-called *structure of causality* of space-time. This structure merely refines our everyday understanding of the causal relationship, specifying exactly which events in space-time can enter into a cause-effect relationship with one another, and which cannot. The causal structure understood in this way is, of course, closely related to the concept of location: a cause here, by means of a certain physical signal (which, however, cannot move in space at a speed greater than light), has an effect there[533].

Heller explains how we can get rid of the temporal sequence of cause and effect, proving that the causal structure in a given space is determined by the Lorentz metric. This defines so-called light cones at all points in space-time,

---

[530] See B.J. Gawecki, *The issue of causality in physics*, op. cit., p. 19.
[531] M. Heller, T. Pabjan, *Elementy filozofii przyrody* (*Elements of the philosophy of nature*), Kraków 2014, p. 125.
[532] Ibid.
[533] M. Heller, *Time and causality*, op. cit., p. 40.

outside of which causal interactions cannot propagate. He writes: "The Lorentz metric, which defines the causal structure, is an additional structure of a given space (or space-time), that is, it does not fit within the concept of space, but must be 'superimposed' on this space"[534].

An appropriately generalised metric (of the Lorentz type) can also be 'applied' to a non-commutative space. If we do this, we get a causal relationship without time and space. As Heller explains:

> "In the notion of generalised causality there is therefore no temporal sequence of cause and effect; there is only a dynamic relationship between states. So we have a trinity: dynamics – probability – causality. One non-commutable geometric structure combines three previously distinct concepts. Is there room for chance in this structure?"[535].

Chance, writes Heller, is an aspect of the overall structure to which we have the right to assign a "small generalised probabilistic measure"[536].

The above considerations make us aware that the whole problem of causality, cause, effect and chance, despite the passage of more than a hundred years, is still an unresolved issue in science. It can therefore all the more be concluded that the step taken by Smoluchowski—leading to the problem's shift from the field of philosophy to the realm of science—was of such importance. His work involved introducing a scientific methodology to causal considerations, enabling a transition from qualitative to quantitative research on causality, which was associated with the introduction of probability calculus into this discourse.

An important issue influencing the process of Smoluchowski introducing probabilistics into the field of causality is the fact that in Polish scientific circles, mainly philosophical ones, at the turn of the 19th and 20th centuries a number of papers appeared on the problems of causality, probability calculus, or the theory of large numbers. Some of them were familiar to Smoluchowski, especially those discussed within the academic community. Some of these works concerned scholars with whom he had personal contacts. The exchange

---

[534] Idem, s. 41.
[535] Ibid.
[536] Idem, p. 42.

of ideas, comments and suggestions can be seen in their correspondence, some of which can be found in the Jagiellonian University Archive[537].

Smoluchowski was undoubtedly interested in this subject, despite its philosophical background. However, he applied to the discourse his own line of thinking—that of a scientist looking for an appropriate scientific methodology to be able to study the problem from the position of a physicist. Scientific inspiration arose as a result of both private and scientific contacts established in a small group of Polish scientists working on causality. The very process of shaping the framework of the theory is interesting, all the more so as sometimes a small suggestion was enough to move this idea in the right direction.

Inspiration from Polish scholars, such as mathematician and physicist Władysław Gosiewski, philosopher and naturalist Władysław M. Kozłowski, philosopher of natural sciences Joachim Metallmann, historian of philosophyWładysław Heinrich (1869–1957) or philosopher Bolesław Gawecki, with all of whom Smoluchowski maintained contact, can be seen in the works *On the concept of chance and the origin of the laws of physics based on probability*, *Notes on the concept of chance in physical phenomena* and *The Self-Study Handbook*. Citing these scholars' philosophical and scientific considerations demonstrates the seriousness of the discussion ongoing at the time in Polish scientific circles. It was undoubtedly a direct stimulus inspiring Smoluchowski to take up the issue of chance and causality.

One of the important publications in the context of Smoluchowski's deliberations on causality is Bolesław Gaweck's book *The Problem of Causality in Physics*. Gawecki dealt with causalism and functionalism throughout his scientific work, arriving at very interesting conclusions. In the above-mentioned publication, discussing various concepts of causality and chance, he conducted an analysis of Mach's functional causality, and—importantly—revealed the sources of Smoluchowski's conception of chance. It cannot be ruled out that Gawecki's work influenced Smoluchowski's decision to take up research on the problem of causality.

---

[537] Sample correspondence with Władysław Gosiewski in the Jagiellonian Library: reference number BJ Rkp. 9415 III vol. 3, with Władysław Heinrich: reference number BJ Rkp. 9420 III vol. 8, with Bolesław Gawecki: reference number BJ Rkp. III.

During Gawecki's work on causality, Władysław M. Kozłowski held the position of a respected philosopher in Polish science. Gawecki became acquainted with Kozłowski's deliberations on causality through his 1906 book *Causality as a fundamental concept of the natural sciences*. He cites it in the footnotes to *The problem of causality in physics*. In time, Gawecki also wrote a biographical book, *Władysław Mieczysław Kozłowski (1858–1935)*[538]. Kozłowski's conception assumed that the notion of dependence is broader than that of causality, which constitutes one form of dependence (causal dependence), while the concept of function is a dependence according to a rule determined quantitatively. A more defining feature needs to be added to dependence to arrive at causality. This feature can be found by considering one of the issues "which runs through the entire development of the philosophical concept of causality"[539]—the issue of the temporal sequence of cause and effect.

Gawecki recalls that in 1872 Mach put forward the idea of replacing causality with functionalism, in which the so-called causal relationship is a sequence of successive phenomena that are functionally related to one another. He cites Mach's earlier outlined concept of function in the mathematical sense, in which the philosopher proposes removing the concepts of cause and effect from science, as vague and ambiguous, and replacing them with the mathematical concept of a function describing the dependence of features of phenomena on each other. The functional dependence of phenomena is supposed to be mutual and reversible[540]. According to Mach, successive relationships in nature can be defined as bilateral and simultaneous dependencies—and can therefore be expressed as mathematical functions. Time can be omitted in causal relationships as all of them—both spatial and temporal—irrevocably come down to the functional dependence of phenomena[541].

Kozłowski's remarks, with which Gawecki had the opportunity to familiarise himself, and which influenced the process of creating *The problem of causality in physics*, are important in the context of these considerations. Kozłowski notes that Mach wants to eliminate the concept of causation and

---

[538] B.J. Gawecki, *Władysław Mieczysław Kozłowski (1858–1935)*, Wrocław 1961.
[539] See W.M. Kozłowski, *Causality as a basic concept of natural science*, op. cit., p. 19.
[540] See B.J. Gawecki, *The issue of causality in Physics*, op. cit., p. 179.
[541] Idem., pp. 179–180.

replace it with two principles: the principle of continuity and the principle of differentiation. The first involves retaining, as far as possible, the acquired habit of mentally combining A and B, but under slightly different conditions. This is the principle of possible generalisation. The second principle tells us to look for differentiating moments in the substitution of the known relationship between A and B, instead of the new $A_1 B_1$[542]. He points out, justifying Mach's point of view, that temporal and spatial relations are useless and create confusion in the concept of causality. As scientific concepts, time and space are "abstractions"—auxiliary hypotheses. The "concept" of time arises through changes and has significance only as long as changes exist. Temperature changes with time—that means it is dependent on the angle of the Earth's rotation. Spatial and temporal relationships are ultimately reduced to the interdependence of phenomena, which replaces all basic relations coming from outside (time, space and causality). Therefore, in place of a "metaphysical" dependence between cause and effect, there will eventually be a "purely logical" relationship between "the conceptual elements determining the fact"[543].

Kozłowski analyses the schema proposed by Mach with a specific example. Breaking away from the temporal and spatial form of phenomena, we realise the relation of causal to functional dependence in the following way: we have a clockwork mechanism in which all the cogs intermesh and there is no hindrance to movement in either direction; by turning any cog left or right we can set the whole mechanism in motion in the appropriate direction. But if each cog has catches (like those on a watch with a mainspring) allowing movement only in one direction, and not allowing it in the other, such a mechanism will be able to be moved only in one direction and moving any cog will set in motion only those following it (in the direction of the designated movement), the cogs preceding it remaining static.

The first mechanism, as Kozłowski indicates, presents functional dependence, the second a causal relationship.

> The first joins the moving parts in such a way that they are all dependent on each other and this dependence manifests in any direction; the second creates a one-way fixed chain (the advent of a cause elicits an effect) and all the subsequent

---

[542] See W.M. Kozłowski, *Causality as a basic concept of natural science*, op. cit., p. 16.
[543] Idem, pp. 16–17.

ones (not in time but in the direction determined by irreversibility) but nothing changes in the previous one, there is no dependence here from effect to cause. i.e. in reverse (in the opposite direction). We can now answer the question (...) whether the mathematic form of a law of nature expresses everything contained in that law. Here of course we do not mean that functional dependence expresses only a quantitative relationship while laws of nature express a relationship between the qualitative contents at play; that substitution of quality with quantity is justified and does not impede every specific interpretation of the law. Mathematical symbols are used in this case like nominative numbers, indicating not only how many but what. What interests us is whether beyond the form of a function itself something implicit lies hidden without which this form would lose all connection with reality and therefore all non-mathematical meaning. Such an assumption exists and represents an indispensable component of all mathematical formulae expressing laws of nature: this is that the dependence relationship exists here only in the direction from certain parameters to others, not two-way, as the mathematical form of the law would lead us to suppose[544].

Kozłowski clearly does not identify with Mach's functionalism. According to his further reasoning, every mathematical function, when used to express a law of nature, contains the implied stipulation of a one-way dependency of parameters. One-way dependence is the nature of causality. Therefore the previous theorem means that every mathematical function, insofar as it expresses physical relations, contains an implied causal theorem. Hence it follows that not only can functional dependence not replace causality, but that it is through it that it acquires meaning in application to real-life phenomena[545].

Kozłowski's thought guided Gawecki's understanding; he also disagreed with Mach's thesis, though he justified his position more forcefully, asserting that the proposed reform consisting of the elimination of the temporal corollary and introducing invertible functions cannot be applied to all types of physical relationships[546]. Not all relationships found in physics can be interpreted only as mathematical functions: apart from Mach's function, there are also essentially irreversible functions, which can be called 'physical' functions where there is an element of time, differentiating one of the terms, when B follows A but A cannot follow B. This happens in a situation

---

[544] Idem, p. 22.
[545] Idem, p. 23.
[546] See B.J. Gawecki, *The issue of causality in Physics*, op. cit., p. 105.

in which the order in which the states of a phenomenon occur after each other is constant, determined by nature itself[547]. Gawecki made a distinction between causalism, operating through physical functions (understood as a certain dependence relationship between the quantities characterising the parts of a given physical phenomenon, following each other in time in a certain defined order) and functionalism utilising reversible functions, which Mach gave exclusivity in science[548].

The irreversibility of succession in this group of phenomena cannot be eliminated. This is not some intellectual invention but stems from the real world we study directly, and this directness is the source of the causal account of phenomena, known as causalism. Not all relationships found in physics in a mathematical form can be interpreted simply as mathematical functions[549]. The main source of error, Gawecki writes, was Mach's prejudice against the causal method as against a metaphysical superstition, which resulted in extreme mathematisation of the relation. Rather than replacing a pre-scientific concept of cause with a scientific concept, Mach entirely ruled out causality, claiming that functionalism is fully sufficient. Mach's critics raised the objection that such a generally formulated concept of dependence as he proposed cannot always be applied. The concept of function is too broad to replace the relationship of causal connectivity[550]. Metallmann summed up Mach's functionalism by stating that determinism is a principle applied to the real world in which something takes place, i.e. the world in time, not the world of ideal relations[551].

In reversible phenomena, 'cause' and 'effect' do not express any content justifying their application in science. To take account of reversible phenomena, the concept of function is sufficient in the mathematical sense. However, in the case of irreversible phenomena, there is no way to eliminate the temporal element, hence it seems expedient to apply to these phenomena a causal method operating through temporally interpreted 'physical functions'.

---

[547] Ibid.
[548] Idem, p. 106.
[549] Idem, pp. 105-106.
[550] Idem, p. 106.
[551] See J. Metallmann, *The problem of chance*, op. cit., p. 91.

Indicating a certain defined direction of the permanent succession of real phenomena, it can be treated as a causal clarification of these phenomena. The general determination of natural phenomena should not be prejudged; after all, we are talking here about the nature of a causal law for physics rather than about the principle of causality. The assumption that every phenomenon is related in a defined way to other phenomena applies both to reversible and irreversible phenomena[552].

Determinism—both in science and in everyday life—expresses nothing more than human faith, deeply grounded and completely justified by previous experience, in the constancy of the laws of nature, which can take the form of either a reversible functional relationship or an irreversible, causal relationship[553]. It is worth warning against, writes Gawecki, because similar mistakes have often been made. As tempting as it may be to think that solving the problem of physical causality automatically entails solving the problem of causality for the real human world, we must regard such a thought as a dangerous illusion[554].

Gawecki built his theory at least from 1913 but his first publication—*Causalism and functionalism in Physics*—only appeared in the *Philosophical Quarterly* (*Kwartalnik Filozoficzny*) in 1921, four years after Smoluchowski's death. However, Smoluchowski became acquainted with Gawecki's concepts, undoubtedly in detail. Bolesław Gawecki's doctoral file, stored in the Jagiellonian University Archive, contains a protocol[555] from a *rigorosum* in physics that was held on January 27, 1914, as the final pre-doctoral exam. The examiners were Władysław Natanson and Marian Smoluchowski. On 30 January 1914, Gawecki defended his doctoral thesis, *Causalism and Functionalism in Physics*, and received the degree of Doctor of Philosophy at the Jagiellonian University. In his doctorate, he dealt with Mach's functionalism;

---

[552] See B.J. Gawecki, *The issue of causality in Physics*, op. cit., pp. 183–184.
[553] Idem, p. 184.
[554] Idem, p. 185.
[555] Minutes of the one-hour *rigorosum* in physics held with Mr. Gawecki on January 27, 1914, in the presence of Władysław Natanson and Marian Smoluchowski with an excellent result, Card 23 of the Jagiellonian University Archive.

Smoluchowski was the promoter[556] of Gawecki's work and promoting him proves that he knew the work itself well.

It is therefore no coincidence that Smoluchowski published his considerations on causality two years later. In 1916 the essay *Notes on the concept of chance in physical phenomena* was published, and in 1917 the paper *Subject, task, method and the division of physics*. In 1918, after his death, an article appeared in German: *Über den Begriff des Zufalls und den Ursprung der Wahrscheinlichkeitsgesetze in der Physik* (*On the concept of chance and the origin of laws of physics based on probability*).

The most important argument, however, is the fact that both Smoluchowski's concept of chance and his causal approach to phenomena are reminiscent of Gawecki's contentions. This author's analysis of Mach's philosophy undoubtedly inspired Smoluchowski to build a conception of chance. Gawecki's arguments led him to a special way of perceiving causality. Taking note of Gawecki's reasons for distinguishing between a causal and functional understanding of causality, Smoluchowski understood that these two concepts need not be mutaully exclusive but, on the contrary—they can be understood dually, in a way in which there is room for both causalism and functionality. The breakthrough here is the differentiation of the perception of causality into epistemic and ontic ways (cognitive and existential), where in the first case the participation of the time factor is permissible, and in the second superfluous.

---

[556] Letter of November 18, 1913, from Dean Jan Łoś of the Faculty of Philosophy of the Jagiellonian University (written on a typewriter and signed with a facsimile) in which he sends professors Witold Rubczyński (1864–1938) and Władysław Heinrich a doctoral dissertation of Bolesław Gawecki along with a doctoral student's note—a request for admission to *rigorosum* examinations. At the Dean's request, there is a handwritten entry made in pen: "Promotion, on January 14, 1914. Supervisor Prof. M. Smoluchowski", Card 12 of the Archive of the Jagiellonian University. Smoluchowski was not a promoter according to today's understanding of the term. The doctoral supervisor, as Gaweckiwrites in his *Curriculum Vitae*, was Prof. Heinrich. Smoluchowski promoted *sub auspiciis Imperatoris*, which means that he had to study Gawecki's doctoral thesis thoroughly. In a book from 1913, the author Kazimierz Kumaniecki writes on p. 100: "According to the regulation of 28 August 1888, only a candidate of philosophy who has not only passed the strict examinations but has also demonstrated that his dissertation, in terms of its scientific value, significantly exceeds the ordinary standard may be considered for promotion *sub auspiciis Imperatoris*".

5.4. Ontic causality

Grasping the ontic aspect of causality is an important step towards the scientific study of certain types of phenomena in nature. We obtain a concept of the notion of causality that allows the application of new mathematical methods such as statistical calculations and probability calculus to the description of an event. Taking a deterministic standpoint, we treat cause and effect as a permanent necessity connected by internal relationships linking partial events. An ontic account of causality enables a description of the physical effects occurring with the aid of probability calculus. This stems from the possibility of determining the mathematical relations connecting partial events to the effect.

By studying the movements of gamboge particles, reacting to the energy transmitted to them by vibrating, Smoluchowski discovers their cause. The observed process is both epistemic and causal in nature. Brownian motion is induced by a cause, namely moving particles of the solution, which as a result of collisions with gamboge particles cause the effect in the form of these particles' vibrations. This case illustrates the causal and epistemic understanding of cause-effect relationships in irreversible physical events. Interpreting philosophically understood causality was not limited to knowledge of its epistemic nature, because Smoluchowski, by applying a research method involving collating the statistics of solution particle displacements, introduced mathematical relations to causality, enabling the calculation of the effect through a scientific method. Presented this way, the issue did not pose a methodological problem due to the mathematical tools the Polish scientist used.

At the same time, Mach's line of reasoning made Smoluchowski aware of the possibility of disregarding time in causal relationships and expressing causality in mathematical functions, with the difference that, unlike Mach, he did not analyse causality in its epistemic aspect, but instead examined its ontic nature. That was the intention of another perception of causality, which was related to a shift in considerations of causality from the philosophical to the methodological space of science, especially of physics. Interpretation of the essence of chance became the tool enabling this intention.

Smoluchowski arrived at the essence of causality's ontic nature through analysing chance as a special type of causal relation. In this approach, a given event depends on chance, which is a function dependent on a variable causing the event. This cause is in line with the law of probability and is primary in relation to the effect, for which we note the immutability of the law of probability. The occurrence or non-occurrence of an event depends on a very small change in the variable constituting the cause. The key issue is the occurrence of the event itself in terms of cause and effect, disregarding the factor provoking the relation, while focusing on the statistical possibility of its occurrence.

This subtle difference shifts the consideration of causality from a philosophical discourse to the realm of empirical research and scientific verification of the event, thus enabling, through the use of probability calculus, determination of the potential for the event to occur. According to Amsterdamski, "an objective interpretation of the concept of probability requires it to be treated as a certain physical concept, as an ontological category"[557]. Wanting to capture the essence of the problems that arose, Smoluchowski asks how it comes to pass that the result of chance can be calculated. How is it possible that random causes have the right effects? How can chance arise if all phenomena should be reducible only to constant laws of nature? Or, in other words—how can the right causes have a random effect[558]?

Chance—according to Smoluchowski—is a specific type of causal reference. This is an ambiguous concept. It can be understood in a variety ways. Lemańska conducts a fairly detailed analysis of the possibility of chance occurring. She cites its characteristic features: 1) chance is the accidental action of many causal factors that are disordered in relation to each other and independent of each other in their action; it does not have its own essential cause, of which it would be a direct effect, but assumes many proper causes, unrelated to each other in action; 2) chance is an event that is actually subject to strict causality, but due to the multiplicity of causes they are not possible to grasp cognitively; 3) an event is random when its cause does not exist. Chance in the relative sense is an event that has no cause within a given frame of reference. Chance in the absolute sense is an event that has no cause in the entire material world; 4) a random event is an event for which there is no explanation, i.e. there is no such law of science the conjunction of which with sentences stating the occurrence of certain initial conditions results in a sentence stating the occurrence of the event in question; 5) chance is treated

---

[557] S. Amsterdamski, Z. Augustynek, W. Mejbaum, *Law, necessity, probability*, op. cit., p. 91.
[558] Idem., p. 29.

as absolutely independent of all laws; 6) an event is random when ot lacks an ontic cause; 7) a random event is one for which we cannot provide a cause[559].

This philosophical *résumé* ofvarious functions of chance brings us closer to its nature but does not touch on the function of chance that interested Smoluchowski, which would enable the application of probability theory to calculate the probability of its occurrence.

> In seeking an approach to chance that would effectively make it subject to mathematical calculations, Smoluchowski analyses various premises; it is generally believed that a particular event y depends on chance when it is such a function of the variable x—called a cause or partial condition (x may also be unknown as to its numerical value, or intentionally ignored)—that the appearance or non-appearance of this event depends on a very small change in the argument x (small in relation to the range of variation of x)[560]. The scientist came to the conclusion that a mathematical law of probabilities $W(y)$, referring to the magnitude of y, can be talked of when $y = f(x)$, in addition to the property that this event occurs in the case of a very small change in the argument x, also has the special property that the distribution of the y-value (within certain boundaries) is independent of the type of dispersion $\varphi(x)$, determining the relative frequency of the *x-value* (assuming that the function $\varphi(x)$ has the correct course)[561].

> In general, Smoluchowski argued, we explain this situation by assuming that although there is a proper causal relationship between a specific cause and effect, the type of relationship is unknowable to us because the phenomenon is too complex. Hence the apparent break from regularity. In this sense, chance should be described as a 'partial' cause unknown to us. The Polish scholar sees a certain affinity of this idea with Meinong's understanding, according to which randomness would mean the factuality of "something unnecessary", though he denies that necessity has to be either internal or external (in relation to a certain set of objective phenomena)[562].

Chance suited to the calculation of probability is therefore distinguished from chance in the broader sense by the essential and characteristic feature that the effect shows a certain regularity with frequent repetition of

---

[559] See A. Lemańska, *Determinism*, op. cit., pp. 11, 12.
[560] See M. Smoluchowski, *On the concept of chance and the origin of laws of physics based on probability*, op. cit., p. 11.
[561] Ibid.
[562] Idem, p. 29.

the phenomenon, independently of a specific type of cause[563]. An important moment of change to the erstwhile reasoning is the shift of attention from the cause to the very occurrence of the event, defined as a chance event occurring in terms of cause and effect. In this situation, it is not important for the researcher to consider either the element stimulating the causal relation, i.e. cause in the Aristotelian sense, or to study the direct relationship between cause and effect. The effect itself is studied, or rather the statistical possibility of its occurrence, focusing on the existential side of causality.

The subtle difference causing a shift in emphasis of the perception of a causal event's occurrence means that we can use a tool, such as probability calculus, to subject the event to scientific verification. Changing the definition of the problem enabled chance to be brought into the realm of physics. Expressing the occurrence of an event through the use of a mathematical formula was of tremendous importance as it enabled a scientific definition of the problem under consideration. This became possible as a result of shifting the focus of understanding causality from the discourse of seeking a cause, to considerations regarding the occurrence of chance.

Chance is that aspect of an event's occurrence that can be approached mathematically. Considering this idea of Smoluchowski's in the example of the decay of the element radium, its essence can be described. The half-life of the radium-226 isotope is 1,599 years. Over this period, the semi-decay of any specified quantity of radium occurs, which means that the quantity of radium atoms is reduced by half during this time. It is impossible to determine and know which atoms will decay, at what time, and for what reason. However, this is no obstacle to determining the time over which half the isotopic atoms will break down. In this case, the most important problem is neither epistemic nor philosophical as we are not asking what the cause of the disintegration is. We are focusing on the statistical aspect of the occurrence of chance, and hence on the possibility of calculating it. The ontic fact of decay becomes the most important thing, as a result of which we achieve the effect in the form of a different number of radium 226 atoms. This causality is obviously of a philosophically causal nature but due to the omission of the element of knowing the cause, we are working in the physical realm of this event's ontic occurrence.

---

[563] Idem, p. 34.

Lemańska presents this issue in a similar form, but with a different philosophical tone:

> Such events as the transition of an electron from one permitted orbit to another, or the radioactive decay of the nuclei of elements, are events that occur with no apparent cause. Since no reason has been found for radioactive decay or the transition of an electron from one shell to another at any given time, it is impossible to determine either when an electron will transition from a higher energy state to a lower energy state, or when and which nucleus will degenerate. Here we are dealing with chance events for which no known cause exists. An electron does not move at a given moment from one level to another due to some specific known cause. From a statistical point of view, the lower level is more desirable than the higher one, so sooner or later the electron will jump, though we cannot predict when this will happen[564].

The claim that there exists no cause in these random events is too far-reaching a statement and it could be agreed that there is no cause, but only in a given frame of reference. The cause is unknown to us, which does not mean it does not exist. Smoluchowski's idea did not assume the existence or non-existence of a cause. The scientist often stated that ignorance of the laws of nature is eliminated from science along with the development of our knowledge. In Smoluchowski's theory, cause is not analysed not because it is assumed to not exist but because in a statistical account of the effect its existence or non-existence has no real significance. Not knowing the cause of a phenomenon does not affect the estimation of the probability of its occurrence.

Smoluchowski argues that if we take a deterministic position, we will treat cause and effect as a constant necessity, linked by internal relations connecting partial events. A lack of necessity can only be talked about in a relative sense, insofar as necessity is not externally knowable, i.e. insofar as some of the causes at play remain undetermined. This statement is an answer to the question about the place of chance in physics. The view expressed by Smoluchowski reduces the essence of chance to incomplete knowledge of the laws acting in nature, or to ignorance of all the causes at play.

The use of probability calculus in this particular situation had much more serious repercussions—it convinced physicists that the language of probabilistics could be successfully used to illustrate certain phenomena in physics

---

[564] A. Lemańska, *Determinism*, op. cit., p. 13.

as a tool for describing events, which had previously been questioned. The direction of research into chance set by the Pole resulted in the creation of a serious branch of mathematics dealing with the application of probability theory in physics research.

According to Amsterdamski, many physicists with a range of philosophical views are approaching the concept of probability as an objective ontological category expressing the dispositions of physical systems. In his book devoted to the philosophical problems of modern physics, Werner Heisenberg puts it in a specific way, taking objective probability as a measure of potential akin to Aristotelian potency. However, Heisenberg emphasises that in physical theorems, in addition to objective probability, the concept of subjective probability should be distinguished, and argues that probabilistic statements in quantum physics contain a particular mixture of objective and subjective probability[565].

Smoluchowski's proposition offers an extremely important and effective mathematical tool subjecting new areas of reality to empirical research. However, the dilemma in modern physics of whether indeterminism is ontological indeterminism or whether there also exists some deeper level of reality according to which quantum phenomena are determined, is one of the main problems for physicists and philosophers, provoking discussions which still seem far from unambiguous resolution[566].

To sum up Gawecki's arguments presenting a fairly unambiguous and distinct theory of causality, it may be wondered why Smoluchowski does not refer directly to Mach's concept in the publications he left behind. Why, as a rule, do we not see direct reference to these views in his work? This is all the more interesting as in many places Smoluchowski expresses parallel views, such as the conviction that until then the problem of causality had been treated as anthropomorphic and unscientific, or polemics towards the Austrian's arguments, such as opposition to his categorical functionalism. Smoluchowski distanced himself from Mach's functionalism, but did not want to enter into discussion with Mach himself.

Mach's functionalism, while referring to mathematical functions, was philosophically postulative and hence quite severely biased. The scientific aims

---

[565] See S. Amsterdamski, Z. Augustynek, W. Mejbaum, *Law, necessity, probability*, op. cit., p. 75.
[566] See A. Lemańska, *Determinism*, op. cit., p. 13.

and methodologies of the two academics focused on different aspirations. For Smoluchowski, who saw the possibility of applying probability calculus to physics, the phenomenological approach proved a dead end in researching the problem. Mach's concept was of more use to philosophy than to physics. Also in the work published after Smoluchowski's death, *On the concept of chance and the origin of laws of physics based on probability*, there is no direct reference to Mach's theory, though there is a general thought in which the researcher specifies the difference he sees in the approach of a physicist and a philosopher to this problem:

> The various philosophical analyses of the concept of probability provide no great clarification in this regard. In general, the philosopher is usually concerned here with something entirely different to the physicist. The philosopher (…) is not in the habit of considering the question more closely as to the nature of objective facts that form the basis of probable propositions. In contrast, the strict natural sciences are not interested in subjective statements and presumptions, justified or not, but are interested in objective or 'mathematical' probability, i.e. the relative frequency of specified chance events occurring[567].

The characteristic Smoluchowski outlines of the physicist's and philosopher's differing approaches in perceiving the occurrence of events does not represent a declaration of a radical separation of the strict sciences from philosophical concepts. For example, Smoluchowski enters into a polemic against the arguments of the Austrian philosopher Meinong[568]. He was cited here because his understanding of chance and cause corresponded in some respects with Smoluchowski's ideas. The Polish physicist may have disagreed with him, as in the case of his thoughts on "a partial cause unknown to us"[569], but he held a dialogue with him, while he remained silent about Mach.

In *Notes on the role of chance in physics*, Smoluchowski refers to Mach's philosophical conception, but only to phenomenalism, while he does not touch on functionalism, which was closer to the topic of his work under

---

[567] M. Smoluchowski, *On the concept of chance and the origin of laws of physics based on probability*, op. cit., p. 30.
[568] Idem, p. 29.
[569] See., ibid., p. 31.

discussion[570]. In Smoluchowski's other works devoted to chance, Mach's concept is also not considered and this cannot be considered a coincidence. In addition to the reasons discussed earlier, in this rather specific situation there is another probable motive—one not based on substance. Knowing the Austrian's character and seeing the atmosphere in which the disputes with Boltzmann took place, Smoluchowski did not seek any kind of confrontation with him. It should be noted that, avoiding pointed discussions with Mach and Ostwald, he did not publish his most important research on Brownian motion on time.

Another incomprehensible and quite intriguing issue in literature on the subject is the fact that although Władysław Krajewski for many years dealt with Smoluchowski's philosophy, published two books on it, and referred to Smoluchowski's works in many publications and articles, in the book *Związek przyczynowy* (*Causal Relationship*) published in 1967, he does not refer to his thoughts and does not indulge in any polemics against the physicist's conception. He refers only once in the book to a work of Smoluchowski's—the most important in terms of his approach to causality, i.e. *On the concept of chance and the origin of laws of physics based on probability*, but this reference concerns only a marginal example which has nothing to do with the Polish physicist's conception of causality[571]. Krajewski tries to answer the question of what ontological categories cause and effect belong to. What do we mean when we talk about an effect (what cause are we looking for)? What do we mean when we talk about a cause (what is the cause of a given effect); is it, for example, a thing, an event or a feature? The problem, he states, is by no means easy to resolve[572].

Among the proposals made by Krajewski to solve this problem are statistical causal laws, in which we are not dealing with an unambiguous dependence of the effect on the cause, but with an ambiguous dependence. Continuing the topic, the philosopher cites Reichenbach, who in the book *Wahrscheinlichkeitslehre* (*The Theory of Probability*) gives examples of such dependencies, such as, for example, flu as the cause of death, a shot as the cause of hitting a target, or bombarding of the nucleus of a nitrogen atom

---

[570] See M. Smoluchowski, *Notes on the role of chance in physics*, op. cit., p. 280.
[571] See W. Krajewski, *Związek przyczynowy* (*Causal relationship*), Warsaw 1967, p. 180.
[572] Idem, p. 9.

with alpha particles as the cause of proton release. In all these cases, the first event produces the second with a certain probability[573]. As can be seen, many elements of Smoluchowski's concept appear in Krajewski's work, but there are no references to, or citations of, his works.

Małgorzata Stawarz takes a different position on the issue in the article *Punkt wyjścia filozoficznych rozważań Mariana Smoluchowskiego na temat przypadku i prawdopodobieństwa* (*The starting point of Marian Smoluchowski's philosophical deliberations on the subject of chance and probability*), stating that Smoluchowski presents the problem from a slightly different perspective in an unpublished paper[574]. According to this author, Smoluchowski differentiates between various understandings of chance depending on the way it is perceived, as the conceptual chaos and difficulties that arise when studying the foundations of probability calculus are caused by the differing perspectives from which the problem is considered by people from three fields: mathematics, physics and philosophy. Smoluchowski's remark contained in the paper is a look at the problem of chance from a meta-scientific position, enabling a departure from colloquial language. The mathematician[575], says Smoluchowski, deals with the formal side of probability theory, calculating the probability of a complex phenomenon on the basis of component, elementary phenomena. In determining what he considers to be random, the physicist is usually guided by some intuition and tests the consequences of probability calculus experimentally. The philosopher, on the other hand, is interested in the psychological side of the subject or considers the problem of how to fit probability within the system of formal logic. However, he disregards the question of what objective conditions external phenomena must be subordinated to in order to explain the application of the concepts of chance and probability[576].

In his essay *On the concept of chance and the origin of laws of physics based on probability*, Smoluchowski confirms this earlier idea—various analyses of

---

[573] Idem, s. 237.
[574] See M. Stawarz, *The starting point of Marian Smoluchowski's philosophical considerations on the subject of chance and probability*, op. cit., p. 84.
[575] Notes on considerations about probability in Smoluchowski's manuscript, found by M. Stawarz, are similar to a fragment of the essay *Notes on the role of chance in physics*.
[576] See M. Smoluchowski, *Notes on the role of chance in physics*, manuscript, op. cit.

the concept of probability do not provide any closer explanation, because the philosopher is drawn towards the psychological concept of probability. The philosopher studies the problem from the epistemological point of view, examining how probable statements could be included in formal logic along with true and false ones[577]. In Smoluchowski's view, this means that in the case of an event occurring, an analysis of its probability in such a way that its determination is useful to physics can only be performed in relation to the conditions affecting the event's occurrence.

Smoluchowski's conclusion, which stems from the differences in the perception and understanding of chance, and therefore also of probability theory, in the perspective of the earlier-mentioned conceptual chaos resulting from an attempt to go beyond the common understanding of the issue and to approach the problem scientifically, boils down to the fact that this perception is determined by the position from which it is formed. For example, the philosophical approach to the problem of probabilistics is epistemological in nature, since the philosopher is mainly interested in the problem of placing probability in the system of formal logic. Epistemologically oriented philosophy is not interested in the practical use of the concepts of chance and probability.

In Smoluchowski's understanding of epistemic and ontic causality, and especially in his search for an answer to the question of the place of chance in physics, the thought can be seen that is contained in the idea of the *theory of utility* (to which the next chapter is devoted), which enables the creation of a theory that could be used to the extent it is needed and to which it would prove useful.

A quote from Amsterdamski may constitute a summary of considerations and disputes on the issue of causality:

> At one time, the most important and exciting topic of this dispute regarding determinism and indeterminism was the problem: necessity or purposefulness; later, although the dispute with teleologism did not become obsolete and did not cease to inspire philosophical reflections, the issue of causality or chance came to

---

[577] M. Smoluchowski, *On the concept of chance and the origin of laws of physics based on probability*, op. cit., p. 30.

the fore; now the main subject of controversy has become the question of whether the laws of nature are unambiguous or probabilistic[578].

Smoluchowski contributed significantly to the above-mentioned evolution of the problem's understanding. This fact has not been properly recognised in scientific publications to date. A few statements, such as that of Metallmann: "I would see Smoluchowski's merit in that in place of the sterile notion of subjective chance, as a negation of determinism, he simply introduced (…) statistical regularity"[579], of the Russian Storczak: "It has transpired that Smoluchowski's deep investigations were necessary in order to finally establish that statistical regularity is an entirely new type of regularity, strictly determined by physical conditions"[580], or of the brilliant mathematician Stanisław Ulam (1909–1984), who, analysing the contribution of physicists to the creation of mathematics, wrote: "In some of the more concrete parts of mathematics—for example probability theory—physicists like Einstein and Smoluchowski have opened certain new areas even before mathematicians"[581], only emphasise how the scientific literature has ignored the Polish physicist's achievements in the development of probability in physics.

---

[578] S. Amsterdamski, Z. Augustynek, W. Mejbaum, *Law, necessity, probability*, op. cit., p. 75.
[579] J. Metallmann, *The problem of chance*, op. cit., pp. 93–94.
[580] Ł.I. Storczak, Дискуссия о природе физического знания, (*Discussion s on the origins of physics*), op. cit., p. 206.
[581] S.M. Ulam, *Adventures of a Mathematician*, Berkeley and Los Angeles, 1991, p. 294.

# Smoluchowski's *utility criterion*

6.1. The principle of economy of thought

6.1.1. Mach—economy of thought

In his book *Zasada ekonomii myślenia, jej historya i krytyka* (*The principle of economy of thought, its history and criticism*)[582], Joachim Metallmann alludes to Richard Avenarius and Ernst Mach, the creators of empiriocriticism, a philosophical doctrine that examines the conditions of the validity of knowledge and advocates radical empiricism. Almost simultaneously, at the end of the nineteenth century, both philosophers developed their own concept of the principle of economy of thought, which they incorporated into two different, independent theories. Analysis of this issue is important because the principle of the economy of thought was an important methodological assumption in Smoluchowski's philosophical concept. It was subjected to repeated analysis[583] and was involved in the creation of *utility theory*. Numerous statements referring to Mach's philosophy can be found in the Polish physicist's works.

Avenarius's concept—in competition with Mach's position—was also known to the Polish physicist, but however interesting from the philosophical point of view, it did not play an important role in his deliberations. Smoluchowski was quite close to the idea of the apperception of habit, as he repeatedly referred in publications to the functioning anthropomorphic

---

[582] J. Metallmann, *Zasada ekonomii myślenia, jej historya i krytyka* (*The principle of economy of thought, its history and criticism*), Warsaw 1914, p. 12.
[583] See M. Smoluchowski, *Self-Study Handbook*, vol. I, op. cit., p. 18.

interpretations of the truths of nature, which were created based on the intentions contained in the apperception of habit; for this reason Avenarius's concept will be presented here, but only as an overview.

In 1876, Avenarius published the work *Philosophie als Denken der Welt gemäß dem Prinzip des kleinsten Kraftmaßes. Prolegomena zu einer Kritik der reinen Erfahrung*[584] (*Philosophy as a conception of the world according to the principle of the minimum expenditure of effort. Prolegomena to a critique of pure experience*), in which he argued that the "soul" is the only organ predestined to cognition. At the same time, he claimed that the soul's ability to perform cognitive functions is limited. Taking into account both complementary assumptions, he came to the conclusion that they are a necessary, but also sufficient, condition for the soul's purposefulness, which it achieves with the least effort, using minimal means[585]. He argued that the process of creating science can be reduced to apperception, which is a special form of cognition. Apperception as self-awareness of perception with understanding, based on possessed knowledge and experience, inspires the soul[586], which from among many sensations chooses what it is accustomed to, and therefore the most common and simplest form of apperception is the apperception of habit[587]. Apperception induces passive or active habituation of the body to conditions and is expressed, among other things, in the fact that "no more sensations are drawn into apperception than necessary"; it is therefore the simplest and most common expression of the soul's desire to perform a task with the minimum effort[588].

The process of apperception consists of the interaction and permeation of two groups of sensations—the apperceiving, which rests in consciousness and acts as an assimilator and an active factor, and the apperceived, i.e. that

---

[584] R. Avenarius, *Philosophie als Denken der Welt gemäss dem Prinzip des kleinsten Kraftmasses. Prolegomena zu einer Kritik der reinen Erfahrung* (*Philosophy as a conception of the world according to the principle of the minimum expenditure of effort. Prolegomena to a critique of pure experience*), Leipzig 1870.

[585] Idem, pp. 1–10.

[586] Avenarius does not mean the soul in a philosophical and theological sense, but as the manifestation of human psychism.

[587] See J. Metallmann, *The principle of economy of thought, its history and criticism*, op. cit., p. 12.

[588] See R. Avenarius, *Philosophy as a conception of the world according to the principle of the minimum expenditure of effort*, op. cit., pp. 11–12.

which is processed and assimilated by the assimilator. All cognition or recognition consists in the mutual permeation of these two groups of sensations attracting, subordinating, and assimilating new content[589]. Apperception is the only form of knowledge, and economical; all knowledge can be reduced to apperception and results from its very nature.

A slightly different concept of Mach's principle of economy of thought, embraced in the assumptions of the process of creating science, is quite important in the context of this discussion on the relationships evident between Smoluchowski's and Mach's views. According to the Austrian philosopher, facts should be described in science as simply as possible so they can be understood with the least mental effort. Mach adds that everything science has learned so far "could in principle be known and without it [the simplest formula – J.G.], it would just cost infinitely more time and effort. In every case, the principle of economy, considering thinking not as a fact, but as a means leading to a certain end, is a teleological principle[590]." Smoluchowski refers to this issue in an almost identical way, but qualifies Mach's thought by asking why we are always looking for the simplest formula. He answers that it is because the fundamental laws of nature must be simple. Some philosophers considered the unity or simplicity of nature to be an obvious truth, while others treated the assertion of the simplest formula of the laws of nature as obvious dogma. However, the sceptic, not believing in the objective reality of the laws proclaimed by physics, will treat them as frivolous, unfounded claims. It has happened in the history of science that various laws, initially adopted with faith in their simplicity, were replaced with more complicated formulae as observations were refined, for example Boyle-Charles' law or Newton's law of cooling[591]. To Smoluchowski, the proposition of economy of thought was very important in the building of science, but was not the fundamental element shaping it.

The Polish physicist shows both a continuation of and inspiration from some of Mach's philosophical findings. By drawing on his conclusions or some concepts from the field of empiriocriticism, especially in the field

---

[589] See J. Metallmann, *The principle of economy of thought, its history and criticism*, op. cit., pp. 21–22.
[590] Idem, p. 2.
[591] See M. Smoluchowski, *Self-Study Handbook*, vol. I, dz. cyt., p. 45

of the methodology of physics, he builds his own concept, the so-called *utility principle*. It happens that he directly quotes some elements of Mach's principle of economy of thought, arguing that we seek simplicity and the most straightforward way of describing physical phenomena. He writes that consciously or otherwise, we divide the whole complex of natural phenomena into component phenomena so that their laws take the simplest form, because only such a form meets the needs of science. After all, the scientific system is only useful to the extent that it helps our mind grasp the whole and satisfies our desire for economy of thought[592]. However, Smoluchowski stipulates that it is specific, that it conforms to the nominalist view, according to which the laws of nature would be a subjective creation of our mind, not something objective, existing in nature[593].

6.1.2. Ockham's razor

The principle of economy of thought has a long tradition, having been mentioned as early as Plato (ca. 424/423–348/347 BC)—not directly, but through certain references and formulations—for example in the Parmenides, where the entire dialogue concerns considerations relating to the One[594].

The problem of economy of thought was addressed by Aristotle in *Physics*. In the first book of *Physics*, he argues with Anaxagoras (ca. 500–428 BC), stating: "it is better to rely on a finite number [of principles], like Empedoclesdoes, than on an infinite number"[595], and in Posterior Analytics we read:"Assume that proof from fewer positions, hypotheses, or propositions, is, *ceteris paribus*, preferable to a longer deduction: for if the conclusion conveyed is equally certain, the shorter proof conveys it with greater rapidity"[596].

The idea of the principle of economy of thought appears in the statements of many thinkers, such as Ptolemy: "We consider it a good principle to explain the phenomena by the simplest hypotheses possible"[597]. Leonardo da Vinci (1452–1519) is credited with *the* phrase "simplicity is the ultimate

---

[592] Ibid.
[593] Idem, p. 46
[594] Plato, *Parmenides. Theaetetus*, trans. W. Witwicki, Kęty 2002.
[595] Aristotle, *Physics*, trans. Robin Waterfield, Oxford, 1999, p. 22
[596] Aristotle, *Posterior Analytics*, trans. Edward Poste, Oxford, 1850, p. 84
[597] J. Franklin, *The Science of Conjecture: Evidence and Probability before Pascal*, Baltimore 2001, p. 241.

sophistication". In *Book III* of his *Philosophiae Naturalis Principia Mathematica (Mathematical Principles of Natural Philosophy)* entitled *The System of the World*, Isaac Newton formulates the following as the first rule: "We are to admit no more causes of natural things than such as are both true and sufficient to explain their appearances. To this purpose the philosophers say that Nature does nothing in vain, and more is in vain when less will serve; for Nature is pleased with simplicity and affects not the pomp of superfluous causes"[598]. Einstein is credited with saying that a theory should be as simple as possible, but not simpler[599].

In the history of philosophy, however, it is William of Ockham (1285–1347), a medieval Franciscan friar and philosopher to whom the greatest merit is attributed for raising awareness of the importance of this essential principle in the methodology of science. The so-called Ockham's razor (or Occam's razor), known under several different names, such as the principle of parsimony, the principle of simplicity, or the principle of economy, defines a rule according to which in explaining natural phenomena one should strive for simplicity, and in choosing explanations one should use the criterion of as few assumptions and concepts as possible.

Controversy surrounding the term 'Ockham's razor' has caused some contemporary philosophers to have a rather contentious attitude towards the medieval thinker. Some sceptics have argued that in Ockham's works, including *Summa totius Logicae (The Sum of Logic)*, there is essentially no such original concept of the author, and that the conception of the idea attributed to him known as Ockham's razor should be consigned to the realms of myth. A publication that presents the topic this way was the book *The Myth of Occam's Razor by William M. Thorburn, issued in 1918, in which the author lists the evidence supporting the above-mentioned thesis.* He writes that Ockham's razor is a modern myth and that there is nothing medieval about it except the general sense of the postmedieval formula *Entia non suntmultiplicanda praeter necessitatem*[600] (Entities must not be multiplied beyond necessity).

---

[598] I. Newton, *Mathematical principles of natural philosophy*, London, 2022, p. 398.
[599] Alice Calaprice in *Einstein in Quotes* (trans. M. Krośniak, Warsaw 2014) claims that this is a variation of Einstein's lecture: "It is difficult to deny that the overriding goal of each theory is to obtain the smallest number of simplest elements without having to resign from adequately reflecting even one empirical data".
[600] See W.M. Thorburn, *The Myth of Occam's Razor*, "Mind" 1918, vol. XXVII 3, pp. 345–353.

In 13-century scholastic texts, phrases such as "It is in vain to accomplish by several means what can be done in a few." (*Frustrafit per plura quod potest fieri perpauciora*) and "Plurality is not to be posited without necessity" (*Pluralitas non est ponenda sine necessitate*) were in quite common use. These maxims, used by Ockham, do not come from him and he was not the only one to use them. The origins of the medieval idea of Ockham's razor may be connected with other philosophers, such as Maimonides (1135–1204) or John Duns Scotus (1265–1308), who lived before Ockham. Also, in *Contra Gentiles*, Thomas Aquinas states: "If a thing can be done sufficiently by means of one, it is superfluous to do it by means of several: for we observe that nature does not employ two instruments where one suffices" (*Quod potest fieri sufficienter per unum, superfluum est si per multa fiat: videmus enim quod natura nonfacit per duo instrumenta quod potestfacere per unum*)[601].

In this context, an obvious remark can be made—philosophical thought, especially a spectacular one, usually hatches over a long time and is often supported by many, more or less brilliant, minds. At the same time, the prestigious appellation of creator or author is often assigned only to one person, not always the most deserving.

The best known formulation of the principle of economy of thought—"Entities should not be multiplied beyond necessity" (*Entia non sunt multiplicanda praeter necessitatem*)— comes not from Ockham, but from the 17th-century German philosopher Johannes Clauberg (1622–1665), and was first applied in 1852 by Sir William Rowan Hamilton. The maxim: "Entities should not be multiplied without necessity" *(Entia non sunt* multiplicanda praeter necessitatem) is not a medieval idea, but was formulated in 1639 by Scottish scholar John Ponce (1603–1661). The Latin legal phrase: "Beings should not be multiplied" *(Entia non sunt multiplicanda)* emerged in 1670 and was associated with Leibniz's nominalism[602]. In *Summa totius logicae*, William of Ockham introduces an important principle of economy of thought: *C'est en vain que l 'on fait avec plusieurs ce que l'on peut faire avec un petit nombre* (*Frustrafit per plura quod potest fieri per pauciora*[603]—"It is futile to do with more what can

---

[601] T. Aquinas, *The Summa Contra Gentiles*, Volume 3, Issue 1, London, 1928, p. 173
[602] See W.M. Thorburn, *The Myth of Occam's Razor*, op. cit.
[603] Guilelmus de Ockham, *Summa Totius Logicae*, Bavarian State Library, 1522, I, 12.

be done with fewer"). This statement, Thorburn argues, is a thought taken from Aristotle's *Physics*.

The above reflections, though highly intriguing in terms of the history of philosophy, are in fact not the most important for the problem under consideration here. They were cited not with the intention of studying the myth of Ockham, but to highlight the importance of the problem of the means and form of shaping scientific thought, and to demonstrate how much this issue has excited philosophers over the centuries. Ockham's contribution to creating the methodology of science is also important as thanks to him the phrase "the principle of economy of thought" took on a new, important meaning, which allowed Ockham's razor to be disassociated from its medieval context in the 17th century and, as the principle of economy of thought, become the basis of modern scientific methodology.

One of this historical contribution's important roles is to show the evolution and continuity of intentions and the importance of the principle of the economy of thought in the development of scientific thinking. It is important to understand that from Plato, through Aristotle, medieval and modern philosophy, to the present day, the way scientific thought is constructed has been heavily influenced by an awareness of the need to apply the principle of economy. Ockham made a great contribution to this notion as he became the keystone of the idea. According to the position he put forward, it was not necessary to introduce new concepts or assumptions to science unless there was a strong basis for doing so, and the simplest theoretical solutions, involving the fewest assumptions, were considered to be the best. The most important task of a physicist was to find the simplest and most concise description possible that applied to the maximum number of physical phenomena.

Ernst Mach, who contributed to the resumption of discussion on the principle of economy of thought, added that science has another requirement of an economic nature—to break down complicated facts to make them as simple as possible. He argued that this is a simplification enabling explanation and comprehension of a problem, but at what level of simplification we stop is a matter of economy of thought on the one hand and of preference on the other[604].

---

[604] See I. Szumilewicz, *Koncepcja przyczynowości u Macha w świetle współczesnego determinizmu* (*The concept of causality in Mach in the world of contemporary determinism*), 'Studia Filozoficzne' ('Philosophical Studies'), no. 127–145, 1959., p. 129.

According to Metallmann, two factors stimulated the discourse on the principle of economy of thought in the second half of the 19th century, namely a biological principle[605] and cognitive-theoretical premises. The first introduced the principle of development to consideration of the structure, physiology and morphology of organisms, in particular the human body. The second led to mental processes subordinated to the principle of development being considered from a teleological perspective, which resulted in the perception of mental phenomena as brain functions and entailed the need to extend the principle of development to psychological phenomena. It was found that all psychological processes take the route of least effort. The more they are united by a certain thought, a certain goal, the more effective a tool they become. Hence they are subject to the principle of development, one of the forms of which is the principle of economy of thought[606].

Mach borrowed a concept from biology to define the nature of the role of thinking, relating it to the concept of adaptation. Thoughts, especially those related to the natural sciences, are subject to transformation and adaptation, like organisms in Darwinism. Scientific ideas transform, expand, fight, compete; some triumph over other, less suitable ones. Ideas should conform as closely as possible to the facts; the wider the field of phenomena, the more sparing and economical the means of representation used must be, so that the material can be mastered with a small expenditure of memory and labour. Hence, learning methods must also be economical in nature[607].

It should be noted that Mach's theorem is not only about a theory's logical incontrovertibility, but also about simplicity and economy. It can be argued that logical incontrovertibility is a necessary but insufficient condition for the value of a theory or law[608]. We cannot therefore talk about the economy of scientific methods, as they are all economical or relatively uneconomical, but we should rather talk about greater or lesser economy, manifested at different stages of the development of science, in terms of its stated objective[609]. The principle of

---

[605] Metallmann introduces this principle by writing: "There are, I believe, two sources from which the principle of economy of thought has sprung. The first is the biological principle", *The principle of economy of thought, its history and criticism*, op. cit., p. 1.
[606] Idem, p. 1–2.
[607] Idem, p. 51.
[608] Idem, p. 56.
[609] Idem, p. 66.

economy encompassed the concept of knowledge more broadly. In *Principles of the Theory of Heat*, Mach argues that the aim of knowledge is to enter into a relationship with reality, to create "picture" of it, and that science should be an economic means to that end. The history of science will therefore be a constant advance towards this ideal; a series of "pictures" capturing reality in an ever more economical way. In the second half of the 19th century, many physicists were involved in the imaging of physical quantities and events. The characteristic voices in this discussion were the thoughts of Hertz and Boltzmann.

6.1.3. Imagery in science

According to Frederik V. Christiansen of the University of Copenhagen, in the famous introduction to *Die Prinzipien der Mechanik in neuem Zusammenhange dargestellt* (*The principles of mechanics: presented in a new form*), Heinrich Hertz writes that the most important problem scientific knowledge should be able to solve is the prediction of future events from past ones. Towards this aim, we strive for images from which we can deduce the 'necessary consequences'. We create for ourselves images or symbols of external objects, and the form we give them is such that the necessary consequences of the image in our minds are always images of the necessary consequences of the nature of the thing concerned. For this requirement to be met, there must be some concurrence between nature and our thought. Experience teaches us that this requirement can be fulfilled, and therefore such concurrence actually exists, when from our previously accumulated experience, which we have been able to deduce from images of nature, we can draw in a short time, as with the help of models, consequences that in the external world would arise only after a comparatively long time or as a result of our own interposition. The images we are talking about here are our ideas about things[610].

Hertz tells physicists something undoubtedly long since discovered by philosophers, namely that no theory can be objective, actually concurrent with nature, but merely a mental image[611].

---

[610] See F.V. Christiansen, *Heinrich Hertz's Neo-Kantian Philosophy of Science, and its Development by Harald Høffding*, 'Journal for General Philosophy of Science' 2006, vol. 37, No. 1, p. 2.

[611] H.W. de Regt, *Ludwig Boltzmann's Bildtheorie and Scientific Understanding*, 'Synthese' 1999, no. 119 (1), p. 115.

CHAPTER 6

Ludwig Boltzmann was one of the creators of the so-called *'Bildtheorie'* (picture theory), which made a significant contribution to treating some theories of science as mental images. Henk de Regt, a philosopher at the University of Amsterdam and a commentator on Hertz's thesis, argues that Boltzmann rejected Hertz's demand that mental images be laws of thought. While Hertz's view of laws of thought was thoroughly Kantian, Boltzmann argued—against Kant—that laws of thought may exist and may even be innate, but could always be modified by education and experience. Boltzmann claimed:

> I therefore wish to modify Hertz's demand and say that insofar as we possess laws of thought that we have recognized as indubitably correct through constant confirmation by experience, we can start by testing the correctness of our pictures against these laws; but the sole and final decision as to whether the pictures are appropriate lies in the circumstance that they represent experience simply and appropriately throughout so that this in turn provides precisely the test for the correctness of these laws[612].

Boltzmann's picture theory assumes that scientific theories are "mental pictures", having partial similarity to reality, and this, according to him, is a foundational element of the philosophy of science. In addition to the epistemological nature of science, attention should be paid, Boltzmann claims, to this neglected aspect of solving a scientific problem, in which explanation and understanding should be considered the central goal of science, and it is picture theory that has important implications due to the nature of scientific understanding of scientific theories. According to de Regt, the discussion of theory was limited to the debate about realistic instrumentalism, which is why Boltzmann's concept can be used as a starting point to develop an analysis of the concept of scientific understanding, which shows that the analysis of Boltzmann's philosophy is not only historical but can be of significant importance for contemporary philosophy of science and the methodology of physics[613].

According to de Regt, Boltzmann distinguishes between two different interpretations of mechanical pictures. Firstly, a physical theory in itself can be regarded as a picture, as an 'Analogon'. Secondly, specific mechanical

---

[612] Idem, p. 116.
[613] Idem, p. 113.

analogies can be used to visualise the consequences of the theory. This amounts to distinguishing between epistemological and methodological roles, which relate to their representative and understanding-generating functions, respectively[614].

The proposals of Hertz, and especially of Boltzmann, of using the imaging of theory as a real application of the principle of economy of thought yielded very good results as a tool for learning about the laws of nature. This method is currently also widely used in many other fields —in medicine, psychiatry, chemistry, and especially in modern physics. However, in light of discoveries now being made in physics and astrophysics, it can be seen that in a highly mathematised science, a specific mathematical equation appeals to physicists more than the best imagery. The role of images in physics, particularly since the quantum-mechanical revolution, has significantly decreased, and although it has not completely disappeared, perhaps we are witnessing another change of one of the epistemological paradigms.

The depiction of certain concepts has always been a simplification of the perception of reality's functioning. The usefulness of this simplification to date favours its use. However, there are indications that a side-effect of this mental shortcut is sometimes such a distortion of the object being depicted that this method of cognition's usefulness is insufficient. For example, in the Big Bang theory imagined to date, the image has been used of a huge amount of the universe's energy being accumulated at a single point, referred to as a singularity. Meanwhile, Penrose constructs a hypothesis in which the Big Bang occurred from the dispersed energy created after it was finally evaporated from black holes, when there was no space, no time, and therefore no dimensionality, only diffused energy. Penrose does not write what diffused energy means in this case, so it is difficult to see how in such a situation one can create an acceptable image of diffusion with no dimensions. Imaging, which is a natural and helpful tool, can narrow or misdirect our perception of reality in certain situations due to the human brain's four-dimensional perception. There is a clear analogy to the macroscopic functioning of thermodynamics in the physics of the 19th century, which at some point became an obstacle to a fuller understanding of reality.

---

[614] Idem, p. 116.

Indirectly, Metallmann's dictum refers to the case described above, as it states that the position of the principle of economy entails a fundamental shift in how the question is asked about the manifestations of knowledge. What we call knowledge, that which serves to grasp and master reality, may be not only true or false, but may also have neither quality, so as to not be subject to the criterion of truthfulness, but to another—an assessment of its utility. In certain cases, applying the criterion of truthfulness does not lead to any result, but leaves us with an unanswered question, or even the impossibility of posing the question itself. It is clear that in these situations, such a point of view is quite absurd. After all, it is impossible to answer the question of whether the centimetre-gram-second (CGS) system of units or the metric system is true, or whether Cartesian coordinates are right, just as it is impossible to answer the question of whether or not an avalanche killing a tourist is moral. Such questions cannot be answered because they are improperly posed, and therefore in themselves ridiculous. The enduring, everlasting importance of the principle of economy lies not in demonstrating the total concurrence of knowledge with the criterion of truth, but in finding a different criterion[615].

In this respect, Philosophy of Science contains an interesting statement written by an Australian philosopher from the University of Pittsburgh—John D. Norton. He devoted the article A Material Theory of Induction to the issue of the functioning of the inductive scheme of inference in the perspective of a material analysis of the theory of induction. However, his reflections relating to economy of thought introduce an intriguing differentia specifica into the considerations raised above.

Norton notes that the most obvious and for a long time the most popular proposition is the hypothetico-deductive model using the concept of simplicity. While many hypotheses or theories may entail evidence deductively in an extended scheme, we unfortunately approve of inductive support only for the simplest ones. The consequence of such decisions is that they function mainly within the orbit of previously accepted interpretations of facts and laws.

Norton argues that our decisions concerning what is simple or simpler depend essentially on facts or laws that in our opinion have precedence over others and it is these facts that dictate which theoretical structures we can use.

---

[615] See J. Metallmann, *The principle of economy of thought, its history and criticism*, op. cit., pp. 156–159.

Invoking simplicity is actually an attempt to avoid introducing theoretical structures ill-adapted to the physical reality governed by these facts and laws. This can be quite clearly seen in the popular example of adjusting a curve, when data presented as a finite number of points on a piece of paper are related as a linear or square equation, which can be adjusted in such a way that we routinely match the marked points to a linear equation because we get a simpler solution that way. Our choice is satisfactory and tacitly accepted to get the desired result. In order to see this, it should first be noted that application of the procedure depends on certain conditions being met. An equation expressing a linear law with only certain options of variables can be created for a law that uses any function that takes our fancy by simply intentionally rescaling one of the primary variables.

This is so natural that the emerging theories are created on the basis of knowledge already embedded in the existing culture of science. The tendency to create theoretical structures guided by the attainment of maximum simplicity and related to the use of equations or rules known to us represents not only an attempt to avoid equations ill-adapted to physical reality but also those ill-adapted to our understanding of that reality. These laws must function in a way that is comprehensible to us, but there exists a tendency to build them based on formulae according to which, in our general imagination, reality functions. Nature is 'mathematical', hence we seek relations of a linear, square, exponential or sinusoidal type which hold true, but we do not know whether they hold true because that is the way nature is or because we have imposed a matrix upon it that meets our demands and expectations. Smoluchowski's utility criterion explains this mechanism. Accepted scientific theories are verified through their current usefulness, in which they prove true. The mathematical analogies used, giving priority to already accepted solutions, dictating the theoretical structures used to create a theory, are valid within the boundaries verifiable by the principle of utility.

Reality is often more complex than we assumed. In the book *Lost in Math. How Beauty Leads Physics Astray*,[616] Sabine Hossenfelder draws quite surprising conclusions regarding the verification of theories created in modern physics. The *utility criterion* is more culturally burdened in today's complex

---

[616] S. Hossenfelder, *Lost in Math*, op. cit. p.

conditions than it would be if science functioned rationally. In difficult conditions, cultural burdens make their presence felt more than we might expect. A hundred and twenty years after the last crisis in physics, as a result of which many scientists assumed the end of the discipline, we are once again faced with a fairly complicated situation. Hossenfelder asserts that in certain branches of physics, no important new data has appeared for decades. The Higgs boson, proposed independently by several researchers in the early 1960s, was the last fundamental particle to be discovered (its detection occurred in 2012). Since 1973, no new accurate prediction has been made that would go beyond the standard model[617]. At a symposium in Vancouver in 2011, the Italian physicist Guido Altarelli put it more bluntly: "It is not time to be desperate yet… but maybe it is time for depression already[618]." This state of affairs creates a specific situation in which the requirements set for theories by the *utility criterion* do not include experimental verification. Faced with a lack of indicators from experiments, physicists are becoming disoriented and are applying aesthetic criteria to the assessment of theories, such as the criterion of beauty[619].

Scholars have long been guided by the criterion of beauty when creating theories, but it has not always been a good determinant. If there is currently a lack of data, and a theory is needed to decide where to look for it, then any mistake in its construction can lead us up a blind alley. Deliberations on how to move forward despite a lack of data and whether to modify the scientific method go beyond the foundations of physics[620]. When new data are insufficient, theoretical physicists rely on their sense of simplicity and beauty in the evaluation of various theories. Beauty is not a scientific criterion, but can be based on experience[621]. If beauty is to determine the choice of scientific theory, this is the best example of creating a *utility criterion* appropriate to the needs of the situation, which would not exist in the same form under

---

[617] Idem, p. 85–86.
[618] X. Portel Bueso, *Supersymmetry Searches at the Tevatron and the LHC*, paper delivered at the 31st International Symposium on Physics in Collision, Vancouver, Canada, August–September 2011, slide 41.
[619] See S. Hossenfelder, *Lost in Math*, op. cit., p. 32.
[620] Idem, p. 66.
[621] Idem, p. 100.

other conditions. Without deprecating the role of the beauty criterion in accepting new theories, we must be aware of how its adoption affects the development of science. There are many physicists who consider this influence to be positive, such as the Nobel laureate Leon M. Lederman, who stated: "We believe that nature is best described in equations that are as simple, beautiful, compact and universal as possible[622]." Previously, Heisenberg asserted that beauty is closely connected with truth: "If nature leads us to mathematical forms of greater simplicity and beauty, (...) we cannot help thinking that they are 'true,' that they reveal a genuine feature of nature"[623]. Paul Dirac, on the other hand, wrote: "A physical law must possess mathematical beauty"[624]. Hence theoretical physicists also use simplicity, naturalness and elegance as criteria to evaluate theories[625]. Nima Arkani-Hamed claims that naturalness is neither a principle nor a law and has been treated as a guide. According to some researchers, naturalness is pure philosophy, although it has helped theoretical physicists achieve successes. Despite this, some have abandoned it in favour of another idea—so-called splitsupersymmetry[626].

Theoretical physicists have much cause for complaint about the laws of physics discovered so far. They especially dislike numbers bearing the hallmarks of unnaturalness. Naturalness has served in the process of developing natural theories since at least the 16th century[627]. Quantum mechanics works fine though many physicists complain of its lack of intuitiveness and elegance[628]. This perception of quantum mechanics has resulted in its lack of acceptance by many theoretical physicists, who believe that being guided by beauty can be justified by acquired experience. But this will be of no use if the new laws of nature being sought are beautiful in a way unknown to us. The support of a like-minded community has a significant impact on how scientists assess the usefulness and prospects of the theories they wish to pursue[629].

---

[622] L.M. Lederman, *The God Particle Et Al*, 'Nature' 2007, no. 448, pp. 310–312.
[623] S. Hossenfelder, *Lost in Math*, op. cit, p. 96.
[624] Idem, p. 39.
[625] Idem, p. 172.
[626] Idem, p. 119.
[627] Idem, 128.
[628] Idem, p. 200.
[629] Idem, p. 244.

In today's theoretical physics, certain practices have been established of creating new laws of nature, which will not be verifiable for a long time. Contact with philosophy could help physicists determine which questions are worth asking, but currently such contact is sporadic. The reliance of theoretical physics primarily on aesthetic criteria and the resulting lack of progress demonstrate the field's inability to repair itself[630].

Citing simplicity or beauty as a result of achieving confirmation in science is implied in the attitude towards the alleged facts prevailing in a given field. If our system represents any kind of cyclical process, we will quickly review the sine or cosine functions, the polynomial expansion of which has infinite possibilities, in order to be able to use them in a justification of the hypothesis at hand. In short, we have no universal schema or universal formal rules that let us define what is simpler or the simplest, which in its essence deserves the term beauty. If we are able to, we choose variables and functions appropriate to the facts that, in our opinion, prevail. These facts are material postulates of inference to adjust the simplest curve[631]. Agreements concerning universal principles defining what is appropriately simple are important but secondary since, in contrast to what Norton asserts, we have an instrumental universal principle enabling definition not of what is simpler or beautiful but what is more useful. That which is more useful contains within itself that which is simpler because while agreement on what is simpler constitutes a postulative statement, discovering what may be more useful constitutes a substantive argument verified in daily life.

Norton sees a cultural anthropomorphism imperative in the functioning formula of the economy of thought, thereby undermining our trust in the assumed principles that enable us to designate what is simpler or simplest. The way we perceive reality, determined by biology and culture, affects the process of creating scientific theories through giving precedence to the ideas of simplicity, although we cannot define it. Meanwhile, to paraphrase Poincaré, not only does nature know nothing of our ideas of simplicity and beauty but—very probably—may "understand" that simplicity entirely differently, as has been proven in pre-Copernican and Copernican astronomy in

---

[630] Idem, p. 316.
[631] See J.D. Norton, *A Material Theory of Induction*, 'Philosophy of Science' 2003, vol. 70, no. 4, pp. 655–657.

which it was assumed that planets' orbits must be in a perfect circle because a planet must move in a perfect, beautiful arc as described by a circle. However, planets are objects that formed in the ancient history of the solar system whose orbits have various shapes and there exists no particular reason why their orbits should be circular[632].

Here an attempt can be seen to refer to the thoughts of Francis Bacon contained in *Novum Organum*, where, among other things, we read:

> The human understanding, from its peculiar nature, easily supposes a greater degree of order and equality in things than it really finds; and although many things in nature be *sui generis*, and most irregular, will yet invent parallels and conjugates, and relatives where no such thing is. Hence the fiction, that all celestial bodies were in perfect circles, thus rejecting entirely spiral and serpentine lines, (except as explanatory terms.) Hence, also, the element of fire is introduced with its peculiar orbit, to keep square with those other three which are objects of our senses. The relative rarity of the elements (as they are called) is arbitrarily made to vary in tenfold progression, with many other dreams of the like nature. Nor is this folly confined to theories, but it is to be met with even in simple notions[633].

Bacon's and Norton's reflections confirm the validity of Smoluchowski's arguments for utility theory. The two thinkers' sceptical positions as to the absolute value of science mean that the foundation of knowledge, which is science, is not—as we would expect—stable in its understanding of reality. Smoluchowski points to the theory of utility, which is a premise systematising our approach to science, providing a stimulus that organises the understanding of knowledge and makes us aware that it is mainly the degree of utility of existing theories that determines their acceptance. The Polish physicist attributed to utility both a methodological disposition and an epistemic value, which constituted the basic criterion for verifying theories.

Utility is a theory that references the anthropomorphism of science, assuming that in the first step of creating elements of science, the most important argument is the effectiveness of their application. If the characteristic of anthropomorphism of knowledge has been a burden through the ages on effectively learning the truth of reality, then in utility theory

---

[632] See *Newton's Dream*, ed. M.S. Stayer, McGill-Queen's University Press 1989, pp. 37–38.
[633] F. Bacon, *Novum Organum 1620*, ed. and trans. Basil Montague, Philadelphia, 1854, p. 347.

anthropomorphism becomes an epistemic attribute enabling verification of the value of accepted knowledge. The principle of economy of thought has been an element contributing to scientific progress, and though it has happened that it has limited that progress, it has above all added to its potential and, importantly, made it more useful. The two concepts—the principle of economy of thought and the theory of utility—constitute a tandem that is the driving force of progress in science.

6.2. The utility criterion
6.2.1. Pragmatism in methodology

Ignoramus et ignorabimus ("We do not know and will not know"), a thought of German physiologist Emil Du Bois-Reymond (1818–1896), raises certain questions that Smoluchowski also poses. Does this mean that man's capacity to know nature has limited and unbreakable boundaries? Do phenomena exist that are always inaccessible to us and that do not influence our experience, which do not concern us because we will never notice their influence? Are we totally indifferent to rays that we will never be able to perceive and whose influence will never be manifest in the phenomena available to us? Are we really losing something more than a beautiful fantasy? Are events that affect our experience partly knowable to us, albeit indirectly[634]?

This conjecture, consistently conducted by Smoluchowski in the form of successively posed detailed questions, leads the author to fundamental questions. On what basis does science work? By what principle do we verify hypotheses and theories functioning within the realm of science? If our knowledge is relativistic in relation to the truths about reality, what criterion makes it possible to accept the currently binding science?

In answer to these questions, Smoluchowski specifies such a criterion—the *criterion of utility*. We do not distinguish—he argues—between true or false theories, more or less likely theories, but between more or less *useful* ones[635].

This bold thesis, enabling the verification of accepted theories, assumes that the theory that is most *useful* is the closest to the truth. *It constitutes a*

---

[634] See M. Smoluchowski, *Self-Study Handbook*, vol. I, op. cit., p. 16.
[635] Idem, p. 51.

driver, supported by the principle of economy of thought, which forces constant verification of the rules of science through the everyday experience of their application. The *category of utility*, often functioning completely unconsciously, *is the basic instrument in the scientist's work in recognising which of the competing theories is more congruent with reality*. It is a natural tool for using the principle of economy, because at the same time it is a verification of the economy of each theory. The more categorically the principle of economy of thought is applied, the easier it is to establish the *usefulness of a theory*. The interaction is two-way as selection of hypotheses and scientific theories made on the basis of their *utility* will simultaneously be subject to the principle of economy. The *utility of theories* and hypotheses enables thorough and clear knowledge of the world of phenomena available to us and their combination into a whole comprehensible to our mind[636].

Doubts about the compatibility of prevailing scientific theories with reality have plagued scientists for years. Gregory W. Dawes, a New Zealand philosopher, notes that modern physicists generally accept the theory of quantum mechanics developed in the 1920s by Erwin Schrödinger (1887–1961), Werner Heisenberg and Max Born (1882–1970). This also includes the intuitively accepted but surprising uncertainty principle, which states that it is impossible to simultaneously determine both the position and velocity of a sub-atomic particle. A physicist could accept this theory, justifying it with its huge success of verifiability, but at the same time be unable to believe it much. The scientist should accept this theory as the best available explanation of the behaviour of sub-atomic particles. They should use this explanation as a premise for further reasoning in the relevant field. However, we cannot conclude that a given physicist, accepting this principle, is fully convinced of its truth. For example, they may consider there to be no evidence that no hidden variable exists that, if known, would allow for an accurate prediction of both quantities. They may consider it to be the best available theory, and therefore the one to work with, without accepting it as true. If acceptance of a theory has occurred with the aim of broadening knowledge, and that knowledge is consistent with reality, it may seem that a theory cannot be accepted that is considered false. When dealing with straightforward situations, such

---

[636] Idem, p. 17.

thinking seems correct but when it comes to scientific theories the situation becomes more complex[637].

Introducing the *category of utility* changes the view of things; a theory's acceptance occurs through a perception of its *usefulness*, which most often involves its acceptance as the best available theory, true, one which should be worked with and which seeks to broaden our knowledge. However, convictions concerning the veracity of theories do not have to be entirely correct as it is not they but the currently understood *utility* of a theory that is of decisive importance. False theories are known which functioned as true for many years, such as the concept of ether[638], or even for centuries—like the theory of geocentricity. Therefore, the question arises whether it is a perception of *utility* ending that creates the search for new theories, or whether a new theory makes us aware of the decline in *utility* of an earlier one. Most probably, there is no clear answer here (although Kuhn proposes one, as will be discussed below), because both the search for the truth about our reality and the search for more efficient theories are influenced by too many different factors.

We trust a theory mot because we believe it to be consistent with reality, but because it is the best explanation available. But the fact that it is the best explanation available means we should accept it and adopt it as a premise in our practical reasoning in every field to which it applies, whether we believe in it or not. The reason for this approach is pragmatic. A theory that bears traditional traits such as value, simplicity, explanatory power, a high degree of testability, and conformity to what we have already accepted, can be a useful tool in advancing our knowledge of the world[639].

This is how science is created—we adopt pragmatically the best possible explanations for phenomena occurring in nature. Dawes notes that these theories, with all the baggage of characteristics defined by the methodology

---

[637] G.W. Dawes, *Belief is Not the Issue: A Defence of Inference to the Best Explanation*, 'Ratio. An International Journal of Analytic Philosophy' 2013, vol. 26, No. 1956, p. 69.

[638] The concept of ether returns in some hypotheses, but in a different form that that functioning in the 19th century. See. L. Kostro, L. Kostro, *Alberta Einsteina koncepcja nowego eteru: jej historia, sens fizyczny i uwarunkowania filozoficzne*, (Albert Einstein's concept of a new ether: its history, physical meaning and philosophical conditions), Gdańsk 1999.

[639] See G.W. Dawes, *Belief is Not the Issue*, op. cit., p. 78.

of science, become a *useful* tool in developing our knowledge of the world. The dependency is two-way: accepted theories must demonstrate their *utility* in order to be included in the realm of science, and at the same time by their *usefulness* they assist in the creation of new theories, which can eliminate current ones from science. This, among other things, has been the reason for the success of science.

In the researcher's work, it is important to assess the value of the hypotheses and theories created. Smoluchowski treats hypotheses as assumptions that seem probable to us in advance, calling them thought experiments. They are adopted on a trial basis as a foundation intended to lead to experimentally verifiable conclusions. On the other hand, he understands theories as the entirety of well-founded hypotheses together with all conclusions related to the phenomenon. For example, the theory of electricity and magnetism is based on several basic hypotheses, partly derived from multi-faceted experiments, such as the principle of electromagnetism and induction, hypothetically adopted by Maxwell[640].

The hypotheses and theories created by scientists do not reflect the reality actually being studied but are merely a part of the 'truth' about nature. Those who naively believe in the reality of physical theories do not understand how a researcher can treat the same subject from opposite points of view, as Helmholtz did for example by developing first a mechanical theory of light absorption and then an electrical theory. This becomes completely understandable if we consider both of them, as well as any physical theory, only as an analogy[641].

In analysing the value of theory in science, Smoluchowski notes that a hypothesis or theory has been "verified" if the conclusions drawn from it are in line with experience. It is only when at least one conclusion is not confirmed that we consider the theory to be "untrue". If "truthfulness" meant compliance with the truth, such a presentation of the matter would have to raise serious doubts. We never know if evidence will ever be found to refute a given hypothesis, even if it seems well-grounded. Even if no evidence were

---

[640] See M. Smoluchowski, *Self-Study Handbook*, vol I, op. cit., pp. 47–48.
[641] M. Smoluchowski, *Several remarks on physical analogies, especially in theories of electric currents, thermal currents and diffusion phenomena*, in: *The Writings of Marian Smoluchowski*, vol. 3, op. cit., p. 242.

found against it, the conclusionstill cannot be drawn that the hypothesis reflects reality because we never know whether other hypotheses are also possible which will lead to the same conclusions[642].

Smoluchowski does not write about the search for arguments confirming the correctness of a theory but about the search for a correctly derived conclusion that does not confirm a theory. An association of Smoluchowski's reflections with Popper's falsificationism is quite obvious. The fact that Smoluchowski's idea emerged several decades before Popper's theory proves the originality of the Polish physicist's philosophy, as well as the fact that concepts of studying falsifiability had appeared in science before a theory was fashioned from them.

Dawes writes that scientific theories commonly involve idealisations or approximations and would also be true under conditions not obtained in reality. Surfaces are treated as if they were frictionless, even though no such surfaces actually exist, and fluids are treated as if they were continuous, even though they are made up of separate molecules. A scientific theory may be accepted as a whole even if the recognition of some of its assumptions is taken nominally as false. Scientists accept a theory that has at least one indisputably false observational consequence. In such a situation this theory cannot be true, at least not as it is, but can be partially true in the sense of making true predictions, within certain domains, under certain conditions, or as restricting chance. However, it cannot be considered a true judgment. In Dawes' deliberation, a natural need appears for Smoluchowski's *utility criterion*, which is a sufficient determinant of a theory's acceptability. Were it not for the *category of utility*, we would surely get lost in assessing a theory's acceptability.

A scientist need not claim to "believe" in a theory but must consider whether there is reason to accept it. Such cases support the distinction between belief and acceptance, without undermining the idea that the acceptance of scientific theories is aimed at expanding knowledge. The scientist may concede that a partial truth is the best that can be achieved at the moment, that an idealised law—as it is, ignoring friction—is the best solution for successful predictions. This does not mean that he has abandoned the goal of "attaining

---

[642] See idem, *Self-Study Handbook*, vol. I, op. cit., p. 49.

truth and avoiding error with respect to the very thing one accepts"[643]. This means that putting forward a new hypothesis and applying the *principle of utility* at the stage of existing knowledge is the best new theory to be had at the moment, in terms of its level of *utility*.

6.2.2. The concept of truth in science

The criterion of truth and acceptability of created theories, despite its departure from the radical falsification of a theory, in relation to what Smoluchowski wrote has essentially not changed; it is still the degree of these theories' *usefulness*. If the *utility* of a prevailing theory is satisfactory, then despite flaws and shortcomings, it still functions in science.

It is important when accepting a hypothesis as a new valid theory to be convinced it is closer to the description of reality than the one previously in force, although we have also had the opposite situation in science. One could mention here for example the attitude towards atomic theory among physicists at the end of the 19th century, the subordination of science to ideological goals in the Soviet Union, or the attitude to the theory of natural selection in some university circles in the USA. If we cannot objectively relate scientific theories to reality without reservation, what makes them credible? What parameter determines that at a given moment we have the conviction that the theories prevailing in science are approved? What is the criterion by which we could assess a theory in terms of its concordance with reality? We cannot objectively establish that accordance since we simply do not know the real laws of nature, we are not in a position to define them and—equally importantly—we do not know if we will ever be able to know them. However, in the longer term, theories that are closer to a true description of reality hold up.

Norton points to reliabilism[644.] In reliabilist calculus, we assume acceptance of the trustworthiness of a given hypothesis or theory that was created by a reliable method. For example, we routinely accept the diagnoses of car mechanics or doctors in the belief that the method they used to obtain the

---

[643] See G.W. Dawes, *Belief is Not the Issue*, op. cit., pp. 69–70.
[644] Reliabilism is a theory by Alvin Goldman (1967) which states that our belief in something is justified if (and only if) it was created as a result of a *reliable* cognitive process. The cognitive process is reliable if (and only if) it leads much more often to true beliefs than false ones.

diagnosis is reliable, although we, and even they, may not understand that method. We may condemn a hypothesis as being ad hoc simply because it is not introduced by the appropriate method.

The theory of reliabilism argues that properly applied science, using reliable methods, enables us to believe in its products. According to Norton, a widespread uncritical attitude can be seen towards hypothetical deductivism[645]. The material stipulation that guarantees inference in terms of accepting results is simply our belief that the method is reliable. This applies not only to the method, but also to the world, in relation to which we are convinced that these methods can work reliably. In a non-cooperative world, no method can be used. For example, there is no hope there will ever be a reliable method to beat casinos, because casinos set up their games in such a way as to prevent us from doing so. We can never have the slightest hope of the credibility of forecasts by readingentrails[646], no matter how expertly it is done. The real world does not allow for predictions by reading intestines, unless they concern the health of the herd from which the specimen was taken[647].

It is usually believed, Smoluchowski notes, that in the case of the principles of physics, science does not distinguish between true and false hypotheses but between more or less probable ones[648]. But what does it mean that a hypothesis can be more or less probable? Apart from the fact that the probability of a hypothesis does not give any guarantee of objective truth, is such a guarantee

---

[645] Deductivism is the process of confirming the validity of a conclusion from a set of premises to which the value of truth has been assigned. Paul Hoyningen-Huene wrote that deductivism is an attempt to develop a position that avoids the difficulties that beset inductivism. It is accepted that theoretical elements enter science at all stages and that inductive generalizations lack proper justification. The basic idea of deductivism is that theories are not built bottom-up from theory-free data, but that they are deductively tested against data. Inductivism and deductivism share the view of scientific explanation and prediction, see J.D. Norton, *A Material Theory of Induction*, op. cit., pp. 652–653. (Czy ten przypis jest prawidłowy? Znalazłem to: http://www.zeww.uni-hannover.de/Theories%20and%20Methods.%20Chapter%203.pdf)

[646] Haruspicy—reading from entrails; a kind of divination consisting in reading from the entrails of sacrificied animals.

[647] See J.D. Norton, *A Material Theory of Induction*, op. cit., pp. 658–659.

[648] Idem, pp. 48–49.

what building a theory is about? What is the point of talking, for example, about the probability of the hypothesis that electrons are rigid bodies[649]?

In the search for answers to the questions posed by Smoluchowski, we find an important remark from Heller—"the concept of truth is on the horizon of science: without this horizon, science would turn into a mosaic of *ad hoc* skills"[650].

The conformity of science's theorems with reality is undoubtedly one of the goals of science: "The truth of thought consists in its agreement with reality. *Veritas est adaequatio rei et intellectus*"[651], Kazimierz Ajdukiewicz (1890–1963) pointed out, expressing this goal of science explicitly. Conformance is not the only goal of science; in many scientific theories we can only talk about the degree of probability of this alignment. According to Smoluchowski, we do not want to know the essence of what lies behind appearances. The task of physics is—as far as possible—to clearly understand the world of phenomena presented to us. It is about examining these phenomena as thoroughly as possible and combining them into a whole that is comprehensible to our mind[652].

Smoluchowski emphasises the importance of the principle of economy of thought posited by Mach, drawing attention to its helpfulness in the methodology of science. The purpose of creating hypotheses is to understand the functioning of nature in the most accessible way, and therefore hypotheses should be treated as a means of describing phenomena in the simplest way possible. All analogies of physics are merely formal analogies, and it is generally the task of theoretical physics to look for such analogies, enabling it to encompass mentally the widest range of physical phenomena. Which hypotheses do we call more probable[653]? It seems to be those that give the impression of being closer to the truth, i.e. those that are simpler and more in line with the concepts we have become accustomed to[654].

---

[649] Initially, hypotheses were created regarding the structure of sub-atomic particles. It is interesting that to this day we do not know the internal structure of the electron.
[650] M. Heller, *Moralność myślenia* (*The morality of thought*), Tarnów 1993, p. 31
[651] See K. Ajdukiewicz, *Problems and Theories of Philosophy*, Cambridge 1973, p. 9.
[652] See M. Smoluchowski, *Subject, task, method and the division of physics*, op. cit., p. 165.
[653] The probability of hypotheses and scientific theories was considered in various later concepts of theoretical scientific approaches (starting with the logic of induction of late neo-positivism in the 1940s).
[654] See M. Smoluchowski, *Self-Study Handbook*, vol. I, op. cit., p. 50.

All speculations as to the actual mechanism of phenomena contain hypothetical elements and there is no phenomenon that could not be explained in an infinite number of ways. We choose theories that seem the simplest to us, but we will never know for sure if they are true. The extreme sceptic will even say that we think with the rules of human logic, but we do not know whether nature, reality, adheres to these rules[655].

Smoluchowski could be accused of underestimating science, and even of a rather specific relativisation of scientific knowledge itself, whichwould not be true however. Being a scientist, an outstanding physicist, Smoluchowski was convinced of the constant development of science and its unquestionable role in supporting almost every aspects of our lives. But at the same time, he maintained a philosophical distance from the treatment of science as a font of certain knowledge.

The convergence of Smoluchowski's and Mach's views in understanding the importance of both the principle of economy of thought and the methodology of creating science means that a number of the Austrian philosopher's ideas were used to establish Smoluchowski's theory of the *utility criterion*. However, there are some subtle differences between Smoluchowski's concept and Mach's principle. Smoluchowski states that in the history of science, it has happened that various laws, initially adopted with faith in their simplicity, have been replaced by more complex formulae as observations have improved[656]. It should be emphasised that the premise of Ockham's razor is widely accepted, although it is a metaphysical assumption and there are no empirical arguments confirming that the world is actually simple. Smoluchowski adds that we seek simplicity not only because it is a need of science, but mainly so that the human mind can accept and understand these laws. A scientific system is useful when it facilitates the mind's grasp of the whole, favouring the principles of economy of thought[657].

Smoluchowski's attitude also differs in regard to the assumption that the laws of nature are a subjective creation of our mind, not something objectively existing in nature[658]. It is also an indisputable fact that the *utility criterion* is

---

[655] See idem, *Subject, task, method and the division of physics*, op. cit., p. 164.
[656] Idem, p. 45.
[657] Ibid.
[658] Idem, p. 46.

a much more complete theory than the principle of economy of thought. Smoluchowski's concept could be considered an approximation, in Popper's understanding, of Mach's principle of economy of thought.

6.3. The principles of utility in philosophy
6.3.1. Utility according to Charles Sanders Peirce and William James

Before Smoluchowski, considerations on the application of the utility criterion appeared in Charles Sanders Peirce and William James's concepts of pragmatism, which assumed utility to be the criterion for the truthfulness of judgments and concepts[659], which—seemingly—is an idea identical to Smoluchowski's concept. According to James, an idea that comes true can be verified. That it is useful because it is true means it is true because it is useful. Both expressions mean exactly the same thing[660]. Truth is the name of an idea that triggers the verification process; utility is the name of a function fulfilled in experience. James argued that "[t]he practical value of true ideas is (...) primarily derived from the practical importance of their objects to us"[661]. This means that the truthfulness of an idea is determined by the practical relationship of truth and utility in relation to the objects to which the idea pertains. Smoluchowski would not agree with such a mutual relationship of truthfulness and utility. He claimed, as previously noted, that we do not distinguish between true and false theories, but between more or less *useful* ones[662].

As a philosophical doctrine, pragmatism was close to Smoluchowski, but he understood *utility* slightly differently than Peirce and James. The American pragmatists assumed the truthfulness of scientific judgements in terms of the effectiveness of subsequent processes. In their view, truthfulness is nothing more than that property of a given judgement that leads to effective actions. The truth of an idea is not a permanent property inherent in it— truth happens to an idea. It becomes true when events make it true. Its truthfulness is

---

[659] See J. Herbut, *Pragmatyzm (Pragmatism)*, in: *Universal Encyclopedia of Philosophy*, vol. 8, Lublin 2007, p. 442.
[660] See W. James, *Pragmatism*, op. cit., p. 163.
[661] Ibid.
[662] See M. Smoluchowski, *Self-Study Handbook*, vol. I, op. cit., p. 51.

actually an event—the process of its substantiation, its validation. Its validity is the process of justification[663].

The reasoning in pragmatism was deductive in nature; the effectiveness of creating science was conditioned by the assumption of the veracity of previous judgements. We thereby arrive at a specific conclusion, which is a new theory, created on the basis of assumed premises that are the hitherto verified judgements. The reasoning in Smoluchowski's theory is inductive in nature; the *utility criterion* is open to any new hypothesis that may replace a previously functioning theory. In Smoluchowski's concept, the new theory is also based on the veracity of previous judgements but it may happen that it contradicts those judgements, though this does not disqualify it as the most important thing is that its *usefulness* is greater than the *usefulness* of the previous theory.

James defines the attitude to the issue of truth as follows: "True ideas are those we can assimilate, validate, corroborate and verify. False ideas are those that we cannot. That is the practical difference it makes to us to have true ideas; that, therefore, is the meaning of truth, for it is all that truth is known as"[664].

The truthfulness of Newtonian theories, according to pragmatism, occurs in the context of events that make them true, which relates mostly to the macro-world on the scale of our planet. The truthfulness of Einstein's theories occurs in the context of events that take place on the scale of our planet but also on a scale far beyond it. The conclusion that results from these comparisons would incline one to abandon Newton's theories due to the much greater concordance of events making Einstein's theory more true. Why then do institutes, laboratories and production companies in many areas of technology base their work on Newton's theories? This happens because the *utility* of Newton's theories is incomparably greater in these fields. This fact does not undermine the pragmatic definition of the truthfulness of a theory since Newton's theory still holds true in terms of the events that make it true. According to Hawking, Newton's theory is a fair approximation of the

---

[663] See W. James, *Pragmatism*, op. cit., p. 161.
[664] Idem, p. 97

theory of general relativity at the boundary of weak gravitational fields[665]. For these reasons, the Smoluchowski criterion does not exclude theories that work well in practice, because they are *useful*.

6.3.2. Relation of the *utility criterion* to the concept of truth

Let's take a look at the notion of the classical concept of truth in Kazimierz Ajdukiewicz's studies. The Polish logician writes: "thought m is true—this means: thought m states that it is so and so, and in reality it is so and so"[666]. On the basis of this definition, Ajdukiewicz cites interesting considerations regarding the objections raised by sceptics, referring to the possession of justified knowledge making it possible to conclude that it is so-and-so. In order to possess such knowledge, it must be justified by some method, i.e. it has to be guided by some criterion. The knowledge acquired on the basis of this criterion will be properly justified knowledge if we know in advance that the criterion we use is reliable and that it always leads to the truth. In order to be persuaded of this, this criterion should be subjected to critical examination by another criterion, which is able to verify the first criterion, and this in turn should be subject to the validation of the next, and so on indefinitely. Therefore, sceptics argue, it is impossible to acquire valid knowledge about anything[667].

In arriving at a definition of truth, there is no other way of ascertaining whether a proposition is true than to put that proposition to the test of an ultimate criterion, the verdict of which is irrevocable in the sense that the verdict of any other criterion must yield to it. Whether or not propositions that stand the test of this ultimate criterion reflect reality we cannot know and—as sceptics have demonstrated—will never be able to know. So actually, when distinguishing truth from falsehood, it is not a question of whether or not a given statement is consistent with reality, but whether it complies with the ultimate criteria. Therefore, if we want to define the concept of truth in a way consistent with how we really use this concept, we should define truth as the compliance of thoughts with ultimate and irrevocable

---

[665] See S. Hawking, *Wszechświat w skorupce orzecha* (*The Universe in a Nutshell*), trans. P. Amsterdamski, Poznań 2018, p. 125.
[666] K. Ajdukiewicz, *Problems and Theories of Philosophy*, op. cit., p. 26.
[667] Idem, p. 19.

criteria. In Smoluchowski's concept, such a verifier is the *utility criterion*, which performs a satisfactory verification of the truthfulness of a theory for the duration of its validity.

Defining truth as conformity with criteria fits into Smoluchowski's concept, but not entirely, as the *utility criterion* does not require a definition of truth in order to function in the methodology of science; it is inspired by the maxim *ignoramus et ignorabimus*—we do not know and will not know the truth. This concept seeks to reconcile a theory having an unspecified concept of truth, located somewhere on the horizon of science, with the way it is *used*.

A similar view to Smoluchowski's is proffered by Tadeusz Kotarbiński (1886–1981), who identified two main tendencies in the approach to understanding truthfulness: classical and utilitarian. According to the classical tendency, 'truly' amounts to 'in line with reality'; in the utilitarian tendency, 'truly' amounts to 'in some respect beneficial'[668]. When it has transpired that we cannot unequivocally determine what is truly consistent with reality, we use a utilitarian perception of reality that also leads to this truth—asymptotically and infinitely, but continuously. There is no need to argue that the term 'beneficial' in this case is synonymous with the term 'useful'.

Since the times of Peirce and James, a growing caution can be seen among philosophers who, in their contemplations, try to create a definition of truth. For example, Ludwig Wittgenstein (1889–1951) in *Tractatus logico-philosophicus* posits a fairly naive, according to Popper, "pictorial" theory of truth—a statement is an image or projection of the fact it is supposed to describe, and is characterised by the same structure (form)[669]. Everything changed when Alfred Tarski (1901–1983) created a theory of truth as a correspondence between a statement and facts, answering the epistemological question of when the statement is true, that is, when it is consistent with the facts.

We try to arrive at the truth by analysing and checking a given statement either directly or on the basis of its consequences. Meanwhile, Popper cautions that there are no ultimate sources of knowledge. Neither observation nor reason is an authority. Intellectual intuition and imagination are important,

---

[668] T. Kotarbiński, *Elementy teorii poznania logiki formalnej i metodologii nauk* (*Elements of the theory of knowledge, formal logic and methodology of the sciences*), Warsaw 1986, p. 111.

[669] See K. Popper *The Growth of Scientific Knowledge*, op. cit., p. 376.

but are not reliable. Although they are essential as the primary sources of a theory, they are still mostly false. Accuracy or precision, unlike clarity, is not valuable in and of itself[670]. Popper's epistemological conclusions concerning the definition of truth constitute premises in favour of Smoluchowski's *utility criterion*.

Returning to the definition of truth, according to Popper, we can use the intuitive conception of truth as a correspondence between a statement and facts, because Tarski's greatest achievement, with real significance for philosophy and the empirical sciences, consists in restoring the correspondence theory of objective (or absolute) truth[671].

In *Logical-philosophical writings of 1933, Alfred* Tarski argues that "for infinite languages the problem of the definition of truth has a negative solution"[672]. He puts forward the thesis that it is not possible to build a correct definition of a true statement for the vernacular or for infinite languages of formalised deductive sciences; it can only be defined for an infinite language in a deductive system. He believes that in reference to colloquial language "not only does the definition of truth seem to be impossible, (...) but even the consistent use of the concept in conformity with the laws of logic"[673].

The language of science, which is an infinite language, cannot define a true statement. In justifying the impossibility of constructing a definition of truth for infinite languages, Tarski references the thesis of Austrian logician Kurt Gödel (1906–1978), who showed that in arithmetic, being an infinite language, it is not possible to construct a strict definition of truth. Gödel's 1931 incompleteness theorem demonstrated that any coherent mathematical system powerful enough to perform what we know as elementary arithmetic must be subject to the constraint that its own coherence can never be proved[674].

---

[670] Idem, pp. 53, 54 and 377.
[671] Idem, p. 377.
[672] A. Tarski, *Pisma logiczno-filozoficzne. Prawda (Logical-philosophical writings. Truth)* vol. I, trans. J. Zygmunt, Warsaw 1995, p. 11.
[673] A. Tarski, *The concept of truth in formalized languages*, in: *Logic, semantics, metamathematics*, Oxford 1956, p. 153
[674] R. Smullyan, *Na zawsze nierozstrzygnięte. Zagadkowy przewodnik po twierdzeniach Gödla (Forever Unresolved. A mysterious guide to Gödel's theorems)*, trans. J. Pogonowski, Warsaw 2007, p. 7.

Gödel's theorem accentuates Russell's aphorism that mathematics is a "subject in which we never know what we are talking about, nor whether what we are saying is true."[675]. This thought is complemented by a statement by Józef Życiński:

> Regardless of the extent to which theses about the existence of perfect Mathematics, also called the formal field or the spirit of rationality, are justified, philosophers of science are primarily interested in small mathematics with its deficiencies and uncertainties. Therefore, the existence of Mathematics does not entail the perfection of mathematical knowledge, just as the perfect supralunar world in Aristotle's physics did not change the imperfections and deficiencies of the sublunar world[676].

Two conclusions can be drawn from the above deliberations: objective truth about reality eludes philosophy and mathematics, and there exists an abyss between that which can be proven and that which is true. Both theories—Tarski's and Gödel's—illustrate a fundamental truth concerning the classically understood truth of statements, representing empirical proof that it is impossible to outline an unambiguous criterion defining the objective truthfulness of a hypothesis or theory.

In the context of Smoluchowski's *usage principle*, this is an argument of key importance as it justifies the thesis that in verifying a scientific theory, the *utility criterion* is more effective than the truth criterion.

According to Życiński, a breakdown occurred in the hope held for twenty-five centuries of achieving certain knowledge of reality and its truths[677], which the researcher encapsulated in the statement: "*Epistēmē* shrinks rapidly, giving way to the omnipotent *doxa*"[678].

Smoluchowski's theory reveals a mechanism in which *doxa* (δόξα), illusory and uncertain cognition, as opposed to truth and unquestionable knowledge (epistēmē), generates scientific progress and provides an answer to the basic question about the essence *of cognition—which of the competing theories is more true? The effectiveness of utility theory* can be seen in decisions regarding the

---

[675] B. Russell, *Recent Work on the Principles of Mathematics*, 'International Monthly', Vol. 4, 1901, p. 84.
[676] J. Życiński, *Język i metoda* (*Language and method*), Kraków 1983, p. 109.
[677] Idem, p. 102.
[678] Idem, p. 109.

choice of finding a solution to a problem being studied. Each individual decision to choose a theory is dictated by the utility benefits provided by it in the broadest and deepest possible spectrum of understanding of this *utility*. The *utility criterion* appears in the deliberations of many scientists. It is not always verbalised, but, for example in Heller, the essence of the theory is palpable:

> The concepts we use have been developed in contact with our immediate (macroscopic) environment. Beyond its (quite fluidly defined) boundaries, they may not only lose their validity, but rather transcend (for example, by virtue of generalisation) into concepts whose content we are unable to imagine. But these concepts should correspond to our current concepts, that is, when applied to the areas we control, their content should pass into the content of our current concepts[679].

Correspondence of a new concept with the prevailing terms is a necessary condition for *use*.

6.3.3. Theory of abductive reasoning and other explanations

Charles Peirce's theory of abductive reasoning, being—according to its creator—the most scientific, was to be a special alternative to inductive and deductive reasoning. Abductive reasoning refers to the process of creating scientific explanations arising through the process of thinking, which for a certain set of facts are developed as their most probable explanation.

In *Metodologia nauk* (*Methodology of the Sciences*), Adam Grobler notes in reference to the Peirce method that in seeking an explanatory hypothesis for a phenomenon, we generally have more such hypotheses at our disposal and so choose the one that provides a better explanation. Grobler calls this the principle of inference to the best explanation[680]. Neither Peirce nor Grobler indicate by what criterion we assess the accepted explanation.

Also, according to Gregory W. Dawes, a key assumption is that we should accept the best potential explanation available to us for any puzzling fact, provided it is in itself satisfactory. Dawes asks why we should accept the best available potential explanation. The goal of science is knowledge, not simply truth. By this principle, science strives primarily not for true beliefs, but for beliefs assumed to be true, ones we have good reason to believe. That we

---

[679] M. Heller, *Time and causality*, op. cit., pp. 36–37.
[680] See A. Grobler, *Metodologia nauk* (*The methodology of the sciences*), Kraków 2006, p. 102.

are committed to expanding our knowledge and then accepting the best available explanations for any quandaries only proves that we are trying to explain them. The principle of practical reason is based on the assumption that if we strive towards a goal, then under the weight of practical rationality we are committed to the best means of achieving that goal. The reasons for accepting the best available explanations are pragmatic and have to do with accepting what is best and available to us[681].

According to the philosopher Gilbert Harman,[682] all explanations are inferences to obtain the best possible explanation. Clark Glymour believes that Harman understood this idea as inferring the best explanation available (by a defined set of logical or probabilistic standards), rather than the best possible explanation. Science and philosophy have always longed for solutions of the second kind. What is needed in science in order to arrive at, if not the best explanation, at least one of the best, is objective criteria for assessing the validity of explanations regarding data, as well as definition of the premises used to find them. There should be a connection between best explanations and truth, at least a "pragmatic vindication" that would guarantee that in the event of an explanation of all possible data of a specified type not being true, then at least the best explanation is true. Speculations about the truth or falsity, as Harman put it, of certain kinds of data or the best explanations are in the realm of things that are not fully verifiable, despite their falsification, while science demands theories that can be trusted without excessive risk.

Peirce, Dawes and Harman use the same logic of thinking as Smoluchowski, but do not get to the heart of the problem, i.e. the criterion specifying which hypothesis, being more probable, provides a better explanation. Hypotheses are created based on facts accumulated through experiments and observations, but the main driver of their creation is the pressure of possible changes improving the *usability* of existing theories.

Peirce did not take the next step and did not even refer to his own "flawed" *theory of utility*, stating that the truthfulness of the judgements and concepts of a given theory decide its *usefulness*. Similarly, Dawes does not define the phrase "best available explanation of any fact" and does not define the principles on

---

[681] See G.W. Dawes, *Belief is Not the Issue*, op. cit., pp. 74–75.
[682] G. Harman, *The Inference to the Best Explanation*, 'The Philosophical Review' 1965, no. 74, pp. 88–95.

the basis of which we accept the best available explanation. Nor does he give reasons why we should accept them. Harman believes, as mentioned above, that data of a certain type, despite the best explanations and despite their falsification, are in the realm of things not fully verifiable, while science, in his view, demands theories that can be trusted without undue risk. However, the question arises—how do we verify these best explanations?

As Glymour writes in *The Logic of Scientific Discovery*, Popper argues that it is best for theories to be bold and as falsifiable as possible. Let us note that for some philosophers of the last century, all possible data of a definite kind can be formalised in a first-order language in a definite vocabulary. Data should be understood both as single sentences and as generalisations in this language, which are arbitrary first-order sentences that can be considered statements about hypothetical empirical laws. Let us suppose that at least within the limits of all observations of a given type, individual generalisations and sentences in the singular of this language represent a set of verifiable or falsifiable statements. In the spirit of Popper's standard, it would seem better that a theory entail more such testable claims rather than fewer, but that is not quite right. More theoretical boldness requires more independently testable claims—at least logically independent. An ideally bold theory to account for some finite collection of data statements would therefore also entail an infinite collection of further testable claims about possible data, a collection that is not finitely axiomatisable in the language of the data[683].

Glymour's considerations, analysing experiences related to the creation and validation of hypotheses and theories in science in the context of Popper's reflections, focus on the formalisation of language representing a set of verifiable statements and their falsification as well as on boldness in breaking with existing theories. They overlook the emergence of an intellectual imperative prompting action aimed at finding correct explanations for data. They do not take into account the search for a mechanism to effectively derive the best available explanation and the best available verification of a hypothesis.

The above arguments are exemplary representations of the process of creating the best possible scientific explanations; they are pragmatic and relate to the adoption of what is best and available to us. A common feature

---

[683] See C. Glymour, *On the Possibility of Inference to the Best Explanation*, 'Journal of Philosophical Logic' 2012, vol. 41, No. 2, p. 462.

of these arguments is the lack of a category according to which theories and hypotheses are verified and accepted. Smoluchowski proposes the *category of utility*, the potential of which allows for the ongoing testing of accepted science. The theory that is most *useful* in a general sense at a given level of science is the one that enables the best possible technological and technical progress and at the same time stimulates further development of science by inspiring the creation of new hypotheses. Hypotheses are created based on facts accumulated by experience and observation; the main driver of their creation is the pressure of possible changes in improving the *usability* of existing theories by changing or replacing them, falsifying them, or approving them as more *useful*. In the face of such unambiguous demands and the inability of satisfying them, the *utility criterion* represents a pillar on which the credibility of science rests.

6.3.4. Thomas Kuhn—analysis of the structure and nature of knowledge

In conducting his reflections on the limits of knowledge in the philosophy of science of his time, Smoluchowski became, in a way, a forerunner of scientific realism. This is a 20th-century view, according to which scientific terms have objective references, while scientific theorems are generally at least approximately true. The key idea of scientific realism is the so-called scientific success argument, which emphasises the effective operation of scientific theories enabling the creation of new technologies, the design and construction of devices, and the prediction of events.

In the book *The Structure of Scientific Revolutions*, Thomas Kuhn analyses the structure and nature of changes in knowledge that have occurred over the centuries and seeks a criterion that would be the *spiritus movens* of changes and even revolutions taking place in science. Efforts aimed at tracing the source of these differences led Kuhn to identify the role of factors significant to scientific research, which he called "paradigms". These are the universally recognised scientific achievements that over a given period of time provide scientists with model perceptions and solutions[684]. Competition among scientists characterises the pre-revolutionary period, which is guided

---

[684] See T. Kuhn, *Struktura rewolucji naukowych* (*The Structure of Scientific Revolutions*), trans. H. Ostromęcka, J. Nowotniak, Warsaw 2009, p. 10.

by something much like a paradigm. Periods occur when two paradigms may even peacefully coexist. Factors such as technological progress, and external social, economic and intellectual conditions play an important role in the development of science[685]. Kuhn holds scientific progress to constitute one of the elements of social life, closely connected to all other elements, of which Kuhn lists several. This historian and philosopher of science a suggests the existence of a functional relationship linking all elements of social existence in terms of their operations and especially of the possibilities of their development. This universal functional relationship is the utility of each of the elements of a community's life.

Kuhn introduces into his deliberations the terms *"puzzle"* and *"puzzle-solving"*, which he believes permit a better grasp of some of the important issues raised. According to him, "puzzles" are a special category of problems that can serve as a test of ingenuity and proficiency in solving other problems. The criterion of a puzzle's value is not that its result is significant or interesting in itself; it is the very existence of the solution that is important[686].

Thanks to the paradigm, the scientific community acquires a criterion for choosing problems that—as long as the paradigm is taken for granted—can be assumed to have solutions. Only these problems are considered scientific, and they are the ones for which solutions are primarily encouraged. One of the reasons the development of normal science seems so rapid is that, within this framework, scientists focus their attention on problems whose solution could only be precluded by their own lack of ingenuity[687]. A new paradigm sometimes emerges even before a crisis has developed or been clearly recognised. Between the first signs of anomalies within the prevailing paradigm being noticed and an alternative paradigm being recognised, largely unconscious processes take place[688].

In other cases, however—for instance Copernicus, Einstein or modern nuclear physics—between the awareness of the old paradigm's collapse and the emergence of a new one, a great deal of time has had to pass. During this period, a scientist will behave more or less in line with the prevailing ideas

---

[685] Idem, pp. 12–13.
[686] Idem, pp. 73–74.
[687] Idem, pp. 74–75
[688] Idem, p. 156.

about scientific activity. He will be like a man groping in the dark, conducting experiments just to see what will happen, searching for phenomena whose nature he cannot guess. However, since no experiment can be performed without some kind of theory, a scientist in a time of crisis constantly tries to form speculative theories. If they gain recognition, they may pave the way for a new paradigm, if not—they can be relinquished relatively easily[689].

The recognition Kuhn writes about is clearly associated with the *utility* of the new paradigm being appreciated. Kuhn writes about utility but does name it. A theory becomes clearer and more convincing when we introduce to it the factor of the *utility criterion*. The scientific community recognises as scientific those problems the *utility* of which is most apparent, in which it is unambiguous or where there at least exist possibilities of verifying it, and this is safest within the functioning of the old paradigm. Only an extreme situation—in which the stunted development of *utility* signals the weakness of an old paradigm—unleashes the potential for it to be changed. Such was the case when the determined epicycles of planetary motion in Ptolemy's geocentric system showed that *use* of this theory did not contribute to the development of thought in astronomy, and the same was true at the end of the 19th century, when physics dominated by thermodynamics did not stimulate development but rather inhibited progress.

According to Kuhn, Popper denies the existence of any verification procedures. He believes that falsification is essential, i.e. the choice of testing procedures the negative result of which necessitates the rejection of an established theory. Kuhn does not agree with Popper's position as, in his view, no theory ever solves all the "puzzles" with which it is confronted at a given time, and nor are the solutions already achieved often perfect. Moreover, it is the incompleteness and imperfection of the existing data-theory fit that, at any time, define many of the "puzzles" that characterise normal science. Khun notes that if every unsuccessful effort to reconcile a theory with facts were the basis for rejecting a theory, then all theories would have to be constantly rejected. On the other hand, if only a serious failure justified rejection of a theory, then Popper's supporters would have to refer to some criterion of "improbability" or "degree of falsification". In formulating it, they would

---

[689] Idem, pp. 156–157.

probably encounter the same difficulties as defenders of various probabilistic theories of verification[690].

The ability to solve problems is neither the sole nor an unambiguous criterion for choosing a paradigm, because scientists will ensure that the data they collect will grow constantly and be treated with precision and in detail. This will lead to a situation whereby some old problems will have to be discarded. Often, a revolution also narrows the scope of interests of a given group of specialists, increases the extent of specialisation and diminishes contact with other groups, both scientific and lay. However, the list of problems solved by science and the precision of individual solutions will constantly grow—at least insofar as any guarantee is at all possible. Could there be any better criterion than the decision of a group of scientists[691]?

It is impossible to disagree with Kuhn's deliberations or the conclusions he draws from them. Similar thoughts result from analysis of Smoluchowski's theory, in which falsification is an important element but at the same time subordinate to *utility*. There have been, and still are, many theories in science that cannot be falsified. Popper's idea that there may be theories in science that might be falsified in the future is correct. However, this solution does not explain why these theories have been accepted by the scientific community. There exist theories that may be falsified in the future, but do not function in science, and conversely—there are also those that will probably never be falsified, and scientists are poring over them.

For example, it is not known how some theories of parapsychology, which today constitute a para-science, will be perceived in the future. In the history of science we have already seen the example of medieval alchemy, which in modern science has been transformed into the serious discipline of chemistry. *The Sceptical Chymist* (1661) by Robert Boyle (1627–1691) is considered a turning point.

The criterion of a theory's falsification is not unambiguous, while Smoluchowski's *utility criterion*, in which the basic determinant of the *use* of a given theory is its *utility*, is indisputable. Decisions on the *use* of a theory are not solely the domain of scientists; it happens that the whole of society participates, through its daily decisions, in the choice of what is most *useful*.

---

[690] Idem, pp. 251–252.
[691] Idem., p. 289.

However, due to where new theories arise, it is the choices of the scientific community that primarily determine their *usefulness*.

Kuhn poses the questions:

> Why would this evolutionary process have occurred at all? What must nature, including man, be like in order that science be possible at all? Why should scientific communities be able to reach a firm consensus unattainable in other fields? Why should consensus endure across one paradigm change after another? And why should paradigm change invariably produce an instrument more perfect in any sense than those known before? …It is not only the scientific community that must be special. The world of which that community is a part must also possess quite special characteristics, and we are no closer than we were at the start to knowing what these must be. That problem—What must the world be like in order that man may know it?… is as old as science itself, and it remains unanswered.[692]

Both the questions posed by the author of *The Structure of Scientific Revolutions* and the considerations conducted here provide extremely serious arguments in support of Smoluchowski's concept. It is clear to see that introducing the *utility criterion* into Kuhn's discourse solves a number of the problems he poses, such as the peaceful coexistence of two paradigms, the issues of new paradigms replacing old ones, or the problems related to Popper's falsification. The solution proposed by Smoluchowski also constitutes an answer to the questions raised, including the fundamental one about finding a criterion that triggers the evolutionary process of scientific development.

The process of creating science is a consequence of the make-up of the human mind, the essence of whose functioning is cognition. The imperative of competition and rivalry imposed on that prompts ceaseless revision and improvement. This imperative forces actions aimed at the best and fullest *use* of everything a person can have at their disposal. These struggles can be seen most clearly in the field of science, in which the epistemological mechanism increases the pressure for development.

The achievement of lasting consensus by the scientific community, which Kuhn ponders, is an understandable mechanism in the exact sciences, because it is not dependent on the establishment of a common position by a specific group of people, as is the case in the humanities. Although there are differences of opinion, they are temporary as the *use* of individual theories

---

[692] Idem, p. 294.

leads—sooner or later—to their verification through everyday *usage*. Everyday practice forces acceptance of those theories that seem most *useful*. Change in the acceptance of a theory generally occurs gradually but can also take place abruptly with the toppling of prevailing paradigms, as in the case of the theories of Copernicus, Kepler, Newton, Darwin or Einstein. According to Poincaré, paradigms are merely adopted conventions that are valid as long as they are *useful*, but when their *usefulness* comes to an end, they are rejected and replaced by new, more perfect, more *useful* ones, which are also mere conventions. As Kuhn himself notes with some surprise, this rule applies not only to the scientific community, but also to the whole of humanity. *Utility* tests not only paradigms and theories, but all choices made in everyday life.

6.4. Scientific realism and the *utility criterion*

6.4.1. Hilary Putnam's scientific realism

American philosopher Hilary Putnam (1926–2016), who argued in favour of scientific realism, saw an explanation of science's success in its constant discovery of truth. The "No Miracles Argument" as Putnam's reasoning is known, was intended as a defence of scientific realism, a scientific philosophy that does not perceive the success of science as a miracle. Therefore, the term "No Miracles" should not be understood theologically but concerns a philosophical stance that interprets the success of science. Putnam wrote that, "the typical realist argument against Idealism is that it makes the success of science a *miracle*. Berkley needed God just to account for the success of beliefs about tables and chairs (and trees in the Quad)."[693] Scientific realism is presented in his two works: *What is Mathematical Truth?* and *What is "Realism"?*, however, as Jacek Rodzeń writes, Putnam did not provide any clear guidelines as to the status of realism itself—whether it is an empirical or even scientific hypothesis, or merely a philosophical thesis[694].

---

[693] See H. Putnam, *What is "Realism"?* Proceedings of the Aristotelian Society, vol. 76, 1975, p. 177

[694] See J. Rodzeń, *Czy sukcesy nauki są cudem? Studium filozoficzno-metodologiczne argumentacji z sukcesu nauki na rzecz realizmu naukowego* (*Are the successes of science a miracle? A philosophical-methodological study on the argumentation from the success of science in favour of scientific realism*), Obi, Tarnów 2005, p. 36.

According to supporters of scientific realism, only an interpretation that explains the successes of science in terms of the concept of the truthfulness of theories prevents us from perceiving it as something otherworldly and inexplicable[695]. According to Putnam:

> a natural account of the way in which scientific theories succeed each other—say, the way in which Einstein's Relativity succeeded Newton's Universal Gravitation—is that a partially correct/partially incorrect account of a theoretical object—say, the gravitational field, or the metric structure of space-time, or both—is replaced by a *better* account of the same object or objects. But if these objects do not really exist at all, it is a *miracle* that a theory which speaks of gravitational action at a distance successfully predicts phenomena; it is a *miracle* that a theory that speaks of curved space-time successfully predict phenomena; and the fact that the laws of the former theory are derivable "in the limit" from the laws of the latter theory has no explained methodological significance[696].

The success of science is not created by an inexplicable miracle, but the philosophical view posed by scientific realism is the only clear and comprehensible explanation of it[697].

Putnam's assertions that older scientific theories are superseded by new, better ones and that the laws of new theories are drawn from the laws of earlier theories, are important assumptions both of the theory of scientific realism and of the "No Miracles Argument". However, in Putnam's concept, a clear lack can be felt of a crucial element of the proposed theory, which is the cause bringing these changes about. Putnam does not raise this question, which is after all a key issue for understanding the emergence of new hypotheses and theories and, as a result, the "No Miracles Argument".

The element missing in Putnam's theory, an accessory forcing change, is Smoluchowski's *utility criterion*. *Utility* represents an original contribution by Smoluchowski to the concept of scientific success. The idea that arose several decades before Putnam, stating that the theory that is most useful is the closest to the truth, complements the American philosopher's theory.

---

[695] See U. Żegleń, *Putnam*, in: *Powszechna Encyklopedia Filozofii (Universal Encyclopedia of Philosophy)* vol. 8, op. cit., p. 577.
[696] See H. Putnam, *What is "Realism"?*, op. cit., pp. 177–178.
[697] See idem., H. Putnam, *Philosophical Papers*, vol. 2, *Mind, Language and Reality*, Cambridge 1975, p. 73.

This simple criterion for assessing a theory's efficiency, or the value of a proposed hypothesis, is effective and reliable.

In his essay After Metaphysics What? Putnam writes about truth in science, defined as internal realism. He argues that truth is an idealisation of rational acceptability and should be identified with idealised justification, not with justification based on current empirical evidence. However, since the conditions for the assertibility of an arbitrarily selected statement are unfathomable, how do we learn them? This is achieved through experience, the author replies[698]. Putnam does not specify a definition of rational acceptability, relating it to the issue of idealised justification of our knowledge of a selected statement. Without rejecting Putnam's thinking, let us note that the source of rational acceptability lies in experiencing the degree of *utility* of the selected statement. We accept statements that are useful in any context of their *usefulness*. The rational acceptability of statements is confronted with constant verification through subsequent experimentally tested hypotheses, falsifiability, efficiency and so on, but the final verifier is the *criterion of utility*.

Would the success of science be real if the motor that is *utility* did not drive scientific research? *The utility criterion* constitutes a natural, active factor in the testing of scientific theories, attributing the qualification of truthfulness to the scientific theories it initiates and verifies. It is like the market mechanism in economics—the driving force of change, at the same time testing the effects of these changes.

We can talk about the usefulness of a theory, as already mentioned, in three ways. A theory is more useful: 1) the simpler and more illustrative its essence is; 2) the greater the area of known phenomena it explains and makes accessible to our mind; 3) the better guide it transpires to be for further research[699]. Smoluchowski's utility criterion—by its nature—initiates the creation of hypotheses and theories, constantly appraising them in an asymptotic approach to the "truth of nature" and over the longer term is the only natural and direct gauge of this.

The scope of influence of the concept of utility operating in our culture is understood in broadest possible sense of the term. It translates into the

---

[698] See idem, *Wiele twarzy realizmu i inne eseje* (*The many faces of realism and other essays*), trans. A. Grobler, Warsaw 2013, pp. 490–491.

[699] See M. Smoluchowski, *Self-Study Handbook*, vol. I, op. cit., p. 51.

application of the utility criterion in the creation of theories and hypotheses in science, thereby forcing a more effective learning and understanding of reality, as well as the use of the utility criterion in everyday life for the full exploitation of knowledge in the development of new technologies. The mechanism of the utility criterion enables the creation of further new theories through the criterion's application as a natural factor imposing and forcing the dynamics of change. Utility is therefore the natural creator of the success of science.

In his book *Czy sukcesy nauki są cudem? (Are the Successes of Science a Miracle?)*, in reference to Putnam's concept, Jacek Rodzeń proposes treating scientific realism as a factor that systematically stimulates metaphysical reflection, rather than as a doctrine to be defended. This realism could be approached as a philosophical research programme that is open to testing, constructive criticism, and modification of the theses put forward[700]. Entering into a discussion with Rodzeń, Stanisław Wszołek does not discredit the idea itself, but raises doubts as to the possibility of testing science without a clearly defined criterion: "It is not clear how such tests would be conducted. In this work we have only a vague suggestion that we should draw upon the history of science and current research practice. However, there are no attempts to outline a strategy for such 'tests'. Moreover, combining the testability criterion with the debatability criterion is quite problematic"[701].

According to Wszołek, Rodzeń justifies the dynamics of scientific development, as well as the miracle of its success, with the theory of scientific realism understood as a philosophical research programme. However, he does not explain what mechanism in the proposed research programme would test the theses, or by what rule they would be modified. Both philosophers lack the mechanism of the Smoluchowski criterion, which resolves the disputed issues using the *principle of usability*, because this principle is a basic link in the proposed research programme. Unstoppable, permanent verification of scientific theories in terms of their current *usefulness* is a test to which all scientific theories are subject. Such verification entails continuous study of the correctness and effectiveness of prevailing scientific theories

---

[700] See J. Rodzeń, *Are the successes of science a miracle?*, op. cit., p. 13.
[701] S. Wszołek, *W obronie realizmu naukowego*, (*In defence of scientific realism*) 'Zagadnienia Filozoficzne w Nauce' ('Philosophical Problems in Science') 2006, No. 38, p. 155.

on an empirical basis and within the realm of the functioning scientific discourse. In his further considerations, Rodzeń places greater emphasis on the importance of the effectiveness of the scientific method, rather on its truthfulness, which fits in with Smoluchowski's views. However, for him, the method's effectiveness concerns the effectiveness of mathematics. This is a true assumption, but a reductive one, as it overlooks a significant part of scientific practice. Smoluchowski's proposition is more universal and more complete as the effectiveness of the *utility principle* covers the entire field of knowledge creation.

In reference to Rodzeń's theory, Zygmunt Hajduk argues that justifying such a programme in terms of empirical testing is possible by referring to meta-scientific data from the scope of scientific research practice—both contemporary and that known from the history of science. We can then reasonably call realism a hypothesis in the empirical sense, subject to meta-scientific testing. This is not a doctrinal position, known mainly from tradition, which is not subject to criticism, but rather a kind of philosophical research programme, open to the entire spectrum of types of scientific success, considered also in historical terms[702]. Hajduk's proposal justifies the need for a philosophical research programme, however it is only the application of Smoluchowski's *utility criterion* that makes it clear that it is a key element of this programme. The proposed principle fulfils the conditions set by Hajduk, allows for empirical testing with reference to meta-scientific data from scientific research practice, but also—which is worth emphasising—allows for testing at the scientific level.

Rodzeń analyses Putnam's argument through Ghins's interpretation. Michel Ghins, a critic of 'traditional' empirical epistemology and a supporter of moderate scientific realism, distinguishes three concepts of scientific success: predictive, progressive (the success of subsequent theories) and methodological[703]. Let us analyse these three notions of scientific success in terms of Smoluchowski's proposed concept.

---

[702] See Z. Hajduk, *Między filozofią przyrody a filozofią nauki* (*Between the philosophy of nature and philosophy of science*), 'Roczniki Filozoficzne' ('The Annals of Philosophy') 2006, vol. 54, No. 2, p. 431.

[703] See J. Rodzeń, *Are the successes of science a miracle?*, op. cit., p. 39.

The predictive success of a theory, according to Ghins, lies in the "correct prediction by theories of observed phenomena"[704]. The correctness of a prediction comes down to the statement that the natural explanation for the success of the theories under consideration is that they are partially true descriptions of the behaviour of observed phenomena.

A predictive understanding of scientific success is an acceptance of the degree of a theory's accordance with reality. Scientific realism takes the view that recognition of a particular scientific theory carries with it a belief in the existence of all entities that meet certain criteria. The basic criterion a theory must meet is its *usefulness*, constituting an ongoing test of the acceptability of adopted theories while assuming their approximate truthfulness.

Progressive success is called the success of growth. This category defines the observable fact that some theories are followed by others that are better in terms of their predictive accuracy and have a wider scope in the domain to which they apply[705].

In the success of progressive growth lies the essence of the *utility criterion*; "better" according to Smoluchowski means "more useful". The greater *usefulness* of a scientific theory, applied technology or technique, is a determinant of what we consider to be better as opposed to what we consider to be worse—less useful.

The third category of Putnam's "No Miracles Argument" is the variant based on the concept of methodological success. It consists in the fact that "the laws of the former theory are derivable 'in the limit' from the laws of the latter theory". This, in Putnam's view, is a kind of methodological principle or directive natural scientists follow in their research[706]. The mechanism triggering the operation of Putnam's methodological directive, resulting in the derivation or improvement of existing theories, is a change in the perception of the usefulness of one—older—theory in relation to another—newer—one. The physicist chooses new theories, retaining what is still useful form old ones in terms of its practical application in techniques and technologies and its theoretical application to theories still in use. At the same time, he selects more predictive theories that foresee the emergence of new solutions

---

[704] Ibid.
[705] Ibid.
[706] Idem, p. 40.

and will allow the development of new technologies not predicted by older theories—he selects those that are simply more useful in the broadest sense of the term. Utility theory describes a simple mechanism functioning as the driving force of progress both in science and in practical, everyday life. As shown above, it clearly explains the operating mechanisms of the identified concepts of scientific success.

In the scientific discussion concerning the "success argument", the critical position is interesting of the American philosopher of science Bas van Fraassen, who presented a critique challenging the arguments of the creators of the concept of scientific success from the position of radical empiricism[707]. He believed that the selection of scientific theories is governed by the same or similar mechanisms that guide biological evolution. This is also a point of view found in the empiriocriticism of Avenarius and Mach, particularly in Mach's principle of economy of thought, but which also fits into Smoluchowski's philosophy and in no way undermines it. Van Fraassen's suggestion can actually be considered to be in line with the *utility criterion*.

Van Fraassen claimed that the success of scientific theories is no miracle. It is not even strange to the scientific (Darwinian) mind. According to him, every scientific theory arises as a result of intense clashes. Only the successful theories survive—the ones which in fact latched on to actual regularities in nature[708]. What does "latch on to" mean in this context? This metaphor indicates that in fact science has accepted theories that met the expectations placed upon them, i.e., according to Smoluchowski, those that proved most *useful*. The moment of so-called latching on of a theory is a decision by which a given theory is selected that fits the currently prevailing level of knowledge or surpasses it, but only to the extent that it is still compatible with that knowledge.

Whether the mechanism of scientific progress should be sought in the theory of natural selection, in biological evolution, is a secondary issue in the context of Smoluchowski's concept. More important is the analogy used that just as in Darwin's theory natural selection consists in the constant competition of individuals of a species, so it is in the case of scientific theories, according to van Fraassen, with the difference being that in relation to

---

[707] B.C. van Fraassen, *The Scientific Image*, Oxford 1980, pp. 39–40.
[708] Ibid.

Smoluchowski's theory, the struggle for the survival of the species is replaced by constant competition in the creation of new theories and the verification of existing ones, and its cause is the *usefulness* of theories already in use.

6.4.2. The *utility criterion* and social need

In the history of science, there have been many situations when the theories created were so far ahead of their time that not only were they not accepted, but they were often fought against, and most often went unnoticed. There are many such cases in history, but for the purposes of this book we will recall the figure of Nicholas of Cusa, a medieval philosopher, theologian and mathematician, who was among the unnoticed. 'Cusanus' was a thinker ahead of his time, an initiator of new concepts in the philosophy of nature, as Ludwik Kostro writes in the article *Transpersonal Cosmic Absolute*.[709] Many years before Nicolaus Copernicus,[710] he claimed that the universe is not geocentric but that Earth moves, like other heavenly bodies, around the Sun. He went further in his hypotheses than Copernicus as he said the universe is made up of an infinite number of planets circling an infinite number of stars. He also acknowledged that the universe has no perimeter, or a so-called centre, as the centre is everywhere.

His philosophical reflections were theological in nature, relating to the concept of God. Through an unusual way of understanding the essence of God, Cusanus came to spectacular conclusions; from the fact that God is everywhere and nowhere, he concluded that there cannot be a centre of the universe. He claimed that all planets rotate around their own axis, and that the rotation causes their flattening[711]. He used the thought experiment method.

This method, which is sometimes attributed to Einstein, has been practised essentially since antiquity, and its outstanding representative was Cusanus, who more than 150 years before Kepler, using a thought experiment, hypothesised that planets orbit the Sun not in circles, but in ellipses. He created a theory describing the inertia of material bodies, which was only later given a real scientific basis by Newton. Though born at a time when Renaissance

---

[709] L. Kostro, *Transpersonal Cosmic Absolute*, 'Studia i Badania Naukowe' ('Scientific Studies and Research'). vol. 2/2012, pp. 73–103.

[710] He died nine years before the birth of Copernicus.

[711] The diameter between the poles is less than the diameter at the equator.

ideas were becoming increasingly prominent, Nicholas of Cusa worked in medieval philosophy[712], which makes his hypotheses and theories all the more significant. Copernicus and Kepler became acquainted with some of his works, hence it cannot be ruled out that he influenced their views. Nicholas of Cusa was not an anonymous person in Europe at that time; he had many supporters, but also opponents. A professor at the University of Heidelberg called him an atheist, though what is more important is that Cusanus's theories arose no special response in 15th-century Europe and did not change anything significant in the knowledge of reality of that time.

There may have been several reasons for this situation, but one of the most important was undoubtedly the inability of information to move quickly. Johannes Gutenberg (1400–1468) invented printing only in the mid-15th century and therefore the works of Nicholas of Cusa were not as widely available as those of Copernicus or Kepler. It is also noteworthy that at the time Cusanus's concept was being developed, there was scant demand for new theories. In 1401, Lorenzo Ghiberti (1378–1455) was commissioned to create doors for the Florence Baptistery—this event is the symbolic dawn of the Renaissance, and also occurred in the year Cusanus was born. His birthplace of Keus is only 1,000 km from Florence, but existing knowledge had to be re-evaluated in order for a demand for new theories to arise. *Utility* is an essential but very subtle form of social need, which in the Middle Ages was more theological than scientific in nature, though this changed due to growing scientific demand. The use of existing theories and hypotheses became insufficient and unsatisfactory. It was not until the Renaissance that civilisational pressure arose to change the prevailing theories, which resulted in the theories of Copernicus, and later of Giordano Bruno (1548–1600), Galileo and Kepler. It is hard to believe that for the fourteen centuries that separated Ptolemy and Copernicus there was no opportunity to transcend the

---

[712] The Renaissance reached both Kues (Cusa) and central Germany much later.

geocentric theory, especially since Aristarchus of Samos (310–230BC)[713], and later Seleucus of Seleucia, who lived in the second century BC, propounded the heliocentric theory even before Ptolemy.

The *use of theories*, and hence their acceptance, is subject to social demand. It took many years for the *utility* of Cusanus's theory to be acceptable within the framework of a certain understanding of the laws of nature. The fact that the theories of Nicholas of Cusa were more in line with reality than those in force at the time was of no importance in view of the lack of those theories' *utility* value at the time.

In *Objective Knowledge*, Karl Popper writes about transpositions taking place between theories—the new theory and the displaced old one. He notes that while the new theory explains what the old theory explained, at the same time it corrects the old theory in such a way as to actually contradict it. The new theory contains the old theory, but only in the form of its approximation. For example, Newton's theory contradicts the theories of both Copernicus and Galileo, although at the same time it explains them due to the fact that it contains the previous ones as approximations. The case is similar with Einstein's theory, which contradicts Newton's, but on the other hand explains it as it contains its approximation[714].

Agreeing with Popper on a new theory's approximation of an old one, particularly concerning using the old theory to explain elements of the new one, and transposing Popper's idea into the Smoluchowski criterion, it should be noted that it is the *usefulness* of the old theory in the conditions of the new one's functioning that determines the degree of this approximation. In other words, it is the outcome of the *utility criterion* in action that enables the functioning in science of an old theory that has been replaced by a new one, as long as it is in some way *useful*. An example is classical Newtonian physics. Its use in everyday life is possible because its degree of imperfection

---

[713] In his book *Brief Answers to the Big Questions* S. Hawking writes on pages 52 and 53 that Aristarchus, on the basis of accurate observations of lunar eclipses, put forward a bold thesis that, in fact, an eclipse is the shadow of the Earth passing across the face of the moon. He created diagrams showing the actual relative positions of the Sun, Earth and Moon. He deduced that the Earth is not in the centre of the universe, but orbits the Sun. He put forward the idea that the stars were not holes in the sky, but other suns, similar to our Sun, but very distant.

[714] K.R. Popper, *Objective knowledge*, op. cit., p. 29.

is acceptable in many fields of technology and science. Illustrating the scale of the problem, Czesław Białobrzeski notes that deviations from Newton's law of gravity and the principles of classical mechanics are barely perceptible; for example in the solar system they become visible only in the perihelion motion of Mercury[715]. This system is still *useful* in the macroscopic world, which is an effective verifier of it. Einsteinian physics contradicts Newton's theory, supersedes it, at the same time approximating it in terms of its *usefulness*. The opposite occurred in the case of Ptolemy's theory, which has been completely ousted from everyday life as *useless*.

Izydora Dąmbska drew attention to the methodological requirements set by the utility criterion for newly created theories. Smoluchowski based the criterion, Dąmbska claims, on the traits that every theory should exhibit. The first is simplicity and clarity, the second—the scope of phenomena explained, and the last—predictability[716]. A scientist chooses a theory that contains all the above-mentioned features primarily because the ultimate task of physics is to find, as far as possible, a simple and concise description. Secondly, it is also important that the new theory apply to the largest possible complex of physical phenomena, covering the greatest possible area of known phenomena. Thirdly and finally, it is important that it is more predictive than other theories and more able to anticipate new phenomena unexplained by older theories[717]. The utility of the new theory should be as extensive as possible and—if possible—total.

Undoubtedly, an exhaustive philosophical and scientific analysis of the *utility criterion* verifiers used could help determine which questions are worth asking and give order to the categories adopted to verify emerging new theories. Unfortunately, the *category of utility*, proposed by Smoluchowski over a century ago, has not been taken note of by the scientific community. What makes the scientist's thought extraordinary is the very idea of the *utility*

---

[715] See C. Białobrzeski, *Podstawy poznawcze fyzki świata atomowego* (*Cognitive foundations of the physics of the atomic world*), Warsaw 1984, p. 24.

[716] The issue of assessing the so-called criteria of "good" theory among 20th-century philosophers of science was raised in many of Ernan McMullin's texts—in particular in the article *Values in Science*, trans. J. Rodzeń, 'Zagadnienia Filozoficzne w Nauce' ('Philosophical Problems in Science ') 1999, vol. XXIV, pp. 7–25.

[717] See I. Dąmbska, *On the metascientific views of Władysław Natanson and Marian Smoluchowski*, op. cit., pp. 8–9.

principle, which relates directly to the value of theories both old and new and is decisive in the choice of a new theory as being more *useful* than the old one. The above considerations show that this apparently banal thought concisely explains the mechanism not only of scientific progress, but also of the transformations taking place in all aspects of human society. In the case of a range of theories examining the mechanism of science's creation that have emerged over the last few hundred years, it complements these theories with an indispensable element, namely a criterion forcing an objective, continuous examination of their usefulness. The *utility principle* fits perfectly into the philosophical discourse that has been taking place since the mid-20th century on the success of science, and its application to Putnam's programme is important in explaining the mechanism driving the development of science. It is an empirical fact that science succeeds in making accurate predictions, building further hypotheses and theories, and devising better systems for controlling nature. However, the discussion around the success of science—the no-miracles argument or scientific realism—has restricted itself chiefly to stating the facts, has not reflected on the search for the mechanism driving this progress, and has not sought a criterion by which subsequent steps in arriving at the truths of nature are verified. This mechanism and verifier is Marian Smoluchowski's *utility* criterion.

6.5. Induction and the *utility criterion*

6.5.1. Hume's inductive scepticism

An opponent of the inductive method was David Hume, who treated the problem of induction in a similar way as he had previously addressed the issue of causality. He argued that we use the inductive method in reasoning through habit and that, in principle, we have a serious problem with justifying the use of induction as there is no evidence for this method's verifiability. Using it in everyday life, humanity has developed certain beliefs that transfer experiences from the past to what is to happen in the future. We assume that induction works on the grounds that it has worked so far, which, Hume argues, does not mean that it will continue to work. Assuming the future effectiveness of induction on the basis of its past effectiveness is itself an inductive inference. This is a vicious circle in which, in order to justify the inductive method's verifiability, we invoke its correctness. In this way we build a tautological definition according to which we conclude that the current effective use of induction means it will continue to be effective. Such reasoning is an attempt to justify the principle of induction by the principle of induction itself.

The philosopher believed that the make-up of the human mind requires a constant search for regularities according to which experiences can be organised. The regularity of nature does not result from things themselves, but is merely an image of our experience and the way we organise our knowledge of reality; therefore the inductive method is a mere projection of our experiences rather than a proven scientific method.

Hume's arguments were intended to provide grounds for rejecting reasoning that does not meet deductive standards. Arguing that there are no grounds to infer from cases known from experience to other similar cases not known from experience, Hume maintained that there is nothing in anything considered in itself that gives us reason to draw an inference extending beyond it. Even if we have observed that two different things are often or constantly related, we have no reason to draw a conclusion about any things beyond those that were present in the experience[718]. The fact that we believe cases of which we have no experience will be consistent with those we know from experience comes only out of habit, which is necessary for our survival. This means that our mind is formed in such a way as to use the fruitful assumption of the repeatability of situations and successive experiences to function efficiently. However, this does not entail unshakable scientific certainty of obtaining the expected result from an experiment.

Renata Ziemińska confirms Hume's sceptical view regarding causality and induction, in which the philosopher argued that our beliefs about facts, both external and internal, have no rational basis, but are based on habit and instinct. We unconsciously interpret the regular sequence of phenomena as a cause-effect relationship, and the regular similarity of perceptions as the identity of the things these perceptions are supposedly based on. Our deep conviction about the existence of a 'self' is also the result of an instinctive interpretation of a succession of perceptions. Humean scepticism about induction is also associated with criticism of causality—no argument can justify

---

[718] See D. Hume, *Traktat o naturze człowieka* (*A Treatise of Human Nature*), trans. C. Znamierowski, Warsaw 2005, p. 223.

our inference about the future based on the past. We do not observe such a causal relationship in experience, nor are we able to prove it. If someone tried to justify this thesis inductively, it would settle the matter[719].

An example is cited in publications of a view taken for granted in antiquity and shaped by induction. A presumption that has always proven true is that the sun rises once a day and sets once a day. According to Hume's sceptical thesis, even such a strong belief need not continue to hold true. It was invalidated by the famous traveller Pytheas of Massalia (350–285 BC). During his voyage, he saw the sun shining at midnight and the sea covered with ice, but for many centuries he was considered a liar.

According to Hume, sceptical arguments are insurmountable on the grounds of reason. The philosopher tries to develop them by adding previously unknown arguments to the sceptical thesis (the dilemma of judgments about facts and of judgments about relations between ideas). He also develops ancient and medieval arguments against causality and induction[720], but this does not change the proven fact that the inductive method creates and gives impetus to the development of science and makes it possible to build new theorems and new scientific theories.

Hume's sceptical argument questioning the method of induction is all the more intriguing as its initiator was an adherent and advocate of the empirical sciences, which, *nota bene*, have flourished due to the inductive method. This was to some extent due to the weakness of scepticism itself, the credibility of which was also questioned, with the sceptical position shown as being as dubious as any popular opinion. This being the case, Hume could have decided to distance himself from scepticism and return to popular and scientific beliefs. But in this situation, Hume shifts instead to another level of reflection and it becomes clear how this sceptic can use the principles of induction and normative discourse in his texts. In this way, scepticism represents more of an epistemological background as a methodological rule of caution and criticism[721].

---

[719] See R. Ziemińska, *Sceptycyzm umiarkowany Davida Hume'a* (*David Hume's moderate scepticism*), 'Przegląd Filozoficzny – Nowa Seria' ('Philosophical Review – New Series') 2011, No. 4 (80), p. 126.
[720] Idem, p. 119.
[721] Idem, p. 133.

However, based on Hume's deliberations, the 19th century saw strong opposition from scholars averse to the use of induction in science. For Smoluchowski, who was among the supporters of this method, there was no alternative path; he treated induction as an indispensable stimulus for the development of science, especially physics. While he preferred the inductive method, however, he also valued deduction, believing that each method should be used when it is more effective.

Practical application of such an effective method as induction poses a philosophical problem regarding the way we convince ourselves that our reasoning is justified, and thanks to which we become sure our conclusion from inductive inference is sound. In deductive reasoning, premises and their truthfulness play a key role, while in inductive reasoning we face the problem of to what extent inductive inference leads to true conclusions.

Some proponents of inductionism determined the truth of conclusions using probability calculus. This was meant to derive from the concept of philosopher Hans Reichenbach. He argued that "the principle of induction provides us with the means of judging the truth of scientific theories. To be more precise, we should say that it serves to decide their degree of probability. The alternatives in science are not truth and falsehood; instead, there is a continuous scale of probability values whose unattainable limits are truth and falsehood[722]." Opponents of the inductive method formulated in this way believed that no satisfactory answer had yet been given to the question of how to measure the probability of the truthfulness of conclusions.

Smoluchowski, a supporter of induction, was among the sceptics of determining the credibility of a hypothesis using probability theory. There was no contradiction in his statement when he argued that "today science usually does not distinguish between true and false hypotheses when applied to the principles of physics, but we talk about hypotheses that are more or less probable"[723]. Smoluchowski's statement that a hypothesis may be "more or less probable" does not translate into its degree of likelihood. In his opinion, we can only talk about probability in relation to statistical phenomena when a large number of analogous cases are analysed. The probability can

---

[722] H. Reichenbach, *Causality and Probability*, in: *Modern philosophy of science: selected essays*, Trans. & Ed. M. Reichenbach, 1959, London, p. 76.
[723] M. Smoluchowski, *Self-Study Handbook*, vol. I, op. cit., p. 49.

be estimated of the birth in Warsaw on a given day of a hundred children, but one cannot speak of the probability of the hypothesis that electrons are rigid spheres. It is also impossible to talk about the probability of causes, because then it should be assumed that the same effect can be brought about by a whole range of different causes. Another type of issue is the question of whether the hypothesis that there is an ocean of liquid water under the icy crust of Jupiter's moon Europa is probable.

A summation in the context of Smoluchowski's considerations on the physicist's attitude to science is Max Born's conclusion in *Natural Philosophy of Cause and Chance*: "Induction allows one to generalize a number of observations into a general rule: that night follows day and day follows night… But while everyday life has no definite criterion for the validity of an induction… science has worked out a code, or rule of craft, for its application"[724]. Popper argues that Born nowhere discloses the content of this code, emphasising there is no logical justification for it, that it is merely a "matter of faith"[725]. Smoluchowski expressed such a view much earlier than Popper. He also did not believe, as Born writes, that daily life does not have a specific criterion for the value of induction as. According to Smoluchowski, it is the *utility criterion* that allows for ongoing verification of a proposed hypothesis or theory resulting from induction.

6.5.2. Peter Lipton's induction and the *utility criterion*

In his work entitled *Inference to the Best Explanation*, Peter Lipton argues that Hume's sceptical argument regarding induction does not, contrary to appearances, lead to its rejection. Drawing on his own traditional argumentation, Hume points out that our observations do not entail our predictions as the governing principle in all our inductive inferences is that nature is uniform and that the unobserved world is similar to the one we have observed. It is therefore essential to demonstrate that nature is indeed uniform. This cannot be inferred from what we have observed, because the claim of uniformity itself incorporates a huge degree of prediction. The uniformity assumed in this way involves the use of an inductive argument,

---

[724] M. Born, *Natural Philosophy of Cause and Chance*, Oxford 1949, p. 7.
[725] See K. Popper, *The Growth of Scientific Knowledge*, op. cit., p. 95.

which would rely on the principle of uniformity, leaving the question itself unanswered.

The basic principle of inductive inference is *'more of the same'*. We believe that strong inductive arguments are those whose conclusions predict the continuation of a pattern described in the premises. Using the so-called principle of conservative induction, we conclude that the sun will rise tomorrow since it has always risen in the past. We would judge as worthless the argument that the sun will not rise tomorrow since it has always risen in the past. According to Lipton, there is no problem in coming up with a viable proof to underwrite the latter argument, created according to the so-called principle of revolutionary induction: "It is time for a change", and which would consequently sanction a paradigm shift. According to Hume's argument, we have no way of showing that the conservative induction he claims we actually use for our inferences will be better than intuitively "wild" principles, such as that of revolutionary induction. Of course, conservative induction has a more impressive track record. Most of the conclusions from the true premises that it has sanctioned have also had true conclusions. Revolutionary induction, by contrast, has been conspicuous by its failure, so anyone wanting to rely on it completely should bear this fact in mind. However, the question of justifying the choice asks not which method of inference has been more successful, but which will be more successful.

The success of conservative induction stems from the fact that within the framework of existing paradigms and prevailing knowledge, functioning theories are supplemented by subsequent hypotheses. These inductions arise within the framework of the applicable laws of science; falsification of their achievements and demonstration of their *usefulness* are often already possible during the inductive construction of a hypothesis. Successful revolutionary inductions that can upend the existing perception of reality obviously occur less frequently as they bring about profound changes in science and a conviction of their greater *usefulness* often takes longer to achieve.

The success of conservative induction seems to be in itself a reason to trust it; its track record is not perfect (it does not aspire to deduction), but is very impressive, especially compared to revolutionary induction. In short, induction will work because it has worked, and this seems to be the only requirement we could ever place on our inductive methods. The verifier of their effectiveness is the *utility* of their effect. Hume's disturbing observation was that this justification is "talking in circles," that is, no better than trying to convince someone that you are honest by telling them you are[726]. Smoluchowski's *utility criterion* overrules this disturbing observation of Hume's, verifying the positive effect of induction and eliminating the tautology Hume suggests.

The two main questions relating to the principles of induction concern description and justification. What principles are used and how can they be shown to be good principles? At first, the issue of description seems to be the priority. However, from a historical point of view, the question of justification is more important. One reason for this is that the question of justification is framed by sceptical arguments that seem to apply to any principles that may explain how we bridge the gap between the evidence we have and the inference we make. But how can we try to justify our principles until we know what they are[727]? Lipton's argument is one that proves the effectiveness of verifying theories through the *utility principle*. The *utility criterion* entitles us to temporarily disregard issues related to a full understanding of the principles that guide us in explaining how to fill the gap between evidence and inference. The conclusion drawn regarding principles does not imply acceptance or otherwise of the induction's effects, as their value depends on their *usefulness*. This situation is known from the history of science, and we currently encounter it regularly.

The problem of justifying induction does not prove there are other inductive standards better than our own. The problem of justification can be posed before we have solved the problem of description, bridging the gap between observation and prediction. Induction can be defined in many different ways, but this would lead to completely different predictions. We have no way of convincing ourselves that our way is better than any other, not because the conservatives' arguments will not persuade the revolutionaries, even if their standards include one stating that an argument repeated over and over again is wrong. If I am honest, I should admit that the rumour I'm talking about

---

[726] See P. Lipton, *Inference to the Best Explanation*, Second Edition, London and New York, 2004, pp. 9–10.
[727] Idem, p. 7.

may carry no weight. We have a psychological compulsion to favour our own inductive principles, but if Hume was right, we should see that we cannot provide a cogent rationalisation of our opinion[728].

The psychological compulsion to favour one's own inductive principles mentioned by Lipton introduces a logical-cultural order to newly created theories, systematically expanding, step by step, the area of our knowledge. This happens until a radical breakthrough occurs—in the form of a revolutionary induction, such as experienced as a result of the theories of Copernicus, Maxwell, or Einstein. In order to be accepted, each of them had to change a number of paradigms related to the perception of reality. The mechanism for making acceptance of such a radical change possible is the new theory's *usefulness*—providing greater possibilities, a broader view of reality and, most importantly, enabling a fuller understanding of the truths of nature.

Due to the results yielded by *utility theory*, Lipton's earlier cited discourse loses its weight; there are opportunities to show that our way is better than any other, because its *utility* is the greatest.

> But verifying a theory's usefulness is not always so simple and obvious. The time it takes for us to come to the conclusion that a new theory is more useful than an old one varies not only between theories, but also for the same theories in different societies. The usefulness of Darwin's theory of natural selection has been continually questioned by various communities since publication in 1859 of the book On the Origin of Species. This theory's acceptance depends not only on changes to paradigms rooted deep within societies, but is also on the fact that the utility of creationism in a given society is sometimes greater than the utility of evolutionary theory.

In a situation where the *usefulness of a theory* can be perceived relatively, when the theory is of a type unverifiable in everyday life, or when its *usefulness* does not depend for every observer on its truthfulness (like the theory that the Earth is a flat disc located on the back of a large turtle travelling through space), the *usefulness* of such theories is determined in a way developed by the scientific community and decided by a scientific milieu. Then, according to Poincaré's thesis posited over a century ago, a fundamental role is played by the decision of the scientist or scientific community introducing the new theory to science. However, the scientific community, as evidenced by the

---

[728] Idem, p. 10.

example of the theory of global warming, may be divided, and then the verifier of a given theory's *utility* becomes time.

While, as Lipton notes, the lack of a satisfactory solution to Hume's challenge and the bleak prospects of finding are not encouraging, the argument regarding inductive principles as a whole is also doubtful. There are at least two reasons. The first concerns the credibility of hypotheses, and therefore of theories created through the principles of induction. This raises the problem of justifying the reliability of inference methods, in terms of their truthfulness. We want our methods of inference to be "truth-tropic" towards the laws of nature. Therefore, a good argument for deduction is that it is an excellent conduit of truth, whereby it is impossible for false conclusions to come from true premises. In the case of induction, such perfect reliability is out of the question. By definition, even a good inductive argument is one where it is possible for there to be true premises but false conclusions[729]. Such an understanding of the degree of reliability and the relationship of a theory to the truths of nature changes when reflection on an issue does not relate to absolute truth but is limited to a part of the truth or to such a truth that is necessary for the theory to be *useful* at a given moment. The magnitude and formula of reliability are demarked by the current *degree of utility* of the theory being tested. If in some respect it is more *useful* than the prevailing one, then the current theory should prove less true than the new one. This is not always the case; it does not hold true when goals other than those applicable to science dominate in creating the theory, or when the theory's greater *usefulness* is falsely perceived. The falsity of an accepted bad theory will verify the *usefulness* of another new theory. A new theory should be more true than its predecessor, as was the case with Ptolemy's theory, which was superseded by Copernicus's assuming circular orbits of planets around the Sun, which was amended by Kepler's theory describing orbits in the form of ellipses, altered in turn by Newton's, introducing gravitational dependence to the movement of the planets. A theory positively verified by its *usefulness* naturally replaces its predecessor. We state that induction worked because it worked, because it was verified by its *utility*.

---

[729] Idem, p. 8.

Popper asks why it is more reasonable to prefer non-fiscal claims over fiscal ones. He writes that on the basis of the pragmatic position, such a question does not arise at all, because false theories can be most *useful*. Most of the formulae used in engineering and navigation are known to be false, although as good approximations they are convenient to use and are trusted by people who know perfectly well that they are false[730]. It is unknown how many formulae used in science are false in Popper's sense.

Therefore, it is no surprise that we experience the *utility* imperative operating more clearly when science encounters obstacles in the form of generally accepted but not fully explicable theories, such as Heisenberg's uncertainty theory or Gödel's incompleteness theorem. Both theories—although not fully understood—function in science because they have been positively verified by the *mechanism of utility*.

Lipton's second point related to the *concept of utility* concerns the very essence of the principle of induction and constitutes a separate philosophical problem relating to the use of the inductive method. It concerns the way we convince ourselves our reasoning is sound and that our conclusion of inductive inference is justified. Premises and their truthfulness play a key role in deductive reasoning. In inductive reasoning, we face the problem of to what extent inductive inference leads to true conclusions.

6.5.3. John D. Norton's inductive inference

A specific operating method of *utility theory* is demonstrated by an analysis of induction presented in John D. Norton's article *A Material Theory of Induction*, in which the author posits the thesis that all effective inductions are local.

Norton argues that in inductive inference we assume so-called inductive risk, consisting in the danger of accepting a high factor of faith in a fixed outcome which ultimately proves false. In science, we try to minimise inductive risk. Inductive risk-control strategies vary depending on the formal theory of induction or the material of the theory. According to the formal theory, we approach the problem in two ways. Firstly, we try to gather as much evidence as possible, because the stronger the evidence, the stronger our inductive conclusions will be. Secondly, we strive to

---

[730] See Popper, *The Growth of Scientific Knowledge*, op. cit., p. 101.

broaden the schemes of inductive inference available to us. These two ways of controlling inductive risk cannot be separated. We limit our exposure to inductive risk by collecting more evidence and at the same time by expanding our inductive schemes.

In everyday life, we routinely accept certain diagnoses, such as those of experienced car mechanics or doctors, on the basis that the method used in their diagnosis is reliable, although we, and often they, have trouble understanding it. We may reject an idea as being an *ad hoc* hypothesis simply because it was not formulated through an appropriate method. Proponents of such an approach argue that properly practised science uses reliable methods and therefore we can trust its results. This attitude, which provides the improvements needed to tame the uncritical nature of hypothetical deductivism, is common. An effective result guarantees us that acceptance of the proposed solutions is motivated by a correct belief that the method is reliable. This conviction applies not only to the method, but also to the world, which we believe to be such that these methods can work reliably[731].

How then can we explain the fact that many theories that came about on the basis of an appropriate method, according to the best scientific practice, have after some time ceased to be believed and have been replaced by other, new ones? This exchange of theories does not occur by accident. The changes are influenced by concrete arguments. Forced changes are most often the next stage of development, creating new and expanding old horizons of understanding the reality around us. However, in every case, the verifier of the theory's acceptance is its utility. A special case is the already mentioned creationism, which, under certain conditions and in a specific community, through a particularly understood utility, resulted in the elimination of Darwin's theory of evolution. However, in the long run, this theory proves more useful and displaces creationism from science.

---

[731] J.D. Norton, *A Material Theory of Induction*, op. cit., p. 658.

According to the material theory,⁷³² all schemes acquire only local meaning and are ultimately anchored in the facts of a given area. Put simply: the more we know, the better we can reason inductively. The above considerations illustrate the dynamics of efforts to control inductive risk. The conclusions drawn are risky because there are known cases of unreliable patterns. Local inductions are characterised by lower risk. We localise induction within a specific field, the material of which postulates achievement of a safer induction. For example, due to the dependence of the tides on the position of the moon, we will hypothesise that the cause of the tides is related to Newton's theory of gravity, which makes us much more confident that this hypothesis is correct. This dynamic consists in transferring inductive risk from the schema to the fact that constitutes the corresponding material proposition. This transfer is an important means of assessing and controlling inductive risk⁷³³.

Both with respect to the elements of the material postulate of the induction carried out, and in terms of controlling the inductive risk, Norton moves within Smoluchowski's *usability theory*. The problems of induction's effectiveness and of inductive risk are transferred to the *usability* of the theories that emerge. Norton's suggestions that in material theory all schemas acquire local significance and are anchored in the facts of a given domain are found in the *theory of utility*, because it is indisputable that the verification of a *theory's utility* is most effective by localising induction within a specific field.

As long as the inductive risk lies in the schema, Norton goes on to argue, we need to evaluate it through a highly problematic assessment of the schema's overall plausibility. We have little chance of getting a clear result, not to

---

⁷³² The concept of the material theory has many proponents. Perhaps the most important of them is Nelson Goodman (1906–1998), who conceived of the "grue paradox", which illustrates the problem that in order to learn by induction, we must distinguish between predictable and unpredictable predicates. Suppose someone noticed that all emeralds ever observed are green and argues inductively to conclude that all emeralds are green. Now suppose we define "grue" as the property of being green until t (say early 2050) and then blue. The inductive evidence we have confirms the conclusion that all emeralds are "grue", as well as the conclusion that all emeralds are green, so we have no reason not to choose either of these conclusions. Many people (though not Goodman) interpret this as a refutation of induction. Goodman sees the problem of Hume's induction as a problem of the accuracy of the predictions we ourselves make (this topic will not be developed further here).

⁷³³ See J.D. Norton, *A Material Theory of Induction*, op. cit., p. 658.

mention how to mitigate this risk. However, when the risk is transferred to a material postulate, in some local domains, the assessment of inductive risk will depend largely on our confidence in that material postulate. It results directly from how much we can trust a given fact and use it in building a hypothesis, and the trust placed in it is verified by *utility theory*. If there is inductive risk, then it is within the programme of its limitation that we should seek evidence relating to the material postulate, even modifying the material postulate in light of the evidence. As a result of these measures, induction will be safer. In short, we can control inductive risk by converting schematic risk into implicit risk, because the latter can be more accurately assessed and reduced[734].

According to Norton, the more we know, the better we can infer inductively, because the more efficiently we establish a theory's *utility*. This idea can be related to Roman Ingarden's remark that every empirical science is an inductive science—an inductive extension, a generalisation, as a result of which more or less probable general statements arise, which can never be fully justified due to the nature of induction[735]. Norton's position assumes that we can limit this risk by giving schemas local meaning and relating the evidence to the material postulate, however, under the *theory of utility*, we do not demand full justification, but the maximum possible *utility*.

If the inductive risk lies in the induction schema, it needs to be evaluated it through a highly problematic assessment of overall plausibility, and therefore we try to gather as much evidence as possible, because the better the evidence, the stronger our inductive conclusions will be[736]. However, the verifier of the reliability of inductive conclusions is their *usefulness*, which may be greater or lesser and which may result in the replacement of existing conclusions with new, more *useful* ones.

After two thousand years of effort, according to Norton, it has not been possible to achieve consensus on the proper classification of induction. An introduction to inductive logic based on simple calculation has become a favourite way of denigrating this way of thinking for centuries. The extraordinary

---

[734] Idem, p. 664.
[735] See R. Ingarden, *Wstęp do fenomenologii Husserla* (*An introduction to Husserl's phenomenology*), trans. A. Półtawski, Warsaw 1974, p. 51.
[736] See J.D. Norton, *A Material Theory of Induction*, op. cit., p. 647.

success of science in learning about the world, achieved through inductive inquiries, has caused problems. How can this success be reconciled with the continued failure to agree on an explicit systematisation of inductive inference? We need to recognise that our failure to agree on a single systematisation of inductive inference is not merely a temporary lacuna, Norton argues. We have failed not because of a lack of effort or imagination, but because we seek a goal that in principle cannot be found. The proper goal is to develop an account of induction in which failure becomes explicable and inevitable, and without the need to deny the legitimacy of inductive inference[737].

In analysing the dilemmas associated with inductive inferences, Norton arrives at a barrier he believes is impossible to overcome. He concludes that the situation described need not be a failure; our lack of success in establishing an explicit systematisation of inductive inference need not be a mere temporary lacuna but rather is a particular point of perception of the evolution of science. With his *utility theory*, Smoluchowski introduces such a sense of understanding the development of science, which occurs through inductive inferences in which subsequent scientific theories must meet the conditions of the *utility criterion*.

For a clearer understanding of Smoluchowski's proposition, we can imagine an abstract triangle, which we can give the working name of the 'triangle of science', at the base of which we have a well-founded set of scientific views and theories the *utility* of which is little questioned. The higher up, towards the apex, the less certain the theories' *utility*, sometimes temporary, inherently impermanent. The movement of theories that takes place within the triangle runs from top to bottom and bottom to top, the lower—the more significant and certain theories are, building the foundation of the field of science. It is mainly on the basis of these that new ideas and new theories are derived by induction, which from the top, from the apex of the triangle, are incorporated into the field of science. The more their *utility* is positively verified, the more certain their position in the triangle is. Theories whose *utility* is poorly verified are located at the periphery of the triangle's apex. Outside the triangle, in the vicinity of the apex, are theories whose *utility* has not been established, which does not mean they do not have any. An example

---

[737] Ibid.

of such a theory, which did not fit into the described triangle for a number of years, is atomic theory, which functioned outside the 'triangle of science' until the end of the 19th century, its *utility* not being confirmed until the 20th century, after which it was finally included within the triangle.

The 'triangle of science' built within the framework human culture is not the only option for understanding nature. Other triangles are possible, built on the basis of different, alternatively understood *utility*. However this one, unique to us, was built within the framework of our culture, in which a key verifier and driver of expansion has been and remains the verified *utility* of a given theory, supported by the principle of the economy of thought and falsifiability.

What is the relationship of the 'triangle of science' to the truth about reality? In simple terms, the same as the relation of a geometric triangle to the infinite surface of the plane on which it stands. This is the relation of the epistemic 'triangle of science', which happens to be adjacent to the infinite ontological surface of the truths of nature.

We have been misled, says Norton, by a model of deductive logic into seeking an account of induction based on universal schemas, whereas an account of induction is possible with no universal schemas. Instead of the current approach, inductive inferences will be seen as deriving their licence from facts. These facts are the material of the inductions; hence it is a material theory of induction. Particular facts in each domain licence the inductive inferences admissible in that domain—hence the slogan: "All induction is local". This is not about advocating any particular system of inductive inference. Competition between well-established systems is futile; each of them can be used along with their attendant maxims on the best use of evidence as long as their use is restricted to domains in which they are licenced by prevailing facts[738].

It could be stated quite perversely that Norton is both right and wrong at the same time. His diagnosis is correct regarding the model of deductive logic seeking a relationship to induction based on universal schemas, which for centuries has imposed a way of constructing the understanding of nature, thus blocking the possible development of a system of induction without universal schemas. However, this is *post factum* wisdom from the perspective

---

[738] Idem, pp. 647–648.

of centuries, which does not answer the question of to what extent deductive reasoning was a necessary intellectual step in arriving at the concept of induction. The historical examples given of more or less conscious use of induction may indicate that the idea of its use matured slowly.

Today, no one any longer questions induction-based methodology. It should be noted, however, that it does not necessarily represent the culmination of the methodology of scientific thought. To believe so would mean the history of science had taught us little. Our requirements for theory to conform with truth are not constant and equal. There are different requirements for facts obtained "locally", for example in relation to the operation of colloids or the operating principles of a compression-ignition engine, than in the case of cosmogonic theories. Induction is not always possible when it comes to processes inaccessible to us, of which we have the most diverse examples, such as the problem of matter-antimatter asymmetry in the universe, M-theory in cosmology, or brane cosmology. Some hypotheses of theories created in these fields go beyond the framework of induction as hitherto understood. It is likely that Norton's accusation against deductive logic that it hinders the development of induction will one day be levelled at inductive thinking itself, which hinders another, more effective way of building theories in these areas of science which we are already reluctantly trying to use. This, in short, is hypothesising on the basis of mathematical constructs. Although in this case the creation of hypotheses is not accompanied by evidence or repeatability, their credibility and usefulness are always determined by their expected and employed *utility*. Theories are accepted that cannot be falsified but which are sufficiently *useful* that they become a discursive part of science. A visible *novum* is the fact that the creation of hypotheses in these areas is not accompanied by classical deductive and inductive methods, as the methodology used surpasses deductive and inductive hypothesis building. Analysing the problem from the point of view of *utility theory*, however, it must be stated that *use* of the inductive method is insufficient for this type of knowledge and we are witnessing the creation of a new, more *useful*, methodology for creating theories.

Norton is also wrong about the futility of competition—the situation is quite the opposite. Competition in inductive inference systems and in the results obtained from these inferences is constant and permanent, forced by *utility*. Less *useful* hypotheses and theories, arising from the operation of accepted inductive systems, are constantly pushed out of the 'triangle of science' by other, more *useful* ones. Hence the extraordinary success of science in learning about the world, achieved through competition between inductive inquiries. Nothing motivates thinking like competition.

Norton's main argument for the local material theory of induction is that no inductive inference schema can be universal and at the same time function effectively. In the case of enumerative induction, in order to ensure its successful operation, universality must be given up and schemas adopted that are obtained only locally, based on the licence of locally derived facts[739]. All efforts come down to the problem of the irresolvable tension between universality and effective functioning. On the one hand, if they are sufficiently general to be universal and still true, axioms or principles become vague and vacuous.

Because of these difficulties, the current material theory of induction is based on the assumption that material postulates refer only to specific areas, that is, to facts obtained locally. As a result, inductive inference schemas will always be licenced locally[740].

Norton's conclusions about the local action of the material theory of induction are confirmed in Smoluchowski's *utility* theory, as *utility* verifies theories more thoroughly and penetratingly in the local scope of their action.

The conditions considered by Norton concerning the irresolvable choice between universality and the efficient functioning of induction in terms of the ability of a theory to entail proofs of truth cease to be a crucial point in the perspective of *utility theory*. The choice between violating universality and the effective functioning of the material theory of induction, based on the licence of locally obtained facts, ceases to be a key problem. In this vital matter, both states of induction ('universal' and 'local') are verified by their *utility*, which is the determinant of their veracity. As noted above, *utility* verifies locally obtained facts *express is verb is*, approving or rejecting them as incapable of entailing evidence of truthfulness.

In his article *Hume, Norton, and Induction without Rules*, Thomas Kelly writes that the reliability of the inductive rule, in contrast to the credibility of the deductive rule, seems to vary depending on the nature of the world in which it is used; even if a given inductive rule is reliable in our world or

---

[739] Idem, p. 652.
[740] Ibid.

in a world that, in our opinion, is our world, there are probably other possible worlds in which the same rule is not reliable. Moreover, it is natural to suppose that not only are there worlds in which the rule is unreliable, but also those in which proof can be obtained of this unreliability of the rule. But when we acquire evidence that the principle is unreliable in the world in which it is applied, it would certainly be unjustified to continue applying the principle. Although there is no doubt about the moves that can be made here, at first glance it seems that this way of thinking runs counter to the idea that there may exist some truly universal inductive principles, or rules that it would be reasonable to apply without regard to the appearance of 'material facts'[741]. According to Kelly, the existence of truly universal rules of induction is unlikely. By linking these rules to the cultural understanding of 'material facts', he indicates no authoritative factor that would put this relativistically perceived reality in order. Such a determinant must exist, as otherwise the existence of our entire scientific achievements would be impossible, writes Kelly. According to Smoluchowski, such a determinant is the *utility criterion*, which in a characteristic way organises the relativistically perceived reality, relating it to the possibility of *using* the effects of the induction performed.

Norton notes that in terms of hypothetical confirmation, the basic principle is a theory's ability to entail evidence of truth. The basic principle itself is in fact rarely used alone, as it is used quite often. According to it, a trivial piece of evidence that a triangle exists in which the angles add up to 180° is not only proof for Euclidean geometry but brings with it the whole baggage of the science that contains that geometry and the whole of modern physics, chemistry, astrophysics and the like. Developing the problem, we can see that everything—apart from an entirely banal version of hypothetical-deductive confirmation—demands additional limits based on facts[742].

The example of the triangle given by Norton is a specific case of the *differentia specifica* of geometric theories, in which *utility* plays an important role. Let's take a look at Friedman's models. A triangle with the sum of angles of 180° in the flat model of the universe that occurs in Euclidean geometry can be replaced by a triangle with a sum of angles less than 180° appearing

---

[741] T. Kelly, *Hume, Norton, and Induction without Rules*, 'Philosophy of Science' 2010, vol. 77 No. 5, pp. 754–764.
[742] J.D. Norton, *A Material Theory of Induction*, op. cit., p. 653.

in the hyperbolic model. A triangle with a sum greater than 180° occurs in the spherical model of the universe. Today we do not know which of these geometries is true in the sense that we do not know which is crucial to our reality. Hawking writes in the book *The Universe in a Nutshell* that it may be Riemann's saddle-shaped geometry[743]. However, in everyday life we use Euclidean geometry and even if Hawking was right, it would not change anything for millions of people. We therefore ask what the quantifier is of the geometry we use. The answer is unambiguous—this determinant is *utility*. The aims towards which a geometry is *used* determine which geometry we apply.

Norton attaches great importance to the issue of a theory's conduciveness to entailing evidence of truth. In the case discussed above, Euclidean geometry is true on the scale of our globe, but whether or not it is true on the scale of the universe we do not know. It is impossible in this case to refer to so-called objective truth; we can at most refer the question for assessment within the framework of Smoluchowski's *utility theory*, which does not radicalise the concept of truth, but merely formulates the way it is understood in nature.

6.6. Summary of considerations on *utility theory*

Utility is a universal category that applies to all aspects of human functioning in the natural world. It determines the value of a theory or concept. As a result of its influence, we come closer to objectively learning the truth about reality. *Utility* tests every theory, hypothesis and premise. It works on a similar principle to the "invisible hand of the market" introduced into economics in the 18th century by Adam Smith (1723–1790)[744], with the difference that rather than in the free market of economics it functions in the non-ideologised space of knowledge. To paraphrase Smith's idea, it can be said that the driving force is that everyone thinks about achieving success in the creation of something new and unique, whether in science, business, production or even everyday life, that will be accepted, and maybe

---

[743] See S. Hawking, *The universe in a nutshell*, op. cit., p. 19.
[744] See A. Smith, *Badania nad naturą i przyczynami bogactwa narodów* (*An Inquiry into the Nature and Causes of the Wealth of Nations*), vol. II, trans. A. Prejbisz, Warsaw 1954, p. 46.

even admired, by society[745]. We act in this way to strive for the creation of something that would be more useful than that which functions now. The usefulness of every idea, theory or hypothesis, every thought, is determined by the *criterion of utility* and society constantly inspires and initiates those actions. The principle of free competition of theories is present here. Every theory functioning in science, or in everyday life, even one which has not been sufficiently falsified, is recognised and used until it transpires that there exists a theory whose *utility* is greater, which provides greater possibilities and—very importantly—enables more effective development of knowledge. The endless decisions taken concerning the choice of verifying utilised theories result in some theories being constantly used in science and technology, others being abandoned, and still others being consigned to history. The *utility of theories* is inherently related to their *use*. It is *use* that determines a theory's *utility*. This idea of Smoluchowski's, banal in content, brilliant in its simplicity, has driven our epistemology for centuries. On the basis of the above considerations, in the context of many dissertations of thinkers involved in the theory, philosophy or methodology of science, it can be seen that Smoluchowski's *utility criterion* introduces its own kind of order to the understanding of that timeless stimulus to which we owe the ceaseless momentum of the development of science.

---

[745] Idem, p. 42.

CHAPTER 7

Smoluchowski's contribution to the development of the kinetic theory of matter

7.1. Ludwig Boltzmann's influence on Smoluchowski's views

It is impossible to understand the scientific development of kinetic theory without taking into account the role of philosophy in its construction. In these considerations, this thesis was supported by a broad analysis of the philosophical motivations accompanying Clausius, Maxwell, Boltzmann, Einstein, and Smoluchowski in the act of successively creating thermodynamics, kinetic theory, and atomic theory.

In 1857, the German physicist Rudolf Clausius presented the first coherent concept of the kinetic theory of gases, which assumed that gases are made up of small particles. A fundamental aspect of Clausius' theory was the equipartition theorem, in which he assumed the total amount of energy in the translational motion of particles to be evenly distributed among a molecule's degrees of freedom[746].

However, it is Maxwell and Boltzmann who are considered the main architects of the kinetic theory of gases. Although their philosophical ideas differed, their contribution to the knowledge and ultimate acceptance of kinetic theory is shared[747]. In addition to Maxwell and Boltzmann, among the creators of kinetics are also two 19th-century scientists: German physicist and chemist August Karl Krönig (1822–1879), who published a paper

---

[746] See H.T. Bernstein, *J. Clerk Maxwell on the History of the Kinetic Theory of Gases, 1871*, 'Isis' 1963, vol. 54. 2, S. 9-45.

[747] See H.W. de Regt, *Philosophy and the Kinetic Theory of Gases*, op. cit., p. 33.

on the kinetic theory of gases in 1856, and Josef Loschmidt (1821–1895), an Austrian physicist who in 1865 presented the first reliable assessment of the size of gas molecules. He achieved this by using experimental data on the density and viscosity of gases together with kinetic theory formulae. He estimated the number of gas particles in one cubic centimetre, which was called the Loschmidt number[748]. Loschmidt is also responsible for the first reliable assessment of the size of atoms of the order of 10^{-8} cm and the calculation of the mass of an atom of the order of 10^{-24} g.

Some aspects of Maxwell and Boltzmann's philosophy, especially those relating to thermodynamics and kinematics, significantly help to understand the changes taking place in Smoluchowski's philosophy, and the influence on the Polish physicist of Boltzmann in particular was indisputable.

In the article *Philosophical Objections to the Kinetic Theory*, John Nyhof emphasises that a breakthrough in Maxwell's reasoning was his conviction by the late 1860s that, from the point of view of kinetic theory, the second law of thermodynamics is statistical in nature. The laws of gas particles published in 1860 introduced kinetic theory to the world of scientific theories. Maxwell proceeded from the assumption that heat conduction through a gas is based on diffusion of energy, that is, on the passage of particles having higher kinetic energy (warmer) between particles endowed with lower energy (colder), and on the fact that the direction of motion of each gas particle is random, and that the energy value of a particle is an average value, and only such a value is strictly determined by temperature, as Smoluchowski recalled in his essay *On the thermal conductivity of gases according to theories and experiments to date*[749]. Maxwell's hypothesis could not be confirmed experimentally as an experimental method of determining the velocity of a particle was not yet known and it was only possible to indirectly calculate a particle's average velocity.

A little later, Boltzmann developed a statistical interpretation of the second law that contradicted its classical understanding. The thermodynamic understanding of the second law did not take into account the potential spontaneous fluctuations of entropy that kinetics allowed. Before the kinetic theory

---

[748] See A.K. Wróblewski, *Historia fyzki*, op. cit., p. 357.
[749] See M. Smoluchowski, *O przewodnictwie teplnem gazów zgodnie dotychczasowych teoryj i doświadczeń* (*On the thermal conductivity of gases according to theories and experiments to date*) in: *The writings of Marian Smoluchowski*, vol. I, op. cit., p. 166.

of matter could be considered a modification of the classically understood second law, the occurrence of variation would have to be proven in some way.

Both Maxwell and Boltzmann were convinced that the possibility of observing an event such as, for example, the flow of thermal energy from lower to higher temperatures was infinitely unlikely. Attempting to correct phenomenologically understood thermodynamics through kinetic theory by means of unobservable phenomena, as well as postulating a change in the understanding of the second law through the occurrence of statistical fluctuations, which unfortunately were considered unobservable, meant that such a molecular theory could be accepted only within the circle of its supporters while for phenomenologists it seemed a metaphysical absurdity[750].

By introducing statistics into his calculations, Boltzmann proved that the second law of thermodynamics has a statistical basis and that entropy is a function of probability distribution. Previously, entropy was understood as a phenomenological and thermodynamic concept. Boltzmann changed the understanding of entropy, proving that there is an opportunity for thermal energy to flow from lower to higher temperatures, something that until then had seemed impossible according to the second law of thermodynamics. He argued that this phenomenon is imperceptible in the macroscopic world, because statistically it is an extremely unlikely process, but the kinetics he postulated allowed for its occurrence.

Contrary to appearances, the term 'kinetic' does not come from Clausius, Maxwell or Boltzmann. When Maxwell wrote about the "dynamical theory of gases" in 1859 and 1866, he never once used the word "kinetic". The Latin terms used then, *vis mortua* and *vis viva*—literally "dead force" and "living force"—were translated into the much less expressive English phrases "potential energy" and "actual energy". In the 1860s, many physicists primarily used the term 'dynamic', but the still relatively little-known word 'kinetic' was appearing. The initiator of this term was the Scottish physicist William Thomson, who in an 1856 discourse before the Royal Institution entitled '*On the Origin and Transformation of Motive Power*,' explained that "any piece of matter or any group of bodies, however connected, which either is in motion, or can get into motion without external assistance, has what is called mechanical energy", he also used the terms: "dynamical energy", "actual energy" and

---

[750] Idem, p. 92.

"potential energy", which were in use in physics at that time. Many years later, in 1893, he recalled in a footnote that shortly after this lecture he had created the term 'kinetic energy', which is now in general use[751].

Smoluchowski changed his philosophical views under the influence of Boltzmann's philosophy, but it did not happen immediately as he needed time to reflect on the kinetic concept. In 1895, when he defended his doctorate, Boltzmann was already a recognised authority in the field of physics, and at the university a pointed dispute was ongoing between Boltzmann and Mach about the form of the methodology of physics and the nature of matter. Smoluchowski witnessed these disputes, but did not address the issues under discussion before he left Vienna. After spending two years at the universities of Paris, Glasgow and Berlin, the Polish scientist returned to Vienna in 1897 and then, under the influence of Boltzmann's philosophy, accepted kinetic theory. However, Smoluchowski would not be himself if he did not have his own concept of how to understand this theory.

In his 1996 article *Philosophy and the Kinetic Theory of Gases*, contemporary philosopher Henk de Regt of the University of Amsterdam analyses the role of philosophy in the development of the kinetic theory of gases. He interprets philosophically the views of Maxwell and Boltzmann in terms of kinetic theory, confirming that the philosophical views of both scientists fundamentally influenced the results of their scientific work.

Maxwell was convinced, as he wrote around 1870, that absolute scientific truth could be achieved, and that the method leading to this goal was to analyse observed phenomena. He advocated for this method over the hypothetico-deductive method as he thought it could yield more certain knowledge. He believed this method could be used to construct "dynamic theories" that would explain phenomena in terms of the operation of material systems that obey the laws of Newtonian mechanics[752]. Maxwell's view was similar to that propounded by realists, who argued that a correct theory provides a true description of reality, or at least strives for a true description, and the successes of a scientific theory give grounds for believing that

---

[751] See H.T. Bernstein, *J. Clerk Maxwell on the History of the Kinetic Theory of Gases, 1871*, op. cit., pp. 207–208.
[752] See H.W. de Regt, *Philosophy and the Kinetic Theory of Gases*, op. cit., p. 39.

something like the entities and structures posited by this theory really exist. Naive realism interprets scientific theories as a literal description of the world.

According to Tadeusz Sierotowicz, realism in science assumes that in order for a theory to be accepted, it must be successful for a sufficiently long period of time[753]. For Smoluchowski, such an understanding of realism—as a theory illustrating our current conception of reality—would be acceptable, except that the length of time for which a theory is accepted determines, in his opinion, the recognition of its *usefulness*. The scholar puts it this way: we do not distinguish between true, false or more or less probable theories, but between more or less *useful ones*[754].

Claims that the theoretical structures of a theory are something like the structures of the real world were not treated by Smoluchowski as an answer but as a poorly posed problem, because—he argued—we must clearly realise that we are not concerned with knowing the essence of a thing hidden behind appearances as there is a threat of falling into philosophical considerations while the task of physics is to learn about the world of phenomena accessible to us as thoroughly and clearly as possible. It is about the most thorough examination of these phenomena and combining them into a whole that is comprehensible to our mind[755]. Absolute knowledge of things is not possible, so one should focus on exploring what the human mind can accommodate and understand.

De Regt notes that natural realism is a form of ontological realism and, combined with a specific epistemological thesis, is opposed to idealism, whose ideas of the external world are created by the mind. He is opposed to scepticism, which assumes any claim about reality to be groundless. It is possible to acknowledge the existence of the external world, but our minds do not have direct access to it. It seems highly probable that Maxwell, on the general ontological level, subscribed to natural realism[756].

Was Boltzmann also a realist? Historians and philosophers of science offer various answers to this question, but most assert that while it can be assumed he was one earlier, the question arises as to whether in later years he rejected

---

[753] See T. Sierotowicz, *Realizm w kontekście nauki* (*Realism in the context of science*), 'Filozofia Nauki' ('The Philosophy of Science') 1997, No. 1 (17), pp. 28–29.
[754] See M. Smoluchowski, *Self-Study Handbook*, vol. I, op. cit., p. 51.
[755] Idem, pp. 16–17.
[756] See H.W. de Regt, *Philosophy and the Kinetic Theory of Gases*, op. cit., p. 38.

or softened his realist stance in relation to the kinetic theory of matter and the atomistic hypothesis. De Regt cites the claim of Engelbert Brody (in 1973) and John Nyhof (in 1988) that Boltzmann never gave up realism, while Glymour Clark (in 1976), Elfried Hiebert (in 1981) and Yehud Elkan (in 1974) claimed that he softened or even rejected his earlier realism. According to de Regt, this controversial difference of opinion is partly due to a failure to distinguish between the various meanings of the concept of realism at different levels. The problem can be solved by distinguishing between ontological realism and epistemological and methodological realism. On the ontological level, Boltzmann's realism stands in opposition to idealism, and there is no doubt that the scholar was a realist. He believed in the reality of the external world, and he firmly rejected idealistic views in which the realities of the material world were not accepted. For example, he dismissed the idealism of George Berkeley (1685–1753), calling him "the inventor of the greatest folly ever hatched by a human brain". He also repudiated as a form of idealism Mach's phenomenalist position, which assumed that only experience builds true science. Boltzmann was a supporter of the ontological theory of mechanicism and materialism, which seem essential in building kinetic theory. He believed the laws of mechanics to be an indispensable basis of natural science, holding that a mechanical model of a phenomenon is necessary to explain it.

> Boltzmann's concept of epistemology assumed the theory of the 'mental picture'. He argued that the task of theory is to construct an image of the external world that exists only internally and must be the starting point for all conception and experimentation. Boltzmann warned against the danger of naive realism and confusing theory with reality by immersing oneself in abstract deliberations[757]. In his lecture On the Methods of Theoretical Physics, he argued that our task is not to find a fully correct theory, but rather an image that is as simple as possible and represents the phenomenon as precisely as possible. Two completely different theories can be imagined, equally simple and equally congruent with the phenomena under study, which, despite their differences, are equally correct[758].

> Smoluchowski confirms this line of Boltzmann's thinking—we choose theories that seem the simplest to us, but we can never be sure they are true. The extreme sceptic

---

[757] Idem, pp. 40–41.
[758] See J. Nyhof, *Philosophical Objections to the Kinetic Theory*, 'The British Journal for the Philosophy of Science' 1988, vol. 39, no. 1, p. 44.

will even say that as humans we think with the rules of human logic and we do not know whether nature and reality adhere to these rules[759]. A hypothesis or theory is considered proven if the conclusions drawn from it concur with experience. When even one conclusion is not confirmed, we consider it false. Putting the matter this way would have to mean that truthfulness would imply actual existence. However, we never know whether evidence will appear to refute a given hypothesis, no matter how well-founded it may seem. Even if no evidence were found against it, the conclusion still does not follow that the hypothesis reflects reality[760].

Boltzmann's understanding of mechanicism is evident in Smoluchowski's papers, but with a slightly different approach to materialism. In his reflections on the subject of cause, chance and probabilistics as applied to physics, the Polish scientist saw that mechanistic explanations in the field of the emerging atomic theory would become increasingly questionable over time and that the approach to the classical understanding of mechanicism would have to be transformed. In his work On the concept of chance and the origin of laws of physics based on probability, he gives an example of such a situation, analysing the rupture of a metal sphere as a result of it being subjected to high pressure from the inside. The place where the shell breaks and the shapes resulting from its disintegration into parts are subject to chance that depends on unevenness in the material created during the casting of the ball. Molecular chance is responsible for the actual microcrystalline structure of the material. The occurrence here of effects caused by random molecular clusters comes down to the transgression of unstable equilibrium states[761]. Although the described experiment is of a mechanical nature (the rupture of a sphere), the actual cause of this event rather than a different one does not fall within the conventional understanding of mechanics.

Returning to Boltzmann, his ontological position can be said to be similar to Maxwell's mature views. The philosophical differences between Boltzmann and Maxwell appear at the level of epistemology and methodology. The issue of a theory's interpretation is epistemological in nature as it involves the question of what can be known about unobservable reality. Boltzmann was opposed to instrumentalists and positivists who, for purely philosophical reasons, did not believe explanations of the concept of a statistical approach to kinetic theory and considered phenomenological thermodynamics self-sufficient[762].

---

[759] See M. Smoluchowski, *Self-Study Handbook*, vol. I, op. cit., p. 16.
[760] Idem, p. 49.
[761] M. Smoluchowski, *On the concept of chance and the origin of laws of physics based on probability*, op. cit., p. 50–51.
[762] See J. Nyhof, *Philosophical Objections to the Kinetic Theory*, op. cit., p. 93.

Smoluchowski comments on the statistical understanding of physical phenomena in several of his works in which he addresses the concept of chance and the origin of laws of physics based on probability, because he treated chance and probability theory as fundamental issues in understanding physical phenomena. He notes that it was Clausius and Maxwell, between 1857 and 1860, who first introduced a mathematical tool to the kinetic theory of gases, in the form of probability calculus. Probability calculus makes it easier—as with the kinetic theory of gases—to reduce laws of physics to the statistics of hidden elementary events[763].

Boltzmann's kinetic hypothesis, which assumed the second law of thermodynamics to have a statistical basis, was confirmed by two mathematicians at the end of the 19th century—the Frenchman Henri Poincaré and the German Ernst Friedrich Ferdinand Zermelo (1871–1953). Poincaré argued that the movements of finite mechanical systems are quasi-periodic. This means that in every finite mechanical system, after a certain time, there will come a moment at which the system will return to a state arbitrarily close to its initial state. Smoluchowski describes this situation in an identical way:

> Zermelo raised a related objection to the kinetic theory in 1896, though different in form, referring to Poincaré's theorem, according to which movements of finite, conservative mechanical systems are "quasi-periodic", because it is always "possible to give a finite period of time within which the coordinates and velocities of all material points approach arbitrarily close to their initial values". This theorem would also imply that entropy would have to return to its initial value over time[764].

The time after which a return to the initial state would occur is extremely long; for example, the return of the system of atoms enclosed in one cubic centimetre of air to its original state (according to Boltzmann) would take many billions of years (today we know that it would exceed the age of the Universe by several times). In various publications, Smoluchowski also repeatedly cites examples and arguments for the reversibility of phenomena. At the turn of the 19th and 20th centuries, Boltzmann's kinetic theory was the subject of heated discussion, both popularising it and leading the Polish physicist to raise the topic fairly frequently.

---

[763] See M. Smoluchowski, *On the concept of chance and the origin of laws of physics based on probability*, op. cit., pp. 27–28.

[764] Idem, *On thermodynamic fluctuations and Brownian motion*, op. cit., pp. 271.

Opponents of kinematics raised the issue of *perpetual motion*, arguing that thermodynamics strengthens our belief in the irreversibility of the phenomena of diffusion, thermal conductivity, and friction. If these phenomena could be reversed, it would be possible to build a *perpetual motion machine* of the second kind, which physics considers impossible. Kinetic theory cannot be correct as it says the reverse process would have to be possible for each of these phenomena[765]. They pointed out that the theorem implies that entropy would have to return to its initial value over time.

Many scientists, as Smoluchowski notes, were persuaded by this argument. Boltzmann tried to explain these contradictions by arguing that the 'reverse' course of the thermodynamic phenomenon is possible, but extremely unlikely, and that in the case of available observations this quasi-period (Poincaré – Zermelo) is an extremely long period—so that we cannot expect to ever observe a phenomenon exhibiting deviations from the normal course. This argument was not convincing as it was not commonly understood why the course of a phenomenon in one direction would be more likely than in the opposite one.

Smoluchowski was one of the first to accept Boltzmann's theory but was quite critical of the mathematical aspect of the Austrian physicist's work[766]. He believed that the general theorems were largely based on intuition but was convinced that it was only a matter of time until mathematical tools emerged that would convincingly prove the truth of kinetic theory's premises. His predictions bore fruit quite quickly, in which process he himself played a significant role. Years later, he recalled: "Today we have come to the conviction that Boltzmann was right in removing the apparent contradiction by introducing the concept of probability; at the same time, we have broadened his ideas and learned almost tangible evidence of their truth"[767].

Smoluchowski saw confirmation of Boltzmann's assumptions in various transformations of nature, especially in physics, but did not find the Austrian physicist's arguments sufficiently convincing. According to Stanisław Loria,[768] Smoluchowski replaced Boltzmann's vague concept of probability with the

---

[765] See idem, *The current state of atomic theory*, op. cit., p. 64.
[766] Idem, *On thermodynamic fluctuations and Brownian motion*, op. cit., p. 272.
[767] See idem, *The current state of atomic theory*, op. cit., p. 65.
[768] See S. Loria, *Marian Smoluchowski and his work (1872–1917)*, op. cit., p. 37.

term mean fluctuation and the concept of the anomalous state. Assuming that a kinetic-molecular theory using statistical methods provides for the possibility of processes departing from the normal course of phenomena in the macro-world, Smoluchowski claimed, the researcher's task is to theoretically analyse the conditions in which they can be expected to occur and to predict and quantitatively reproduce a hypothetical picture of their course, if only the basis of a logic-defying example of clearly irreversible phenomena. By pouring a layer of black poppy seeds into a box and a layer of white poppy seeds on top of it, and then shaking the box, we will cause the black and white seeds to mix. Theoretically, it would be possible to perform such movements as to separate the black and white seeds back out, but in practice we will always just see progressive mixing until there are more or less equally distributed black and white seeds[769].

The diffusion of initially separated gases, for example oxygen and nitrogen, occurs in a similar way. It takes place on its own and thermodynamics proves that entropy increases during the process. Thermodynamics rules out the reverse—the automatic separation of air into oxygen and nitrogen. According to kinetic theory, it is possible, though extremely unlikely. It is easy to prove that the even distribution of oxygen and nitrogen in one cubic centimetre is roughly 10^{19} times more probable than their complete separation; it is therefore not surprising that in practice we observe only the former situation[770].

Breaking the stereotypical perception of reality presented a problem not only for readers who were not specialists, but also for many physicists, according to whom the reversibility of phenomena was, above all, incompatible with the principle of entropy. Smoluchowski compares human short-sightedness to the situation of flowers that wake up in the spring under the influence of rising temperatures and during their short life probably consider it dogma that "the climate of the universe goes from a colder state to a warmer state." They will never know that autumn and winter will return one day[771].

---

[769] See M. Smoluchowski, *Atomistyka współczesna. Odczyt wygłoszony na ogól. posiedzeniu XI zjazdu lekarzy i przyrodników w Krakowie* (*Contemporary atomic theory. A lecture given at the general meeting of the 11th Congress of Physicians and Naturalists in Kraków*), printout from 'Pamiętnika XI Zjazdu Lekarzy i Przyrodników' ('Commemorative journal of the 11th Congress of Physicians and Naturalists'), p. 9.
[770] Idem, p. 38.
[771] See M. Smoluchowski, *The current state of atomic theory*, op. cit., p. 66.

The atomic-kinetic point of view confirmed the principles of thermodynamics, as long as the observation time turned out not to be too long. However, regarding long-term, cosmic phenomena, the opposite conclusions resulted from these two theories. Smoluchowski related the basic difference between thermodynamics and kinematics to entropy, claiming that on the basis of empirical thermodynamics, the entropy of the Universe is constantly increasing, and therefore the Universe must pass into a state of deadness over time, in which all potential energy will turn into heat and all temperature differences will equalise. The kinetic theory argues to the contrary—after the stage of deadness, new life will appear again, because all states in the eternal procession of entities return with time[772].

The initial value of entropy at the dawn of the Universe was most likely extremely small and has been constantly increasing ever since. In *The Road to Reality*, Roger Penrose considers the arguments behind this hypothesis. Referring to the possible "heat death of the Universe" posited by Lord Kelvin[773], he states that the Universe would have to expand forever, with ever-increasing entropy, or it would come to its end, that is, a state in which entropy would reach its maximum and cease to grow, and the Universe would be "dead", disturbed only by fluctuations. Both scenarios raise the question of how the Boltzmann hypothesis of entropy returning to its "extremely small" minimum would be possible in either state of the Universe described above.

Penrose seeks a way out of this seemingly contradictory situation. Due to the second law of thermodynamics and the constant increase in entropy, the beginning of the observable Universe, which today we call the Big Bang, should be characterised by extremely low entropy, but is this really the case? Penrose puts it this way: "What was extraordinarily special about the Big Bang was actually its great uniformity (…). We must try to understand why this corresponds to a very low entropy, and how it provides us with a Second Law that is relevant to us here on Earth in the familiar form that we know"[774].

According to the scientist, it is likely that entropy would have been extremely high at the time of the Big Bang, though that would rule out

---

[772] Ibid.
[773] See idem, *Lord Kelvin*, op. cit., p. 5.
[774] R. Penrose, *The Road to Reality. A Complete Guide to the Laws of the Universe*, London 2004, p. 705.

the second law of thermodynamics. There is no unambiguous position of physicists on this important issue:

> Observations lead to the conclusion that the entropy of the Universe is increasing, although it does not necessarily follow from this that it was minimal in the past. Some researchers see evidence in the background radiation and structure of the Universe that the past was a hot, homogeneous chaos. Others, however, on the basis of the consequence of the second principle and contrary to observations and tradition, put the past of the Universe in a minimum of entropy. Are such diametrical discrepancies in the resolution of such a fundamental problem possible?[775]

Penrose does not resolve this dilemma, but ends the discussion with the above question, which has been asked since Boltzmann's time yet remains unanswered. However, in the final part of his reflections, he leans towards the concept of a constant increase in entropy in our Universe, from the minimum at the time of the Big Bang to the maximum, heading towards infinity, in the singularity of the ending Universe. According to him, among the support for such a position is the second law of thermodynamics, which states that later states have higher entropy.

Penrose's contemporary dilemmas show what a monumental step Boltzmann took and what philosophical and physical insight Smoluchowski demonstrated. The constant increase of entropy was supposed to result from the simple fact that in an isolated system it is more likely to rise than fall as there are many more ways for it to increase than decrease.

The above problems changed some of the philosophical concepts favoured by scientists in the late 19th century. The phenomenological concept of the structure of matter was relegated to obscurity and replaced by the kinetic-atomic view. Classical *thermodynamics*, known as *phenomenological thermodynamics*, treated solids, liquids and gases as being continuous, with no molecular structure. It dealt with macroscopic thermodynamic phenomena. In its scientific research, it advocated relying on pure experience, on the description of facts, rather than on explaining phenomena.

Empiriocriticism, the *consequence of whose assumptions was the rejection of building theories through hypotheses, claimed that* there are no separately

---

[775] H. Korpikiewicz, *Koncepcja wzrostu entropii a rozwój świata*, Poznań 1998, p. 81.

existing physical and mental phenomena—they are one and the same and can be understood in various ways, hence scientific research should be based on pure experience. It sought *to eliminate the opposition between idealism and materialism, putting forward a theory of the psychological nature of impressions (the so-called theory of pure experience), to which it reduced all reality.*

Smoluchowski disagreed with such a way of thinking as this assumption led to abandoning the category of truth in intellectual activities. These views implied philosophical beliefs he was opposed to. *Empiriocriticism's exclusion* of hypothesis from the process of creating scientific theories was unacceptable to him. He argued that hypotheses are essential in science, and that the dynamic *progress of science is made possible by theoretical considerations being formulated into hypotheses and then subjected to falsification.* The life of the classically understood[776] philosophical concept of the energetcists, represented by Wilhelm Ostwald, ended in a similar way. According to Smoluchowski, scientific description should be formulated in line with knowledge; we should not limit ourselves only to experience as knowledge cannot be based solely on empiricism.

Among physicists, Smoluchowski admired the logical-critical minds, such as André Marie Ampère (1775–1836), Kirchhoff, Hertz and Duhem, accustomed to purely mathematical thinking, who tended towards sober abstract hypotheses involving mathematical relationships (included here are the laws of physics enabling all the formulae of basic theoretical physics to be expressed mathematically). Insofar as we deal only with quantities subject to direct observation, science is an expression of so-called phenomenalism. This includes relationships expressed in the form of ordinary equations, such as the law of light refraction, the laws of electrolysis, differential equations, equations of hydromechanics, electrodynamics, diffusion and thermal conductivity. Naturalistic minds of the geometric-imaginative type will look for support in hypotheses that appeal to models of physical phenomena, to analogies. This category includes the atomic-kinetic, electron, emission, and wave theories, the electromagnetic theory of light, the mechanical theory of acoustic phenomena, and the theory of gravity. This is the prevailing type

---

[776] It should be noted that the natural philosopher of the early 21st century does not radically dissociate himself from monistic concepts.

among physicists, especially of the English school. Examples are Faraday, Maxwell, Kelvin, and Rutherford[777].

The fundamental questions of late 19th-century philosophers regarding realism in science still remain without clear answers. Tadeusz Sierotowicz notes that today we often talk about naive, structural, essential, constructive or critical (scientific) realism. Outlining the various shades of realism in science and interpreting it as one of the possible philosophies of science, it can be stated that, despite the existing differences, "what unites realists is the belief that the changes taking place in the field of science can be interpreted as progressive and that the sciences make it possible to know the world beyond its external empirical manifestation"[778]. It is worth noting here that the discussions around scientific realism do not essentially concern the existence of external reality, but the quality of our knowledge. The issue of the quality of knowledge is expressed in this context in questions such as: "Do our terms refer to external entities? Do our theories express relationships between external entities[779]?"

Perhaps these relations even appear different, because—as de Regt writes in *Philosophy and the kinetic theory of gases*—sense impressions are direct manifestations of external objects in our minds. Relativity of knowledge is therefore an epistemological claim, implying that our knowledge of the external world can never be absolute[780].

7.2. Atomic-kinetic theory at the turn of the 19th and 20th centuries

Atomic science remained at the stage of speculative philosophy from the time of ancient Greece with elements of a scientific atomic theory emerging only in the 19th century. The start of these changes occurred in 1805, when John Dalton (1766–1844) explained the simple numerical rules observed in the formation of chemical compounds through the combination of unchanging atoms into certain groups, molecules and particles[781]. Dalton showed that

---

[777] See M. Smoluchowski, *Self-Study Handbook*, vol. I, op. cit., pp. 52–54.
[778] See T. Sierotowicz, *Realism in the context of science*, op. cit., p. 28.
[779] Idem, p. 29.
[780] See H.W. de Regt, *Philosophy and the Kinetic Theory of Gases*, op. cit., p. 38.
[781] See M. Smoluchowski, *The evolution of atomic theory*, op. cit., p. 17.

each chemical compound contains the same amounts by mass of the same elements, and that this can be easily explained if we assume the atomic structure of matter.

At the end of the 19th century, phenomenological views on the structure of matter dominated the beliefs of European physicists. Atomic science was considered obsolete, a theory doomed to oblivion as an unscientific fantasy. The rejection of this theory by most of the scientific community at the end of the 19th century seemed to be permanent. Marian Smoluchowski outlined in a few words the general intellectual mood of the era of atomic theory's fall. There existed, in his view, very important factual arguments at the time against atomic theory and dissuading from that course. However, scholars of the exact sciences, hard-headed people determined in their pursuits, did not succumb to the general mood[782].

Those doubting the legitimacy of atomic theory cited as their chief argument the inconsistency of everyday observations with the theory's assumptions. They were convinced that if the indefinite and statistical nature of phenomena were to dominate in nature, then both these features must be applicable to observable phenomena in the macroscopic world and should consequently be perceptible in everyday life. In the observable world, there is no way to confirm characteristics of nature that would prove the accidental or random status of matter. It could even be said that explaining physical events with the aid of a mathematical tool such as the calculus of probability is contrary to everyday life experience and so any theory based on such mathematics is mere intellectual speculation. Another hurdle hampering acceptance of atomic theory (mentioned earlier) was the reversibility of natural phenomena assumed in kinematics.

The dispute between energeticists, representing a phenomenological perception of matter, and advocates of the conception of atomic theory, was not superficial; it concerned both an understanding of the essence of science and paradigms forming the foundation of scientific thought. The approach to science had to undergo a change, which, Smoluchowski claimed, should be based on experiment as well as on the theoretical aspect—on mathematics, with particular emphasis on the developing calculus of probability.

---

[782] Idem, p. 63.

A fact representative of the situation at the time was that when in 1895 Ernst Mach was named professor of philosophy at the University of Vienna, where Boltzmann had worked since 1893, taking the floor during a lively academic debate on the values of atomic theory, he cut the discussion short with one statement: "I don't believe that atoms exist." In subsequent academic disputes, Mach asked Boltzmann several times: *"Eines haben Sie gesehen?"* (Have you seen one?)[783].

Smoluchowski noted a situation characteristic of the climate in science at the time, when, following the publication of Boltzmann's work *Vorlesungen über Gastheorie (Lectures on gas theory)* in a German scientific journal, the following statement appeared: "Kinetic theory, as we know, is as flawed as various mechanical theories of gravity, in particular it wrongly understands the principle of the conservation of energy; however, if someone really wants to learn about it, let them take up Boltzmann's work"[784].

The emergence of key evidence for kinetic-atomic theory was initiated by Robert Brown (1773–1858), a Scottish botanist researching in 1827 microscopically fine particles floating in a liquid and making small, irregular movements visible with the use of a high-magnification microscope. He called them molecular movement[785]. Brown was not the first person to have made such observations. In 1784, a Dutch doctor, Jan Ingenhousz, had also noticed the effect. However, Brown was the first to conduct systematic research[786] on small pollen grains suspended in liquid. The Scottish botanist became convinced that the irregular oscillating movements were made by tiny particles of organic or inorganic substances.

Smoluchowski describes this fact in the following way:

> The name "molecular movement" comes from the English botanist Brown, who in 1827 noticed through microscopic research that tiny pollen grains, suspended in liquid, vibrate and make irregular movements reminiscent of the movements of a swarm of mosquitoes or an army of ants; upon closer research, he became convinced that this is a general phenomenon exhibited by any small particles of an organic or inorganic substance when they are in such conditions. This same

---

[783] See J. Bernstein, *Einstein and the existence of atoms*, op. cit., p. 864.
[784] See M. Smoluchowski, *The current state of atomic theory*, op. cit., p. 61.
[785] See M. Smoluchowski, *The evolution of atomic theory*, op. cit., p. 21.
[786] See J. Bernstein, *Einstein and the existence of atoms*, op. cit., p. 865.

phenomenon had actually already been noticed by other scholars before Brown, like Needham 1750, von Gleichen 1764, but they did not take the matter up more closely. A slightly different idea to our current view is associated with the name Brown, however, as by molecule, a concept not as clearly crystallised at the time as today, he meant only tiny moving particles; the name meant just the observed fact with no thought as to its explanation[787].

No thorough research was conducted on the phenomenon for a long time although every naturalist observed them in the course of their microscopic work because the movement itself intrigued researchers. Various ideas constantly appeared to explain the essence of this phenomenon but there was no work focused on understanding the movements' cause. It transpired that the correct explanation of the nature of the phenomenon was a key argument for accepting kinetic-atomic theory.

The contradictions accompanying the observations of various authors, as well as the complexity and inconsistency of the theoretical explanations, according to Smoluchowski, contributed to the phenomenon being widely ignored. When any research was sought to be conducted, contradictions appeared during its course, which discouraged their continuation. An example may be the big differences achieved in the observed velocities of particles. Smoluchowski writes:

> The real core of the issue, however, remained untouched: whether the observed phenomenon corresponds quantitatively with the requirements of atomic-kinetic theory. In this regard, there seemed to be a fundamental contradiction. The velocity of particles, given for example by F. Exner, stands at more or less 0.0003 cm/sec (for rubber particles of 0.001 mm diameter in water). Meanwhile, accepting, in line with kinetic theory, that the average energy of liquid molecules must equal the kinetic energy of these particles, they would obtain a theoretical velocity of 0.4 cm/sec, so over a thousand times greater.[788]

The discrepancies were extremely large and inexplicable.

Apart from laboratory problems, Smoluchowski mentions the climate prevailing among academics to whom the notion of a molecular-kinetic composition of matter was alien, writing:

---

[787] M. Smoluchowski, *On thermodynamic fluctuations and Brownian motion)*, op. cit., p. 297.
[788] Idem, p. 298.

> Contradictions between the observations of various authors as well as the diversity and inconsistency of the theoretical explanations were surely the reason why the phenomenon was generally ignored, such that we find no mention of it in any of the extensive works or physics textbooks (with the one exception of Lehmann's *"Molekularphysik"* (*Molecular Physics*)) until recent years. The foreignness to the general scientific community of the notion of its molecular-kinetic essence is evidenced by the characteristic fact that when, in the last decade of the last century, the school of energeticists and phenomenologists (Mach, Ostwald, Zermelo etc.) decried atomic-kinetic views as naïve, unscientific beliefs, there was nobody, even among proponents of this view, who would highlight Brownian motion as obvious evidence of thermal molecular motion[789].

Scientists inclined towards atomic theory lacked the courage to publicly air their views, as Smoluchowski saw for himself.

According to Smoluchowski, the hypotheses arising around Brownian motion constituted an extremely interesting chapter in the history of physics. He cites their history in his work *On thermodynamic fluctuations and Brownian motion*. Their history proves, he claims, that a known and often-observed phenomenon was forgotten and found no acceptance or interest on the part of official science. Ultimately, they show how a true theory provided the key to understanding and more closely researching the phenomenon of Brownian motion. Through molecules, a concept not yet clearly defined at the time, these small, moving particles were understood. The name merely denoted an observed fact, with no attempt to explain it[790].

Two examples illustrating the scale of tensions arising in academic circles testify to the impact of scientists from the energetics school on intellectual attitudes among physicists. The first concerns Ludwig Boltzmann and his conception of the theory of gases. It was particularly fiercely contested, which may have further exacerbated Boltzmann's depression, becoming, as already mentioned, a factor in his suicide. The Austrian physicist's theories were treated by scientific opponents as a compromising manifestation of speculation. They probed contradictions in his arguments, seeking discrepancies between experience and kinetic theory.

---

[789] Ibid.
[790] Idem, p. 297.

The second example concerns Marian Smoluchowski, who, after developing equations to describe Brownian motion, hesitated to publish them. The delay in publication was due, as he said himself, to the author's caution as he awaited more precise experimental results, but also to fear of the reaction of a scientific community disinclined towards atomic hypotheses. Smoluchowski wrote that

> there prevailed in science an excessively critical current, it could be called: cowardly sober. It is not easy to clip the wings of the human mind, but he who could not keep himself from speculations at least held back from declaring them publicly. I remember how I myself for a long time hesitated and procrastinated over announcing my contributions to kinetic theory. This current primarily turned against the most powerful theory that science had produced to date, i.e. atomic theory[791].

However, despite resistance, kinetic theory systematically developed and the logical consequence was that Boltzmann's hypothesis on the "reversibility of phenomena" had to be accepted. Ice in a glass of water, according to Boltzmann and in line with kinetic theory, can cool by itself and simultaneously warm up the water in the glass but this is such an improbable phenomenon that in practice it is ignored[792]. This hypothesis was unacceptable to most physicists.

The randomness of events in nature, their reversibility, as well as doubts relating to the second law of thermodynamics and to the constant increase in entropy, caused Smoluchowski to become philosophically close to pragmatism, with *utility* as a criterion of truth, and to conventionalism with its approach to chance. *The view of conventionalists that all scientific theorems and theories are conventional in nature, and that attitudes adopted towards perceived reality are variable and tend to evolve, found real application in Smoluchowski's scientific practice.*

The period from 1900, when Smoluchowski became acquainted with Franz Serafin Exner's article *Notiz zu Brown's Molecularbewegung* (*Notes on Brownian molecular motion*), until 1903, when he finished his own research on Brownian motion, culminating in the development of a hypothesis and building proof in the form of mathematical models describing its nature, was a time that shaped his views in the field of atomic-kinetic theory.

---

[791] See idem, *The current state of atomic theory*, op. cit., p. 62.
[792] See idem, *The evolution of atomic theory*, op. cit., pp. 19–21.

CHAPTER 7

It is no accident that the terms 'kinetics' and 'atomistics' are used interchangeably in this work as Smoluchowski used the names of the two theories in the following way:

> Perhaps I may take the opportunity to explain parenthetically why I use these names to some extent as synonyms, although the correct meanings of the terms: "atomic" and "kinetic" theory are different. As we know, the former atomic theory of Dalton has been enriched since the times of the formulation of the principle of conservation of energy, by the additional supposition that atoms and molecules are in constant motion, that heat is a store of kinetic energy and the measure of the kinetic energy of this internal motion is what we call temperature. Since the time of Robert Mayer and Helmholtz, therefore, atomic theory has fused with kinetic theory into a unified whole[793].

The work of Einstein and Smoluchowski became a turning point in research on Brownian motion. The results obtained proved the important role of randomness in nature. Smoluchowski proved that the random collisions of molecules are a cause of the bizarre zig-zag movements that the French physicist Louis George Gouy (1854–1926) compared to the movements of ants around an anthill. He demonstrated that the space occupied by a molecule among the whole collection of other molecules is random and determined by chance circumstances. Hence it cannot be assumed that the molecules of a gas or liquid are distributed regularly and evenly. Theodor Svedberg, the Swedish Nobel laureate who confirmed experimentally conclusions on random unevenness, achieved this by counting ultra-microscopic particles in colloidal solutions.

The hypothesis of Einstein and Smoluchowski defined Brownian motion as chaotic shifts of colloidal particles suspended in a liquid or gas, caused by collisions with liquid particles; the more intensive, the less the liquid's viscosity, the smaller the size of the molecules of the solution, and the higher the temperature of the liquid. The random wandering of emulsion particles is caused by their bombardment by water molecules, which are much smaller, numerous, and fast-moving. The bombardment of particle by molecules is on average the same from all sides. However, if the particle is small enough, it happens that the number of molecules colliding with it from one side will be different at some point (greater or smaller) than the number of molecules

---

[793] See idem, *Today's state of atomic theory*, op. cit., p. 63.

striking from the other. As a result, the particle receives a stronger impulse every so often on the side marked by the impact of a larger (at a given moment) group of molecules. The described movements of particles, according to Smoluchowski, occur in the suspension not so much through the bombardment made by the liquid particles as through the fluctuations occurring in the density of particles in the direct vicinity of the suspension particle.

Smoluchowski raised this issue in response to Swiss botanist Carl Wilhelm von Nägeli, who argued that the size of a solution's molecules relative to the size of suspension particles is so small that they are in no position to effectively cause the movements of suspension molecules. The effect of one collision with a molecule is extremely small, just $\Delta v \sim 10\text{-}3$ μm/s. In a 1906 article, Smoluchowski explained this problem—Brownian motion is generated not by individual particles but by the occurrence of spontaneous fluctuations of solution molecules. He mentions these in a 1904 paper, calling them "systems of a swarming nature".

In another of von Nägeli's objections he claimed that solution particles, striking suspension particles from all sides simultaneously, cannot cause oscillations as they are simultaneous collisions. Smoluchowski's response testifies to in-depth consideration of the problem. That the solution particles collide with suspension particles 10^{20} times per second from various sides does not mean that a suspension particle should remain motionless as this is the same error of understanding as if a person playing a game of chance (for instance rolling dice) believed that he will never bear a greater loss or ever achieve a greater gain than the stake for one throw. We well know, states Smoluchowski, that good and bad luck do not usually entirely balance out, that the longer the game lasts, the greater the average sum either won or lost. Smoluchowski points out that the difference in collisions, positive or negative, can run to 10^8 or 10^{10} collisions persecond[794].

Without getting into the mathematical structures, it is worth emphasising that the essential idea on which the theory of Brownian motion rests is the assumption that suspension particles (at a certain temperature) behave to a certain extent the same a gas molecules and the average kinetic energy of

---

[794] Idem, *Zarys kinetycznej teorii ruchów Brownai roztworówmętnych* (*Outline of a kinetic theory of Brownian motion and turbid solutions*), in: *The Writings of Marian Smoluchowski*, vol. I, op. cit., pp. 495–497.

translational motion performed by the centre of mass of such a particle is equal to the average energy of the translational motion that a gas particle possesses at that temperature[795]. This corresponds to the basic assumption of the kinetic theory of gases and is a direct consequence of Maxwell's energy equipartition principle. The thermodynamic principle of energy equipartition assumes that the available energy a molecule (for example gas) possesses is evenly distributed in all its possible uses (the so-called degrees of freedom)—regardless of whether that is a degree of freedom related to the vibrations of a particle, rotational energy or translational motion.

Much later, in the *Self-Study Handbook*, in the section entitled *Methodological guidelines for students of individual sciences*, Smoluchowski summarised this breakthrough moment in physics. He stated that kinetic theory, having awakened from a temporary lethargy, showed new vitality, not only in predicting a range of previously unknown phenomena (heat transfer in rarefied gases, the transpiration of rarefied gases, radiometric forces) but also resolving the dispute with thermodynamics. He drew attention to the fact that in Brownian motion, taking place in microscopically small particles suspended in liquids or gases, we have a visible example of particle movements.

7.3. Controversy around the Brownian motion breakthrough

Research on Brownian motion, gas fluctuations and opalescence effectively changed the understanding of a number of designations of theoretical terms functioning in 19th-century science and philosophy. The theoretical formulae derived from this were confirmed by the careful measurements of Perrin and his colleagues, notably Stefan Dąbrowski, and at the same time yielded one of the most accurate methods known to date of determining the number of molecules. In his essay *On thermodynamic fluctuations and Brownian motion*, Smoluchowski writes that immediately after the announcement of the discovery, the theory was experimentally confirmed:

> Among the experimenters who undertook this research, Svedberg, Perrin and their colleagues should be mentioned above all. The latter scientist in particular

---

[795] See idem., M. Smoluchowski, *O fluktuacjach termodynamicznych i ruchach Browna (On thermodynamic fluctuations and Brownian motion)*, op. cit., p. 300.

has succeeded in verifying with great accuracy the theoretical formulae relating to Brownian motion, as well as to the distribution of particles in a gravitational field, so that such measurements can conversely be used to calculate the Avogadro number, that fundamental constant in kinetic theory[796].

The importance of these studies lies partly in the fact that they shed light on and old and contentious problem as to the essence of suspensions, colloidal and crystalloid solutions, demonstrating the singularity of the fundamental laws and the gradual nature of the transition from one category to another. Thus, they gave a new impetus to research on colloidal solutions, which today arouses the interest of chemists. The most important thing for the principles of physics, is that they proved a certain superiority of kinetic theory over thermodynamics[797].

Smoluchowski supplemented the theory of Brownian free particle motion with a detailed analysis of the impact external forces will have on the process. He studied the following cases: the influence of a constant force, such as gravity, the influence of forces from a reflecting wall (repulsive forces), the influence of forces from an absorbing wall (attractive forces) and the influence of an elastic (restitution) force. His detailed analysis and insightful interpretation of the results obtained shed a beam of bright light on the relationship of atomic theory to thermodynamics[798].

According to Bernstein, who was initially an adherent of atomic theory but, disappointed with the solutions proposed at that time, became an extreme phenomenological positivist, Ernst Mach claimed that the role of theoretical physics comes down to an economical description of observed facts[799]. For the rest of his life, he did not change his mind, remaining opposed to atomic theory. He was not, of course, the only brilliant mind in history whose scientific intuition failed him. However, there have been few similar cases that have resolved fundamental issues of science. Through his stubbornness, he had a negative impact on the development of physics in the late 19th century.

---

[796] Idem, pp. 299–300.
[797] See M. Smoluchowski, *Themes and issues in today's physics*, op. cit., p. 215.
[798] See idem., *The evolution of atomic theory*, op. cit., pp. 20–23.
[799] See J. Bernstein, *Einstein and the existence of atoms*, op. cit., p. 864.

The aggressive position of Mach and Ostwald contributed to Smoluchowski's postponement of the decision to publish *Zur kinetischen Theorie der Brownischen Bewegung und der Suspensionen (On the kinetic theory of the Brownian molecular motion and of suspensions)*[800], which was a summary of research on Brownian motion that came out only in 1906. The paper was published in the prestigious German scientific journal *Annalen der Physik*, the same periodical in which Einstein's articles were published, but a year later. It was Einstein's publication that prompted Smoluchowski to decide to make the results of his research public. Zenon Klemensiewicz claims that in mid-July 1905 an issue of *Annalen der Physik* was published containing Einstein's first work on the movements of suspended particles. Even if Smoluchowski did not come into contact with it in Lviv, which was possible given the holiday season at the time, he almost certainly read it in Cambridge, where he later stayed. This must have prompted him to publish his own research results on Brownian motion, which were already explicitly covered in Einstein's second work, printed at the end of March, when Smoluchowski was leaving Cambridge[801]. His stay in Cambridge was probably the reason Smoluchowski did not publish his work after Einstein's first article in May, but only in 1906.

Adjacent to these facts is an episode important for this history, significant for Smoluchowski's biography, and momentous for the history of science. It was his writing in September 1903 (the date is important) and publishing in 1904 in a *Festschrift* (commemorative book) containing publications by 125 physicists from around the world, issued to celebrate Boltzmann's sixtieth birthday, the article *On irregularities in the distribution of gas molecules and their influence on entropy and the equation of state*. He writes there about observations of inhomogeneities in density (density fluctuations). The book came out a year before the publications of Einstein, who read Smoluchowski's article immediately after its publication and before writing his own. Stachel, Director of the Boston University Center for Einstein Studies, points out the

---

[800] M. Smoluchowski, *Zur kinetischen Theorie der Brownschen Bewegung und der Suspensionen; von M. von Smoluchowski, (On the kinetic theory of the Brownian molecular motion and of suspensions; by M. von Smoluchowski)*, 'Annalen der Physik' ('Annals of Physics') 1906, vol. 326, No. 14, pp. 756–780.

[801] See Z. Klemensiewicz, *Marian Smoluchowski*, op. cit., pp. 96–97.

visible influence of Smoluchowski's paper of 1904 on Einstein's 1905 paper on Brownian motion.

Stachel showed that Einstein, as a paid reviewer, received the Boltzmann commemorative book and reviewed three other papers from it for a supplement to *Annalen der Physik*. Hence the almost certain supposition that he had read Smoluchowski's work. The above facts should be emphasised because they are not widely known, even among physicists[802].

Smoluchowski had been working on the issue of Brownian motion since 1900. As he writes in *On thermodynamic fluctuations and Brownian motion*, he had been convinced of the molecular-kinetic essence of Brownian motion since 1900[803]. He obtained his results in 1903, three years before the date of his publication, as he described in more detail in *outline of the kinetic theory of Brownian motion and turbid solutions*:

> several years ago, I developed for this phenomenon a kinetic theory, which seemed to me most probable; I have not yet published the results, wanting to check them further against more precise experimental measurements. Meanwhile, however, the discussion on this subject was reopened by two theoretical works by Einstein, in which the author calculates the displacement of fine particles that must arise as a result of molecular motion and concludes from the agreement with observations of Brownian motion that they are kinetic in nature[804].

This discovery, so momentous for the further development of physics and attributed to both Einstein and Smoluchowski, has been accompanied from the start by some ambiguity regarding recognition for explaining Brownian motion. Two facts—researchers establishing that Einstein had read Smoluchowski's article before publishing his own works, and certain ambiguities and doubts arising during the reading of the two scientists'

---

[802] J. Stachel writes similarly in *Einstein on Brownian Motion*, in: *The Collected Papers of Albert Einstein*, vol. 2, *The Swiss Years: Writings, 1900–1909*, ed. J. Stachel, Princeton 1989, p. 215–216. In 1904 he published an article on density fluctuations in gases that has several features in common with his later work, as well as with Einstein's work on Brown motion. Einstein may have read this paper, which appeared in *Meyer, S. 1904*.

[803] See M. Smoluchowski, *On thermodynamic fluctuations and Brownian motion*, op. cit., p. 299.

[804] Idem, *Outline of a kinetic theory of Brownian motion and turbid solutions*, op. cit., p. 490.

papers—prompt us to pore over four key articles, i.e. the article by Marian Smoluchowski of July 1903[805] and two articles by Einstein, of May[806] and December 1904[807], and to satisfy the natural research curiosity that requires us to determine whether Einstein used hints contained in Smoluchowski's article, and if so, to what extent. The fourth article is a paper written on July 9, 1906[808], by Smoluchowski. More than a hundred years after this event, determining the right answers to these questions remains an extremely difficult, though inspiring, research task.

The Polish scientist's discovery was epochal in nature, worthy of a Nobel Prize. Kerker writes about this in his article *The Svedberg and Molecular Reality*. Despite the success of kinetic molecular theory in calculating the properties of gases, no direct evidence of the existence of atoms and molecules was obtained, and the theory suffered from Mach's attacks, and especially Ostwald's. The reality of the existence of atoms was one of the key problems to be solved at the turn of the century. The early 20th century saw a breakthrough leading to the current era of science, and it is generally accepted that Brownian motion was the key test of this. The importance of the subject is confirmed by the exceptional recognition of the discovery evidenced by the fact that as many as three Nobel Prizes have been awarded for work related to it[809].

Today, we can only speculate as to what the actual reasons were for Smoluchowski postponing publication. Did his failure to announce his

---

[805] July 1903—M. Smoluchowski, *On irregularities in the distribution of gas molecules and their impact on entropy and the equation of state* (*Über Unregelmässigkeiten in der Verteilung von Gasmolekülen und deren Einfluss auf Entropie und Zustandsgleichung*), trans. B.J. Gawecki, Warsaw 1956.

[806] 11 May 1905—A. Einstein, *Über die von der molekularkinetischen Theorieder Wärme geforderte Bewegung von in ruhenden Flussigkeiten suspendierten Teilchen* (*On the movement of small particles suspended in a stationary liquid required by the molecular kinetic theory of heat*), op. cit.

[807] December 19, 1905—A. Einstein, *Zur Theorie der Brownschen Bewegung* (*Theory of Brownian movement*), op. cit.

[808] 6 July 1906—M. Smoluchowski, *Zur kinetischen Theorie der Brownschen Molecularbewegung und der Suspensionen* (*Outline of a kinetic theory of Brownianmotion and turbid solutions*).

[809] See M. Kerker, *The Svedberg and Molecular Reality*, op. cit., pp. 191–192.

research results immediately after obtaining them or to present their importance in the ongoing dispute over the 'truth of atomism' stem from fear of an excessively aggressive reaction from the academic community, or did Smoluchowski have to grow into the realisation that his equation was historic and that he could close the debate about the truth of the atomic hypothesis that had been going on for a century? To some extent, as he himself writes, he was also anticipating experimental support for his theory, which would have contributed to gaining confidence in its validity. Time has shown how mistaken Smoluchowski was, and it could even be suspected that this reason was not entirely true, invented *ad hoc* as, by presenting it, he contradicted himself somewhat. Firstly, for experimental confirmation of the theory of Brownian motion to be achieved, it would have to have been sufficiently well known in scientific publications for experimental physicists to take it up. Smoluchowski repeatedly propounded the methodological thesis that all experiments acquire scientific significance when they are preceded by sound theoretical preparation, for which in this case it would have been necessary to know the theory that the experiment was intended to confirm or disqualify. After Popper, we call this activity in short the falsification of a theory. Secondly, time has shown that proving the validity of the theory of Brownian motion following its publication by Einstein and Smoluchowski was no simple matter. That process took several years, over a dozen scientists participated in it, and it only really ended with Perrin's publication *Les preuves de la réalité moléculaire* (*Evidence for molecular reality*) in 1911, six years after Einstein's work was published.

As is usually the case, there were at least several reasons for not publishing, but most likely—as Smoluchowski's own later statements indicate—the three described above were key, although Smoluchowski mainly mentioned two of them, i.e. the atmosphere prevailing at the time in the scientific academic community, and awaiting empirical support for the theory.

Einstein did not work in the scientific community, and did not directly experience what happened to Boltzmann at the University of Vienna when he tried to convince the community of kinetic-atomic theory, something in which—involuntarily—Smoluchowski participated. Prior to Einstein's and Smoluchowski's papers, for the majority of physicists there were no convincing arguments to support atomic theory. Mach even claimed it was not possible to prove the existence of atoms at all and, due to the impossibility of falsifying[810] such a hypothesis, the concept of the existence of atoms could not be considered an element of science. Despite the unequivocal conclusions from research on Brownian motion and the resultant specific assumptions of atomism, he did not change his opinion.

While both Einstein's works explain the essence of Brownian motion and are treated as the first publications that scientifically prove kinetic-atomic theory, they undoubtedly differ greatly from each other. In *The case of Brownian Motion*, Roberto Maiocchi writes that at first glance, the May article has little in common with the phenomenon discovered by Brown. It does not even mention all the previous research on the phenomenon and it seems that the problem considered here is that of abstract mechanics, i.e. calculating, according to the kinetic theory of heat, the motions to which bodies suspended in fluid are subject. That these movements may coincide with Brownian motion is certainly present in Einstein's thinking, but explaining this phenomenon does not appear to have been his primary goal[811].

The May paper was written as if under the influence of a revelatory thought, hastily, and focused on the issue of diffusion. Most probably, it was written as a continuation of a thought contained in Einstein's doctoral dissertation, *Eine Neue Bestimmungder Moleküldimensionen* (*On a new determination of molecular dimensions*), which he completed on April 30, 1905. In it, he combined the technique of classical hydrodynamics with the theory of diffusion to enable the creation of a new method for determining molecular size and the Avogadro number. John Stachel states: "Einstein's method is well suited to determine the size of solute molecules that are large compared to those of the solvent. In 1905 William Sutherland published a new method[812] for determining the masses of large molecules that shares important elements with Einstein's."[813]. This method concerned the molecular theory of diffusion

---

[810] Falsificationism is a set of methodological procedures associated with the name of Karl Popper, although some ideas can be found in the works of William Whewell, Charles Sanders Peirce or Claude Bernard.
[811] See R. Maiocchi, *The Case of Brownian Motion*, 'The British Journal for the History of Science' 1990, vol. 23, No. 3, p. 263.
[812] W. Sutherland, *A Dynamical Theory of Diffusion for Non-Electrolytes and the Molecular Mass of Albumin*, 'Philosophical Magazine and Journal of Science' 1905, series 6, vol. 9, pp. 781–785.
[813] See J. Stachel, *Einstein's dissertation on the determination of molecular dimensions*, in: *The Collected Papers of Albert Einstein, Volume 2*, op. cit., p. 171.

developed by Walther Nernst on the basis of an analogy made by Jacobus 'Henry' van't Hoff (1852–1911) between solutions and gases, and Stokes' law on hydrodynamic friction[814]. It also bears a trace of Smoluchowski's article.

In his paper *On the Movement of Small Particles Suspended in a Stationary Liquid, Required by the Molecular-Kinetic Theory of Heat*, Einstein studied the diffusion process of suspended particles subjected to a constant force and which are not initially distributed uniformly in the liquid, assuming that the cause of such diffusion was molecular agitation. He was thus able to determine the coefficient of diffusion D and the mean square value of the displacement along direction a of the particles in time $A.t$[815]. Einstein, as he wrote himself, was not sufficiently familiar with the problem of Brownian motion to be confident in forming hypotheses, even in terms of the correctness of the molecular-kinetic conception of heat[816]. Einstein was not as well versed as Smoluchowski in the issue of Brownian motion, although he was interested in the topic, as evidenced by discussions and correspondence with Michele Besso[817]. On the one hand, the reservations presented in the May article confirm that when writing he did not feel very confident on this topic and had not considered the problem in depth. There is a perceptible lack of certainty in the conclusions drawn and in-depth understanding of the subject, which is confirmed by the author himself. Einstein acknowledges—firstly—that the data on Brownian motion available to him are so imprecise that he could not form an opinion about this phenomenon. He writes: "It is possible that the motions to be discussed here are identical with the so-called 'Brownian molecular motion'; however, the data available to me on the latter are so imprecise that I could not form a definite opinion on this matter"[818]. However, on the other hand, as Cichocki mentions, it can be noted that:

---

[814] Ibid.
[815] See R. Maiocchi, *The Case of Brownian Motion*, op. cit., pp. 264–265.
[816] See A. Einstein, *Über die von der molekularkinetischen Theorie der Wärme geforderte Bewegung von in ruhenden. Flussigkeiten suspendierten Teilchen* (*On the movement of small particles suspended in a stationary liquid required by the molecular kinetic theory of heat*), op. cit., p. 549.
[817] Michele Besso (1873-1955)—an engineer, and a close friend of Einstein from the period when he worked at the Federal Polytechnic Institute in Zurich, and then at the patent office in Bern.
[818] A. Einstein, *On the movement of small particles suspended in a stationary liquid required by the molecular kinetic theory of heat*, op. cit., pp. 549.

Einstein mentions Brownian motion only once in his work. In the 'Introduction' he states that perhaps the irregular movements of the particles he will write about further are the movements discovered by Robert Brown, but imprecise knowledge does not allow him to take a position on this matter. However, the structure of the work and its content contradict this statement. They indicate that Einstein knew perfectly well that his considerations related to this phenomenon. For example, the ultimate goal of the work is clear—to determine the magnitude that would characterise the above-mentioned movements, i.e. that which most pained experimenters[819].

We are therefore unsure whether it is merely doubt as to the correctness of the hypothesis presented that caused the inclusion of this objection in the text:

> If it is really possible to observe the motion to be discussed here, along with the laws it is expected to obey, then classical thermodynamics can no longer be viewed as strictly valid even for microscopically distinguishable spaces, and an exact determination of the real size of atoms becomes possible. Conversely, if the prediction of this motion were to be proved wrong, this fact would provide a weighty argument against the molecular-kinetic conception of heat[820].

On reading Smoluchowski's 1904 article (which in Stachel's view is basically confirmed), Einstein—having become acquainted with the deliberations on irregularities in the distribution of gas molecules—noticed suggestions indirectly related to Brownian motion and became interested in it. He understood the essence of the theory and the importance of the discovery but was not fully convinced of the validity of the hypothesis. Although the article he read interested him in the theory of Brownian motion, he focused more on a problem that was closer to him, that is, on diffusion.

This duality of the situation has been recognised by many authors; for example, in *The Case of Brownian Motion*, Maiocchi states that the first verification, developed by Einstein in the summer of 1905 one the basis of what he had worked on in the spring, had nothing to do with the formula for λx. This confirms once again that Einstein was not principally interested in

---

[819] B. Cichocki, *Albert Einstein – praca o ruchach Browna z 1905 roku* (*Albert Einstein – the 1905work on Brownian motion*), 'Delta' 2005, No. 6, p. 4.

[820] A. Einstein, *On the movement of small particles suspended in a stationary liquid required by the molecular kinetic theory of heat*, op. cit., pp. 549.

research on Brownian motion. His concern was rather with the coefficient of diffusion D. Einstein's reasoning was rather tortuous and also turned out to be mistaken in certain aspects. From attempting to calculate how the coefficient of viscosity of a fluid varies due to the addition of particles in suspension, he arrived, by calculation, at the verification of the formula of the coefficient of diffusion. From this formula he arrived at the verification of the value of the Avogadro number. It is rather significant that Einstein considered the true originality of his previous work to be the possibility of a new means for determining N through the study of the particles suspended in a fluid[821].

In this respect, Einstein's December publication is interesting, being the result of seven months of reflection and, importantly, preliminary confirmation by some scientific circles of the theory's validity. In December, Einstein overcame his uncertainty, left behind the issue of diffusion and focused on a more important matter, from which he had originally distanced himself. The change in tone of the December article is telling; the problem had been thought through and Einstein has become confident as to the validity of his hypothesis. The article refers to experimental physicists confirming his earlier theses—a physicist from the University of Jena, Heinrich Siedentopf (1906–1963), and a professor at the University of Lyon, Louis Gouy, who by direct observation had become convinced that the so-called Brownian motion is caused by irregular thermal movements of the liquid's molecules.

Roberto Maiocchi notes that the fact Einstein took no account of the existing experimental work on Brownian motion is borne out by the complete absence of even the slightest discussion in his work regarding the new quantity introduced (new as regards the history of Brownian motion). On the contrary, Smoluchowski, who for some time had been doing theoretical research analogous to Einstein's and who was well acquainted with the relevant references, highlighted in his work published in 1906 the important novelty of the notion of displacement. This enables one to get round the serious objection to the kinetic explanation, based on the great disparity between theoretical velocities and 'observed' velocities. In fact, the theoretical velocity with which it is assumed the suspended particle moves is not in any way measurable by observational procedures. What we see, Smoluchowski pointed out, is only the mean velocity of the particle, struck 10 to 20 times a second, each time

---

[821] See R. Maiocchi, *The Case of Brownian Motions*, op. cit., p. 266.

in a different direction. Its centre will describe an unpredictable zig-zag path made up of straight lines much shorter in length than the size of the particle. Its displacement becomes visible only when the geometric sum of these lines is raised to an appreciable value. "Exner and Wiener"—Smoluchowski added in a note—"also stressed that they were not able to take into account the very small zig-zags but they did not realise the importance of this observation"[822].

In view of the various reservations raised in the May article, it was necessary to write a second paper that would dispel any doubts that arose in the first. Hence the explanation:

> The following paper will amplify in some points the author's own paper mentioned above. We will derive here not only the translational movement, but also the rotational movement of suspended particles, for the simplest special case where the particles have a spherical form. We will show further, up to how short a time of observation the result given in that discussion hold true[823].

Maiocchi points out that in the last months of 1905, Einstein finally addressed the problem of Brownian motion directly. The article containing his research confirms how the previous work was free from any concerns about comparison with the available experimental studies: "Shortly after the appearance of my work on suspended particles in fluids from the point of view of the molecular theory of heat, Mr Siedentopf of Jena communicated to me that he and other physicists—first Professor Gouy of Lyons—had come to the opinion, through direct observations, that the so-called Brownian motion is caused by the irregular thermic movement of the molecules of the fluid"[824].

In spite of this, Einstein's epistemological attitude had not changed from that evident in his previous work. He decided to investigate the problem of particles suspended in a fluid using a "more general" method than that used previously. He did so partly in order to better show how Brownian motion is related to the fundamental principles of kinetic theory, and partly also to be able to develop, alongside the formula for the Brownian translation motion already published, a formula for the movement of rotation. He thus again obtained the formula for λx and the formula for the Brownian motion

---

[822] Idem, p. 264.
[823] A. Einstein, *Theory of Brownian movement*, op. cit., p. 371.
[824] R. Maiocchi, *The Case of Brownian Motion*, op. cit., p. 267.

of rotation, i.e. for the rotary motion of the particle, which is caused by the molecular collisions. This he did by a line of reasoning that, more than the previous one, directly links the conclusions to the fundamental principles of the kinetic theory of gases. Nevertheless, as in the previous work, Einstein did not bother about the relationship between these formulae and the empirical data: "I do not want to make a comparison here between the scarce experimental material at my disposal and the results of the theory, I leave this comparison to those who deal with the subject from an experimental point of view"[825].

Most probably, Einstein's reflections following publication of the May article prompted him to write the December paper. Its publication had to take place quickly as Smoluchowski may have been at a more advanced stage of work, or even had a solution ready, as later transpired to be the case. The fact that Smoluchowski's hypothesis relating to the environment of gases is also true for liquids and thus solves the problem of Brownian motion, which the Polish physicist did not want to address directly in his article, was confirmed in a 1906 paper: "The kinetic theory states that bodies suspended in gases must demonstrate movement of the Brownian motion type, but much faster even than those we observe in liquids"[826].

Einstein made a claim that is quite significant. He believed that not only the qualitative properties of Brownian motion, but also the order of magnitude of the paths made by the particles, correspond completely to the results of the theory. But what are these experimental results whose "order of magnitude" corresponds "completely" to that of the displacements provided for by the theory? Einstein gives no answer to this question. Smoluchowski is more specific here, stating that the formulae developed in the works of experimenters already available at the time must necessarily be related to Exner's results, which he considered to be evidence in support of the new theory[827]. This idea (later taken up and applied to the study of molecules in solution) is also present in Smoluchowski's work. The Polish physicist points to a sure cause of error in the theory: "To treat the suspended particles by

---

[825] Ibid.
[826] M. Smoluchowski, *Outline of a kinetic theory of Brownian motion and turbid solutions*, op. cit., p. 507.
[827] See R. Maiocchi, *The Case of Brownian Motion*, op. cit., p. 267.

the same standard as the molecules with which the kinetic theory of the gases dealt, meant having to transfer to the particles all and only all of those characteristics which the theory attributed to the molecules, over and above the ambiguous suggestionswhich could come from experimentation"[828].

It is understandable that Smoluchowski so thoroughly substantiated the details of the theory of Brownian motion when he became acquainted with Exner's work. In his article *Notiz zu Brown's Molecularbewegung* (*Notes on Brownian molecular motion*), Exner described a series of microscopic observations of particles of rubber of various sizes suspended in water at a variable temperature. The basic aim of the experiment was to determine what numerical relationship existed between the temperature of the fluid, the sizes of the particles, and the velocities of moving particles. The observed changes were recorded manually on a blackened photographic plate and the reproductions then enlarged by projection. The path of the particles' movement was measured on these enlargements. The velocity was determined by dividing the length of the path by the observation time. The value thus measured is, in Exner's opinion, that of the velocity with which the suspended particles are moved. The curve obtained turned out to be approximately a straight line. However, Exner's conclusions were not unambiguous. The researcher stressed that the kinetic model he used was not very accurate as it neglected, for example, internal molecular forces and friction and therefore supposing the behaviour of suspension molecules to be analogous to that of fluid molecules was insufficient; he also did not think his experiment proved anything against the kinetic theory. He therefore advocated continuation of research towards proving the kinetic concept: "Nevertheless, one can always consider that, if one starts from the hypothesis of a relation between the movement of the molecules in the liquid and that of the suspended particles, the visible movements and their corresponding measured quantities will be of real use for the clarification of the internal movements of a liquid"[829].

In *The Collected Papers of Albert Einstein*, Stachel writes that Exner explored a completely different way of applying the kinetic theory of heat to Brownian motion, which assumed an equipartition of energy between the molecules of

---

[828] Idem, p. 265.
[829] See F. Exner, *Notiz zu Brown's Molecularbewegung* (*Notes on Brownian molecular motion*) 'Annalen der Physik' ("Annals of Physics') 1900, vol. 307, No. 8, pp. 843–847.

the liquid and suspended particles. He calculated the velocity of particles on the basis of observations that he interpreted as giving the average velocities of the suspended particles, obtaining results inconsistent with contemporary estimates of particle velocities[830].

The results of Exner's research did not speak unequivocally for or against the kinetic explanation but provided an impetus for Smoluchowski to conduct his own research. From that time, he tried gradually to develop the theory, first by considering the probable arrangement of particles, then the phenomenon of movement caused by a combination of free paths, and finally the mechanism itself of the movements of particles caused by the collisions of surrounding molecules. He was also the first to point out that similar phenomena, though governed by slightly different quantitative laws, must occur in gases[831].

Exner's work, Maiocchi writes, highlighted the difficulties of the kinetic programme, which is why some historians interpreted his idea as a negative "crucial experiment". Smoluchowski changed the interpretation of Exner's results, using them to support the kinetic explanation. His method of obtaining this result is based on some rather obscure moves. First of all, it is necessary to ascribe a new meaning to the values measured by Exner, who believed he had measured the velocity of the particle, which is impossible. The distance the researcher measured in the time interval t was not actually the path followed by the particle, but the "displacement"—the distance between the points of departure and arrival. Exner's results are thus comparable with Einstein's and Smoluchowski's theory only if the "velocities" are reinterpreted as "displacements made in the unit of time"[832].

Years later, Smoluchowski recalled:

> several years ago, I developed for this phenomenon a kinetic theory, which seemed to me most probable; I have not yet published the results, wanting to check them further against more precise experimental measurements. Meanwhile, however, the discussion on this subject was reopened by two theoretical works by Einstein, in which the author calculates the displacement of fine particles that must arise as

---

[830] See J. Stachel, *Einstein on Brownian Motion*, op. cit., p. 209.
[831] See M. Smoluchowski, *On thermodynamic fluctuations and Brownian motion*, op. cit., p. 299.
[832] See R. Maiocchi, *The Case of Brownian Motion*, op. cit., p. 268.

a result of molecular motion and concludes from the agreement with observations of Brownian motion that they are kinetic in nature[833].

In 1953, Stanisław Loria succinctly outlined the two paths followed by Smoluchowski and Einstein. Einstein starts from thermodynamics, using the concept of osmotic pressure and the famous Boltzmann theorem. From the general theory of fluctuations around the equilibrium state, he arrives at a calculation of the average value of Brownian particle displacement in one second. Smoluchowski takes the opposite route—he starts by solving the stochastic problem and, following the intricate movement of a particle, seeks to answer the question of the probability with which a particle moving in a straight line, with sections of equal length, forward and backward, will find itself after a certain number of steps at a specified point[834]. The method used by Smoluchowski, as the theory's later development showed, demonstrated the advantage of being microphysical in nature, in the full sense of the word. One can agree with Loria's assessment in terms of the differences presented in the two authors' ways of calculating Brownian motion, but in terms of the paths leading to the problem's solution, there remain questions to be considered.

Smoluchowski refers to Einstein's theses, highlighting the theory's lack of experimental confirmation, which—as he wrote earlier—prompted him to suspend publication, and which was later proven experimentally by Svedberg and Perrin among others. He wrote:

> In this area, therefore, experiments cannot resolve the issue raised; even more interestingly—the same also applies both to Brownian motion and related phenomena. Whether we follow Einstein's argument based on the notion of osmotic pressure or apply the method of direct calculation which I have given, we arrive at the same final formulae for the displacement of particles, for their separation by gravity, and for the diffusion coefficient, regardless of whether the above-mentioned hypothesis is adopted[835].

---

[833] M. Smoluchowski, *Outline of a kinetic theory of Brownian motion and turbid solutions*, op. cit., pp. 490–491.
[834] See S. Loria, *Marian Smoluchowski and his work (1872–1917)*, op. cit., p. 19.
[835] Smoluchowski M., *O pewnem zagadnieniu kinetycznej teoryi roztworów* (*On a certain problem of the kinetic theory of solutions*) in: *A commemorative book to celebrate the 250th anniversary founding of the University of Lviv by King Jan Kazimierz*, Lviv 1911, p. 209.

In many publications, Smoluchowski's opinions confirming the correctness of Einstein's calculations are treated by physicists as evidence of the his discovery coming first. Such an interpretation is made, for example, by Brush, who notes that after Einstein's publication, Smoluchowski's article appeared, in which he confirms the results obtained and indicates that they completely agree with his earlier discoveries, which he made in a different way. Smoluchowski confirms the concordance of the results given by Einstein with his own, and although this is undisputed, to conclude on this basis that the Polish physicist confirms that Einstein made the discovery first is an overstatement. Nowhere did Smoluchowski ever unequivocally state this.

Smoluchowski's attitude stemmed from his surprise at Einstein's publication; he only confirmed what was undeniable—that his results were correct, because they were. Loria notes: "For three years, Smoluchowski's thoughts stubbornly revolved around the problem of the alleged contradiction between thermodynamics and kinetic theory. By probing ever more deeply the intricate complex of difficult and detailed topics involved, he saw connections that had thus far been hidden from other, less astute experts in these theories"[836].

For three years, Smoluchowski studied the nature of Brownian motion, and in 1903 finally cracked the problem. In the hostile atmosphere prevailing around atomic theory at the time, the problem for him was publishing the discovery. The paper included in Boltzmann's commemorative book was to be the first step on the road to making the results of his work public. At first, it probably did not occur to him that he had not foreseen all the implications of his idea as he would not have expected that someone would read his work and, guessing his intentions, make his discovery public, not sharing Smoluchowski's objections to revealing a theory that was not fully developed, and without mentioning where, how, or through whom he had found inspiration.

For a scientist who had achieved success after several years of research work, becoming the only theoretical physicist with in-depth knowledge of Brownian motion, taking the position of someone supplementing this spectacular discovery with important additions must have been an unpleasant experience, especially as he must have seen in many publications appearing after 1905 increased discussion on proving the Einstein-Smoluchowski theory.

---

[836] S. Loria, *Marian Smoluchowski (1872–1917)*, op. cit., p. 804.

This discussion usually referred to Einstein—a theoretical physicist—and a whole galaxy of experimental physicists: Exner, Svedberg, Perrin, Zsigmondy, Gouy, Max Seddig (1877–1963), George G. Stokes (1819–1903), Siedentopf, Kerker, Wiener, and Victor Henri (1872–1940). Unfortunately, Smoluchowski's name rarely appeared in this group, which is strange as some of the aforementioned researchers used his work, but which is also to some extent understandable upon closer inspection of the situation.

Nevertheless, nowhere has there been any contemporary discourse on Smoluchowski's numerous comments on Einstein's publications. It is as if the issue did not exist, but the Polish physicist had a number of doubts, which he presented in a fairly veiled way, though clearly enough to provoke a few questions and provide a few answers. In his work *Outline of a kinetic theory of Brownian motion and turbid solutions*, he wrote:

> I have found in Einstein's formulae some of my own results and his end result agrees completely with mine although we used an entirely different method. Hence, I offer my reasoning, especially as my method seems to me transparent and therefore more persuasive than Einstein's method, which is not free from criticism. I add to that the discussion of other theories and the factual material accumulated by earlier researchers, which I believe is highly persuasive of a kinetic interpretation of these phenomena. At the end of my paper, I include a few comments on the so-called colloidal suspensions theory related to this subject[837].

What does Smoluchowski mean by "I have found in Einstein's formulae some of my own results"[838]? Shouldn't this sentence end with a question mark that Smoluchowski did not use? What does "I include a few comments on the so-called colloidal suspensions theory related to this subject"[839] mean? Is it not a suggestion to indicate an important point of proof that Einstein does not mention in his work? The same overtone can also be detected in the statement "my method seems to me transparent and therefore more persuasive than Einstein's method"[840].

---

[837] M. Smoluchowski, *Outline of a kinetic theory of Brownian motion and turbid solutions*, op. cit., pp. 490–491.
[838] Idem, p. 490.
[839] Ibid.
[840] Idem, pp. 490–491.

Had Smoluchowski guessed that Einstein had become acquainted with his 1903 paper and did he have the impression that he had described a theory of Brownian motion in the May article, as if in passing, without going into its essence in detail? After all, he focused on the problem of diffusion, without particularly stressing the issue of Brownian motion, as if this issue had involuntary occurred to him while reading Smoluchowski's text, and yet the formulation of a theory of Brownian motion soon became an extremely important discovery in the history of physics. Einstein's May article gives the impression of not being fully thought out, of being written hastily, in a hurry, but so effectively that "from the calculation of the displacement of small particles that must arise as a result of molecular motion, he concludes that the truncated Brownian motions are consistent with the thesis of their kinetic nature"[841], which is a key argument of the theory.

Smoluchowski referred several times to Einstein's paper, indicating differences between the two scientists' work on the topic. He highlighted the difference in the starting point for theoretical deliberations. He wrote that Einstein's solutions were more abstract, having resulted from his general research in the field of statistical mechanics; he was also the first to provide a theoretical formula determining the displacements of particles suspended in a liquid, but left open the question of whether the theoretically predicted phenomenon corresponded to what was known as Brownian motion[842]. Smoluchowski repeatedly returns to his discovery and compares it with Einstein's; a certain note of regret is discernible in his argumentation that having such an advanced theory, properly developed mathematical formulae, and conclusions, he refrained from completely and unambiguously publishing them, all the more so as he was aware of the significance of the discovery of scientifically justified Brownian motion in the process of proving the truth of atomic theory.

---

[841] Idem, s. 490.
[842] Idem, s. 490–491.

7.4. Comparison of the works of Einstein and Smoluchowski

An attempt to compare the development of the theses from Smoluchowski's articles of 1903 and 1906, and Einstein's papers of 1905 raises a number of questions, which—as it transpires—are difficult to answer unambiguously. Similarly, it is difficult to answer the question of why Einstein never reacted, either directly or indirectly, to the numerous comments on his work contained in Smoluchowski's articles. Therefore, rather than seeking answers that would inadvertently bring summary statements into the ongoing discourse, it is better to focus on asking better-founded questions.

Let us cite three characteristic elements of Einstein's approach to his discovery, outlined by Stachel, which will be helpful in attempting a comparison: (1) he based his analysis on the osmotic pressure rather than on the equipartition theorem; (2) he identified the mean square displacements of suspended particles rather than their velocities as suitable observable quantities; and (3) he simultaneously applied the molecular theory of heat and the macroscopic theory of dissipation to the same phenomenon, rather than restricting each of these conceptual tools to a single scale, molecular or macroscopic[843].

Let us start the analysis with an unexplained issue Smoluchowski raises, which is the problem of Einstein's use of Stokes' law in the proposed calculations: "If the dimensions of the sphere M are large compared to the free path of the surrounding particles, we can use the ordinary Stokes' formula to calculate the resistance"[844], however, "Einstein does not at all take into account (…) particles so small that they are not subject to Stokes' formula"[845].

Smoluchowski assumes that Einstein's use of Stokes' formula results in him omitting small particles that are not subject to the formula and concludes that in the study of Brownian motion this important element remained beyond investigation. Meanwhile, according to Maiocchi, not only did Einstein not omit them, but he applied Stokes' formula to them. Hence, in his opinion, there is a common feature in the criticism of Einstein's theory and of Perrin's experiments—the problem of applying Stokes' law to ultramicroscopic

---

[843] See J. Stachel, *Einstein on Brownian Motion*, op. cit., p. 210.
[844] M. Smoluchowski, *Outline of a kinetic theory of Brownian motion and turbid solutions*, op. cit., p. 505.
[845] Idem, 506.

particles. The application of this law was as fundamental to Einstein's work as it was to Perrin's reasoning. However, Paul Langevin (1872–1946), who was unable to do without Stokes' law, expressed doubts about the legitimacy of using this formula, and Henri raised a specific question: "Can we really apply Stokes' law to the displacements of granules of 1 μ diameter in water? It may be that this law is not applicable to such small granules". Moreover, he identified the 'illegitimate' use of Stokes' law as the probable cause of the observed disagreement between Einstein's formula and his own data[846].

Outlining Einstein's line of reasoning in his May article, Bogdan Cichocki also notes the application of Stokes' law to the movement of excessively small granules:

> Einstein starts his considerations with an analysis of the phenomenon of osmotic pressure. Let us divide a vessel into two parts, placing in it a semipermeable membrane through which the solvent can penetrate, but the dissolved substance cannot. If there is then pure solvent in one part and a solution in the other, there will be a pressure difference between the two parts called osmotic pressure. The same phenomenon occurs when there is a suspension instead of a solution. Einstein notes that this phenomenon does not exist when there are no suspension particles. Nor does it when there is no fluid. Therefore, irregular particle movements (caused by the presence of the fluid!) are the direct cause of osmotic pressure. He then goes further, stating that a suspension can be viewed as a gas made up of atoms and that the same tools can be applied that lead, for example, to the ideal gas law for sufficiently dilute gases. In this way, he derives (…) van't Hoff's law, but Einstein does not mention it (as he was not keen on citing anyone).
>
> In the next step, he considers a situation in which a constant force K acts on the suspension particles, e.g. in direction x. Then the concentration of particles n in the equilibrium state will depend on the position x. This relationship is given by the so-called barometric formula. Einstein derives this from the principle of minimum free energy and (1), where (…) he notes (and this is his third step) that this equality must be consistent with the condition of dynamic equilibrium.
>
> This would lead to a concentration of particles in the direction of the force. Then, however, as a result of the diffusion phenomenon, a flow of particles will occur in the opposite direction. In the diffusion process, the number of particles flowing per unit time through a unit area equals $D\frac{\Delta n}{\Delta x}$, where D is the diffusion coefficient. In

---

[846] See R. Maiocchi, *The Case of Brownian Motion*, op. cit., p. 277.

a state of equilibrium, the effects of both processes must cancel each other out and equality must therefore occur $\frac{K}{6\pi\eta a}n - D\frac{\Delta n}{\Delta x} = 0$. (2). Comparison of the above formula and the so-called barometric formula gives a very important result for the diffusion coefficient of the suspension: $D = \frac{RT}{N_A}\frac{1}{6\pi\eta a}$.

Einstein used this result in his doctoral thesis to determine the Avogadro number. In this paper, however, he went in a different direction. Namely, he stated that the process of the concentration of particles equalising (i.e. the diffusion process) occurs as a result of irregular movements of the suspension particles. It is possible to attempt to describe these movements using probability calculus. He assumed that the movements of separate particles are independent. Then, that the displacements of a selected particle at non-overlapping time intervals are also statistically independent. In making these assumptions, he used magic words like "obvious," "simple," "easily shown," etc. Today, we know that this is not so obvious and that these assumptions require more thorough justification. However, if we accept them, it can indeed be quite easily shown (Einstein did it in three moves), (…) arriving at a formula that indicated how the intensity of Brownian motion should be "measured". Experimental confirmation of this formula became a turning point in the process of "bringing atoms to life". Except that Einstein did not mention a word about Brownian motion when discussing this result. However, an inquisitive reader of the final part of his paper may notice that, by a strange coincidence, there is a calculation based on the formula (3) of the average displacement of a 1 |μm particle in one second under the conditions of a typical experiment[847].

The application of Stokes' law to microscopic particles in the works of Einstein and Perrin represents an extension of the application of the law far beyond the recognised limits of its applicability. Thus, the object under examination, having little in common with the object originally intended to be examined, is ultimately subjected to a principle that does not apply to it. Such thinking resulted primarily in the fact that:

> First of all, Stokes' law could only be considered as being proved experimentally for particles of some millimetres in radius, and to apply it to orders of magnitude of one micron was an extremely rash move. Secondly, the law presupposes a sphere which moves in a continuous medium, whereas in the kinetic model the surrounding medium is discontinuous and does not operate on the sphere with continuous forces but with irregular collisions. Thirdly, the velocity with which

---

[847] B. Cichocki, *Albert Einstein – the 1905work on Brownian motion*, op. cit., p. 5.

the particle moves in the fluid appears in the law, but the research on Brownian motion had by now made it clear that the velocities which can be measured (...) are not at all the velocities of displacement of the particles. Besides, there were two further problems in the case of Perrin's measurements: he applies the law to a cloud of particles of the emulsion which is slowly being deposited inside a capillary tube, but the law has been defined for the single particle, not for a cloud, and it presupposes the existence of an indefinite fluid, not of a portion of fluid enclosed in the narrow cavity of a capillary tube[848].

The discrepancies highlighted by Maiocchi are supplemented by Smoluchowski's text:

> It was understood that the numbers Exner considered the measure of particle velocities in no way correspond to the velocity of actual movement; they are chance shifts resulting from the geometric composition of a great number of small deflections possessing all possible orientations in space. (...) the displacement achieved over a unit of time must be a unit of a significantly lower order than the actual velocity of movement, occurring in an immeasurably intricate, zig-zag path[849].

At the very beginning of his May paper, Einstein refers to Brownian motion[850] and in his December work he writes about the movement of particles suspended in a liquid, as postulated by the molecular theory of heat[851]. However, the mathematical arguments presented in the May article concern mainly the relationship between the diffusion coefficient and temperature, while the December paper is written differently. The paper *A New Determination of Molecular Dimensions*, created in 1904 and 1905, directed Einstein's attention to the problem of diffusion and he focused on that problem in the May article. Stachel captures the essence of that article in the following way—Einstein derived the diffusion equation from an analysis of the time-dependence of the particle distribution, calculated from the probability distribution for displacements. This derivation is based on his crucial insight into the role of Brownian motion as the microscopic process responsible for diffusion on a macroscopic scale. The solution of the resulting

---

[848] R. Maiocchi, *The Case of Brownian Motion*, op. cit., pp. 277–278.
[849] M. Smoluchowski, *On thermodynamic fluctuations and Brownian motion*, op. cit., p. 299.
[850] A. Einstein, *On the movement of small particles suspended in a stationary liquid required by the molecular kinetic theory of heat*, op. cit.
[851] Idem, *Theory of Brownian movement*, op. cit.

diffusion equation, combined with its expression for the diffusion coefficient, gives an expression for the mean square distribution, λ as a function of time, an expression that, according to Einstein, could be used experimentally to determine the Avogadro number N[852]. Einstein's December paper was entirely devoted to the analysis and calculation of Brownian motion.

It is impossible, without entering into speculation, to answer the question of why Einstein applied Stokes' law to microscopic particles. Any attempt to find an answer merely provokes an unauthorised auction of conjectures.

Moving on to another field of reflection, Smoluchowski writes:

> I will not enter into a discussion here on the very ingenious reasoning with the aid of which Einstein arrived at his formulae, however I believe that both methods used by him rely on indirect reasoning (e.g. neither transferring the laws of osmotic pressure to particles M suspended in a fluid and calculating the velocity at which they diffuse through the liquid, nor applying Boltzmann's theorem (on the influence of potential forces on the statistical composition of mechanical systems) to the non-potential force which is the resistance experienced by particles M in the movement through the medium, is entirely beyond reproach), which does not seem entirely persuasive. In each case, concurrence with the direct method used here, which better explains the mechanism of the whole phenomenon, should be seen as a desirable confirmation of both means of calculation[853].

What were Einstein's applied ideas supposed to be that Smoluchowski did not want to talk about and which he found unconvincing?

Smoluchowski answers this question much later, in his 1914 paper *On thermodynamic fluctuations and Brownian motion*, in which he summarised his deliberations on Brownian motion in relation to the theory of thermodynamic fluctuations. He also addressed Einstein's concept, proving that the study of Brownian motion in terms of osmotic pressure is an indirect and not entirely convincing notion. Let us analyse this problem sequentially in an attempt to falsify Smoluchowski's thesis.

In his *1904 paper On irregularities in the distribution of gas molecules and their influence on entropy and the equation of state*, Smoluchowski writes:

---

[852] See J. Stachel, *Einstein on Brownian Motion*, op. cit., p. 212.
[853] See M. Smoluchowski, *Outline of a kinetic theory of Brownian motion and turbid solutions*, op. cit., p. 506.

> Boltzmann showed that van der Waals' claim that the internal cohesion forces cancel each other out everywhere and create only a constant 'internal pressure', calculated as if the mass was evenly distributed, is justified only on the assumption of the sphere of action is large compared to the mean distances. Well, it seems to me that the applicability of this method needs to be limited a little further still, namely to a case in which the sphere of attraction is large compared to the area within which there are still discernible differences in density[854].

Cichocki writes that Einstein:

> Einstein starts his considerations with an analysis of the phenomenon of osmotic pressure. Let us divide a vessel into two parts, placing in it a semipermeable membrane through which the solvent can penetrate, but the dissolved substance cannot. If there is then pure solvent in one part and a solution in the other, there will be a pressure difference between the two parts called osmotic pressure. The same phenomenon occurs when there is a suspension instead of a solution. Einstein notes that this phenomenon does not exist when there are no suspension particles. Nor does it when there is no liquid. Therefore, irregular particle movements (caused by the presence of the fluid!) are the direct cause of osmotic pressure[855].

According to Einstein, the so-called osmotic pressure results from the pressure exerted by the dissolved substance, separated from the pure solvent by a membrane, i.e. acting on a wall that is permeable to the solution, but not to the substance dissolved in the solution. We therefore have osmotic pressure, which must act on the solution in order to prevent the flow of the solvent with the solute through the semi-permeable membrane that separates solutions of various concentrations. The cause of osmotic pressure is the difference in concentrations of chemical compounds in solutions on either side of the membrane and the system's desire to equalise them. This difference can also be defined as the degree of disorder of electrolyte molecules, which is a determinant of the degree of entropy of an electrolyte in solution. The entropy is greater the higher the degree of disorder of the electrolyte's molecules.

The osmotic pressure described by Einstein is measurable in a situation *in which the mass is not distributed evenly* and the pressure's value appears on the membrane separating the electrolyte from the pure solution. According to

---

[854] Ibid., *On irregularities in the distribution of gas molecules and their impact on entropy and the equation of state*, op. cit., p. 66.

[855] B. Cichocki, *Albert Einstein – the 1905 work on Brownian motion*, op. cit., p. 5.

Einstein, it appears on the boundary of the partial volume V^*, when a situation arises that electrolyte molecules are dissolved in the partial volume V^* of a fluid of total volume V despite there being no membrane. Smoluchowski called this 'artificial osmotic pressure', which *nota bene* is a particular case of internal pressure arising in a situation where mass is not distributed evenly.

In Exner's work, there is no fundamental difference between a dissolved particle and a suspended particle. Einstein came to a similar conclusion, but rather than emphasising the equipartition theorem, he took osmotic pressure and its relationship to the theory of diffusion and to the molecular theory of heat as the starting point of his analysis of Brownian motion; according to this theory, a solute molecule differs from a suspended body solely with regard to magnitude, and it is not apparent why the osmotic pressure of a number of suspended particles is not the same as that of the same number of solute molecules. According to this theory, the solubilized molecule differs only in the size of the suspended body, and it is not known why a certain number of suspended bodies should not have the same osmotic pressure as the number of dissolved molecules. On the other hand, Einstein pointed out, according to the "classical theory of thermodynamics", suspended particles—as macroscopic objects—should not exert an osmotic pressure on a semipermeable wall. Before Einstein, no one seems to have recognised that this contrast provides a touchstone for the kinetic theory. His choice of a suspension to study the relations between the thermo- dynamic and atomic theories of heat amounted to a radical reversal of perspective. Usually the legitimacy of microscopic explanations of thermodynamic results was at issue. In this case, however, the question centered on the applicability of a thermodynamic concept-osmotic pressure–to the suspended particles[856].

In the paper *On thermodynamic fluctuations and Brownian motion*, Smoluchowski shows that the osmotic pressure formulae applied work well in the description of thermodynamic fluctuations that occur in a solution with uneven distribution of particles, hence applying artificial osmotic pressure to the calculation of diffusion is justified, but for the calculation of Brownian motion it is an indirect and problematic tool.

---

[856] See J. Stachel, *Einstein on Brownian Motion*, op. cit., p. 209.

The nature and direction of the changes occurring in the course of spontaneous processes in an isolated thermodynamic system intended for the calculation of the effect of diffusion action is similar to calculations of the fictional action of osmotic pressure. Both processes take place, in line with the second law of thermodynamics, if the thermodynamic system moves from one state of equilibrium to another. Smoluchowski puts this idea of Einstein as follows: According to Einstein, a conclusion of the existence of diffusion phenomena can be taken as a starting point, linking it with the laws of osmotic pressure, as a result of which we get a duality of views on the phenomenon of diffusion since on one hand we explain it, quite rightly, as the result of microscopic Brownian motion and on the other we consider it a result of "osmotic pressure", which in this case is more of a fictitious macroscopic notion. This assumption enables a calculation of the velocity of diffusion movement if we rely on the generally accepted hypothesis that diffusion can be reduced to the fictitious action of osmotic pressure, which shifts particles from places of greater concentration towards places of lower concentration[857].

Smoluchowski signals the possibility of the situation described by Einstein—with an even distribution, the potential energy U is small in relation to the kinetic energy E. We may be presented with such constellations, where, for example, all the molecules are in one half, and the other remains completely empty, with U experiencing a significant change. However, the number of such cases of extreme density is so negligible that we need not consider it here[858]. This is the situation upon which Einstein builds his fundamental argument in the May article.

Hence Smoluchowski's thesis—that studying Brownian motion in terms of osmotic pressure is an indirect and not entirely convincing concept—has merit.

The physicist Robert Alicki notes:

> However, the question arises as to why Einstein made such an identification, which is not obvious. Osmotic pressure occurs when an area of a solution is bordered by a semipermeable barrier, which is not the case in the problem of diffusion. It can therefore be speculated that Smoluchowski's idea of 1903 helped Einstein adopt

---

[857] See M. Smoluchowski, *On irregularities in the distribution of gas molecules and their impact on entropy and the equation of state*, op. cit., pp. 309–310.
[858] Idem, p. 637.

that assumption. Smoluchowski showed that local fluctuations in the density of gas reduce entropy (now described as the "microscopic function of entropy" 'S') and also increase the momentary free energy. As Smoluchowski writes in paragraph 5, "The importance of this 'microscopic' entropy function lies in the fact that gas can 'by itself'' perform the work $T\,(S\_0 - S')$..." and later he writes about Maxwell's demon[859]. In other words, the fluctuations themselves, without the need to introduce a semi-permeable barrier, can generate forces (of the osmotic pressure type) that move a Brownian particle against the force of resistance, performing work. Later, Smoluchowski withdrew from this and today we also do not consider that work, but the image itself of a random force generated by fluctuations remains[860].

Another issue, to which Smoluchowski devotes a great deal of attention in a 1904 article, is entropy. In § 3 of the paper he writes:

> Some interest may be aroused by a variation in the usual notion of entropy, based on the molecular structure of gas discussed here. If we applied a 'macroscopic' formula for entropy (...) (per unit of mass) to calculate the total entropy of a unit of mass from the entropy of individual parts (...), we would have to take into account that the relative number of such parts of the volume, where the density ρ is increased or decreased (...) denotes the normal number, i.e. per volume v with uniform distribution. (...) The 'microscopic' entropy of individual parts of the volume is therefore higher or lower than the normal value in an even density distribution; but the mean value is lower and at the same time depends significantly on the dimensions used to calculate the parts of the volume (consisting of a number of particles v)[861].

Let us compare Smoluchowski's text with an extract from Einstein's article in which he develops his conception as follows: Let z gram-molecules of a

---

[859] "Similarly, Maxwell's 'demon' would violate the second law of thermodynamics though in our case it would not pay attention to velocity but to the density of molecular swarms. Maybe instead of that—as one of my friends noted—an equally perfect one-way valve could be applied (...) it would not even have to be unusually small to create appreciable densities. In order to achieve this result, an additional contribution—admittedly not taken account of in more detail here—would be differences in velocity and temporal irregularities." M. Smoluchowski, *On irregularities in the distribution of gas molecules and their impact on entropy and the equation of state*, op. cit., p. 61.

[860] Robert Alicki – professor of physics, lecturer at the University of Gdańsk. The quoted statement was sent to the author of this book *via* e-mail on February 4, 2019.

[861] See M. Smoluchowski, *On irregularities in the distribution of gas molecules and their impact on entropy and the equation of state*, op. cit., pp. 58-59.

nonelectrolyte be dissolved in the partial volume V^* of a liquid of total volume V. If the volume V^* is separated from the pure solvent by a wall that is permeable to the solvent but not to the dissolved substance, then this wall is subjected to the so-called osmotic pressure. But if instead of the dissolved substance, the partial volume V^* of the liquid contains small suspended bodies that likewise cannot pass through the solvent-permeable wall, then according to the classical theory of thermodynamics we should not expect—at least if we neglect the force of gravity, which does not interest us here—that a force be exerted on the wall; because according to the customary conception, the "free energy" of the system does not seem to depend on the position of the wall and of the suspended bodies, but only on the total masses and properties of the suspended substance, the liquid, and the wall, as well as on the pressure and temperature. To be sure, the energy and entropy of the interfaces (capillary forces) should also be considered in the calculation of the free energy; but we can disregard them since the changes in the position of the wall and the suspended bodies considered here shall proceed without changes in the size and condition of the contact surfaces[862].

Recall that entropy is a thermodynamic state function, determining the direction of spontaneous processes in an isolated thermodynamic system, and is a measure of the degree of the system's disorder. According to the second law of thermodynamics, if a thermodynamic system passes from one state of equilibrium to another, without the participation of external factors, its entropy always increases. On the other hand, osmotic pressure—as described above—is the pressure that separates solutions of different concentrations, which is a direct result of the difference in concentrations of the chemical compounds in the two solutions.

Let us compare three issues described in different language, but referring to the same problems, in Smoluchowski's 1903 article and Einstein's May 1904 article. The first is the noticeable relationship between the magnitude of entropy and the magnitude of osmotic pressure. The orientation of the diffusion problem towards entropy or osmotic pressure depends on the research tools used, but in both cases the results indicate the degree of electrolyte disorder.

---

[862] See A. Einstein, *On the movement of small particles suspended in a stationary liquid required by the molecular kinetic theory of heat*, op. cit., pp. 549–550.

The second issue concerns the research methods proposed by the two scientists. Smoluchowski wants to study entropy by isolating individual volume parts from the total volume. Einstein proposes studying osmotic pressure by isolating the partial volume V^* from total volume V. The scholars' ideas of separating small volumes are similar; in both cases a partial volume is separated from the solution's volume and studied in relation to the overall volume.

The third problem is the concordance of the two physicists' thoughts on the subject of the forces at work in a solution. Smoluchowski writes that "the internal *cohesion forces cancel each other out everywhere and create only a constant 'internal pressure', calculated as if the mass was evenly distributed*[863]", and Einstein that "*a force be exerted on the wall; because according to the customary conception, the 'free energy' of the system does not seem to depend on the position of the wall and of the suspended bodies, but only on the total masses and properties of the suspended substance, the liquid, and the wall*"[864]. How does Smoluchowski's "*internal pressure*" differ in the above case from Einstein's "free energy"? Is the application of osmotic pressure not similar to the application of entropy? The similarity of the ideas and terms is difficult to underestimate. The important difference is that Smoluchowski's text appeared a year before that of Einstein, who *nota bene* could have read Smoluchowski's paper straight after its publication.

If Einstein had read Smoluchowski's article, of the two paths of proof suggested by the author, he was more convinced by the one that led to osmotic pressure, perhaps because he had already worked on it in 1903 with Besso. The idea of using osmotic pressure and indirectly van't Hoff's equation as well as other equations determining standard enthalpy and entropy in his proof was an assumption that enabled Einstein to formulate a conception of diffusion and Brownian motion. He used thoughts from the paper *A New Determination of Molecular Dimensions*, but the source of the idea lies in Smoluchowski's suggestion, as evidenced by Einstein's original intention to arrive at the average value of a particle's Brownian shifts in one second from the general theory of fluctuation around a state of equilibrium, using

---

[863] M. Smoluchowski, *On irregularities in the distribution of gas molecules and their impact on entropy and on the equation of state*, op. cit., p. 66.

[864] See A. Einstein, *On the movement of small particles suspended in a stationary liquid required by the molecular kinetic theory of heat*, op. cit., pp. 549–550.

the concept of osmotic pressure and Boltzmann's famous theorem. This is contained in Smoluchowski's proposals in the paper *On irregularities in the distribution of gas molecules and their influence on entropy and the equation of state*, in § 3, 4 and 8. The application of formulae for osmotic pressure was indispensable in the calculation of diffusion but not so helpful in calculating Brownian motion, while the use of Stokes' law was controversial for calculating the motions of particles with a diameter of 1 μ, as, according to many physicists, it does not work for calculating their displacement of such small particles.

In the May 1905 paper, Einstein wrote:

> But from the standpoint of the molecular-kinetic theory of heat we are led to a different conception. According to this theory, a dissolved molecule differs from a suspended body in size alone, and it is difficult to see why suspended bodies should not produce the same osmotic pressure as an equal number of dissolved molecules. We will have to assume that the suspended bodies perform an irregular, even though very slow, motion in the liquid due to the liquid's molecular motion; if prevented by the wall from leaving the volume V^*, they will exert forces upon the wall exactly as dissolved molecules do. Thus, if n suspended bodies are present in the volume V^*, i.e., $\eta/V^* = v$ in the unit volume, and if the separation between neighbouring bodies is sufficiently large, there will correspond to them an osmotic pressure p[865].

Smoluchowski states: "If we were to apply a 'macroscopic' formula of entropy to calculate the total entropy of a unit of mass from the entropy of individual parts of the volume, we would have to take into account that the relative number of such parts of the volume, where the density is the number of such parts v by volume in which the density ρ, is increased or decreased"[866].

Einstein saw the author's intention, transposing the contents of his argument from gases to liquids; the idea was not new to him as he had encountered it in van't Hoff's analogy between solutions and gases. He also made certain assumptions about Brownian motion, not being entirely convinced of its legitimacy. Smoluchowski underscores the similarity of the action in the two media: "Deliberations concerning Brownian motion were based on the analogy of particles suspended in a liquid medium with gas particles, an

---

[865] Idem, pp. 550–551.
[866] M. Smoluchowski, *On irregularities in the distribution of gas molecules and their impact on entropy and the equation of state*, op. cit., p. 60.

analogy which is expressed quantitatively in the fact that the kinetic energy of the translational motion must be the same in both cases"[867]. Einstein's intuition therefore naturally followed the suggestions of Smoluchowski, who, when writing an article about gases, was actually thinking about Brownian motion in fluids, which he had just finished working on and for which he already had results. Einstein interpreted these intentions, but not fully as he was focused on problems related to diffusion, which were more important to him. Hence in the May article he placed more emphasis on the idea of osmotic pressure. Smoluchowski's reflections on entropy caused Einstein's thoughts to move towards diffusion and osmotic pressure (which was closer to him), only noting the remarks about the essence of Brownian motion, of which he was not entirely convinced.

It can be speculated that Einstein intuitively felt that Brownian motion may be important evidence, but was not convinced and lacked scientific certainty as he did not have properly conducted research and was more interested in the problem of diffusion. So at the beginning of the May article, he writes about the structure of Brownian motion, then raises a number of question marks and queries his own idea, writing about his doubts as to the truth of the thesis. The December article conveys certainty about the thesis and the argumentation no longer contains his previous doubts.

In a way, the situation of the two scholars is paradoxical. Neither was entirely convinced of the correctness of their theses, except that Smoluchowski abstained from publishing despite having already worked through the problem, while Einstein published his paper, making it more intuitive as he had not conducted any targeted research.

In the introduction to his paper *On irregularities in the distribution of gas molecules and their influence on entropy and the equation of state*, Smoluchowski calls for research into the influence of unevenness in the local distribution of gas molecules on deviations from the mean values of velocities of individual molecular gas particles. He bases his argument on mental speculation, in which the distribution is studied of molecules in an ideal gas in a specific volume. The presumed volume of gas contains a normal number of molecules, evenly distributed. This state may be subject to deviations, resulting

---

[867] Idem, *On thermodynamic fluctuations and Brownian motion*, op. cit., p. 328.

in a specific average positive or negative percentage deviation from the normal density. Smoluchowski writes that assuming, after Mayer, that there are 6×10^{19} molecules in 1cm³ of gas, the average deviation for 1cm³ will be $1/2 \times 10^{-10}$ of the normal density, which was impossible to capture experimentally given the capabilities of the time. However, even for microscopically small sizes it would be imperceptible under the microscope. It would be possible if "swarming systems"[868] were created from individual molecules (because we are dealing with gas, in the case of liquids Smoluchowski talks about fluctuations). The microscopic entropy of individual parts of the volume is sometimes greater and sometimes smaller than the normal value in even density distribution[869].

Einstein states at the very beginning of his paper On the Movement of Small Particles Suspended in Stationary Liquids Required by the Molecular-Kinetic Theory of Heat: It will be shown that, according to the molecular-kinetic theory of heat, bodies of microscopically visible size suspended in liquids must, as a result of thermal molecular motions, perform motions of such magnitude that these motions can easily be detected by a microscope. It is possible that the motions to be discussed here are identical with the so-called 'Brownian molecular motion'[870].

Smoluchowski's correct interpretation of the mechanism causing the movement of molecules, and Einstein's omission of this problem by not describing the mechanism causing Brownian motion, suggest a difference between the two scientists in terms of research and experiments conducted in this field. Smoluchowski is aware that single particles of water or other solvent are not able to induce the movements of particles, due to too great a difference between their size and the dimensions of the particles, and he is convinced that individual molecules must create swarms of particles or fluctuations that could effectively induce the particles' movements. Einstein does not mention this important condition of Brownian motion.

Let us analyse another issue; Smoluchowski writes in § 4:

---

[868] Smoluchowski calls these systems fluctuations.
[869] See M. Smoluchowski, *On irregularities in the distribution of gas molecules and their impact on entropy and the equation of state*, op. cit., p. 59.
[870] See A. Einstein, *On the movement of small particles suspended in a stationary liquid required by the molecular kinetic theory of heat*, op. cit., p. 549.

Let us compare this with the kinetic definition of entropy given by Boltzmann. By this definition, entropy is a negative value denoted by the function $H = \int f \log f \, dudvdw$, where f is the number of molecules having velocity u, v, w. If, following Boltzmann, we formulate this expression for a larger volume, without taking into account the changes that the density undergoes within its parts, (…) then we get the "macroscopic" entropy; however, if we take into account the local density variability in f, we must assume: $H' = \sum \Delta x \Delta y \Delta z \int \log f \, dudvdw$, where N should be replaced by $v(1+\delta)$ whereby the sum extends to all parts of the area and integrate, which leads to the same result[871].

Further, in another place (in § 4 and 5) Smoluchowski writes:

The H function was presented by Boltzmann as a logarithm for the probability of the correct velocity system. However, the H' function could be expressed with slight modification as the logarithm of the probability of simultaneous distribution of velocity and density. It is equal to the sum of the logarithms of these two probability ratios, and in relation to density one can reason in a manner completely analogous to that concerning velocity (…) our considerations have only theoretical significance, as a small contribution to Boltzmann's interpretation of entropy as a concept in the field of probability; differences in density also exert in other respects (…) a visible influence, namely in relation to the equation of state[872].

Let us note that in the case of research on the state of entropy, we are dealing with stochastic processes, while in the case of research on osmotic pressure—with statistical processes. Approaching the problem of diffusion through researching osmotic pressure is to narrow the description of a specific situation of the phenomenon compared to a description of diffusion by means of entropy, which is fuller and more universal. An unquestionable speculation that arises involuntarily is the thought that after reading Smoluchowski's article, Einstein made a kind of transcription of the concept described based on entropy into a description based on osmotic pressure, which was natural in the light of the research he had previously conducted with Besso, and in connection with the article *A New Determination of Molecular Dimensions*. Nevertheless, the question constantly returns as to

---

[871] M. Smoluchowski, *On irregularities in the distribution of gas moleculesand their impact on entropy and the equation of state*, op. cit., pp. 59–60.

[872] Idem, pp. 60–61.

why Einstein never addressed the doubts Smoluchowski raised regarding the use of osmotic pressure.

The above considerations and comparisons of texts are speculative, but they reveal quite significant ambiguities, especially concerning who was responsible for this key discovery for early 20th-century physics. In summary, it can be seen that from Smoluchowski's various comments there emerges a picture of Einstein's conceptions and calculations not being very clear or entirely consistent, which the author obviously does not write about explicitly. Einstein approached the problem chiefly from a mathematical perspective and did not fully explain the essence of the discovery. The Polish physicist pointed out that "Einstein's deliberations are of a more abstract nature as they stem from his general research in the field of statistical mechanics"[873], adding "but for now he has left the question open as to whether the theoretically foreseen phenomenon corresponds to what was known as Brownian motion"[874]—this is without doubt a telling remark.

In Smoluchowski's work from 1900 onwards, i.e. from the time the Polish physicist became acquainted with the results of Exner's research, we find digressions and reflections referring to Brownian motion. In his paper *On thermodynamic fluctuations and Brownian motion*, Smoluchowski writes explicitly: "The author of this paper has been convinced of the molecular-kinetic essence of Brownian motion since 1900, when he learnt of its existence from Exner's work and from that time has gradually sought to resolve that theory, first by considering the probable arrangement of particles, then the phenomena of movement caused by a combination of free paths, and finally the mechanism itself of the movements of particles caused by the collisions of surrounding molecules"[875]. Meanwhile it is difficult to find similar comments related to the work on this problem in Einstein's papers. Stachel notes that Einstein probably became aware of the controversy around the molecular theory of heat as early as in his student days, when reading the work of Mach, Ostwald, and Boltzmann. When he finished reading Boltzmann's *Gastheorie (Gas Theory)* in 1900, in which Boltzmann, presumably reacting to a dispute with Ostwald and Helm, suggested he was isolated in his support

---

[873] Idem, *On thermodynamic fluctuations and Brownian motion*, op. cit., p. 299.
[874] Ibid.
[875] Ibid.

of the kinetic theory; Einstein was firmly convinced of his theory's validity[876]. Perhaps earlier thoughts, Sutherland's suggestions, and inspiration from Smoluchowski's article, resulted in the May paper introducing the theory of diffusion and suggestions concerning Brownian motion.

In physics, for a hundred years, a narrative has been built up in which all the glory for discovering the essence of Brownian motion falls on Einstein and this happens despite the fact that during the period in which both these papers were published it was claimed that the contribution of both scientists to the discovery was at least equal, and even that Smoluchowski's study was more thorough, better justified, and approached the problem in a more insightful way. It was assumed that the discovery would bear the names of both scientists. However, from the very outset it was Einstein who became the chief beneficiary. There are many examples confirming this fact; for Perrin the discoverer was Einstein, and his publications *Über die von der molekularkinetischen Theorie der Wärme geforderte Bewegung von in ruhenden Flüssigkeiten suspendierten*[877] of 11 May 1905, *and Zur Theorie der Brownschen Bewegung*[878] of 19 May 1906 were crucial, as Perrin writes in *Atoms*. On May 19, 1927, during a speech at the ceremony to award the Nobel Prize to Svedberg, Henrik Gustaf Söderbaum said: "As we have recently heard, Einstein evolved a theory for this so-called Brownian movement which was then developed to a high degree by the now late Smoluchowski"[879]. In a frequently revised study of the history of physics—*Kulturgeschichte der Physik (A Cultural History of Physics)*, by Károly Simonyi, Smoluchowski's name is completely omitted. In *The Collected Papers of Albert Einstein* (1989), Stachel gives Einstein credit for the discovery. He states: "Some of the consequences of his work were of great significance for the development of physics in the twentieth century. Einstein's derivation of the laws governing Brownian motion, and their subsequent experimental verification by Perrin and others, contributed significantly to the acknowledgment of the physical reality of atoms by the then still

---

[876] See J. Stachel, *Einstein on Brownian Motion*, op. cit., p. 207.
[877] A. Einstein, *On the movement of small particles suspended in a stationary liquid required by the molecular kinetic theory of heat*, op. cit., pp. 549–560.
[878] Idem, *Theory of Brownian motion*, op. cit., s. 371–381.
[879] Quote from a speech given by H.G. Söderbaum, Secretary of the Royal Swedish Academy of Sciences, at the ceremony to award the Nobel Prize to Theodor Svedberg on May 19, 1927.

numerous skeptics"[880]. In *A Brief History of Time*, Hawking discusses the proof of the atomic structure of matter, but only mentions Einstein's publications and achievements[881]. In his essay *Marian Smoluchowski and his work (1872–1917)* Stanisław Loria recalls a book by Percy Williams Bridgman (1882–1961), *The Nature of Thermodynamics*, in which—as he writes—there is no mention of Smoluchowski's research[882]. He also cites a paper, *Stochastic Problems in Physics and Astronomy*, by Indian astrophysicist Subrahmanyan Chandrasekhar[883], who writes: "It is somewhat disappointing that the more recent discussions of the laws of thermodynamics [e.g. P. Bridgman etc.] contain no relevant references to the investigations of Boltzmann and Smoluchowski. The absence of references, particularly to Smoluchowski, is to be deplored since no one has contributed so much as Smoluchowski to a real clarification of the fundamental issues involved"[884].

Smoluchowski's name is rarely to be found in the studies of contemporary researchers. An exception to this is Roberto Maiocchi, who mentions Smoluchowski quite often, but at the same time gives credit for the discovery solely to Einstein. Nowadays, if Smoluchowski is mentioned at all it is only marginally, or—as with Charlotte Bigg—he is overlooked. In two articles by this writer[885], interesting accounts are given of Jean Perrin's reflections on experimental research on Brownian motion, the part in his success played by his assistant Dąbrowski, and the effects of the French physical chemist's research on the development of physics. Although her analyses of the French researcher are insightful, they lack any record of Marian Smoluchowski's contribution to this important discovery. Bigg even states, "To this day, Einstein and Perrin's Brownian motion work is taught to physics students and represented in the way Perrin suggested. In physics, Perrin's image now has only historical relevance as a particular moment in the history of the

---

[880] See J. Stachel, *Einstein on Brownian Motion*, op. cit., p. 206.
[881] S. Hawking, *A Brief History of Time*, op. cit., p. 67.
[882] See S. Loria, *Marian Smoluchowski and his work (1872–1917)*, op. cit., p. 7.
[883] S. Chandrasekhar, *Stochastic Problems in Physics and Astronomy*, 'Reviews of Modern Physics' 1943, vol. 15, No. 1, pp. 1–89.
[884] Idem, pp. 6–7.
[885] C. Bigg, *Evident Atoms: Visuality in Jean Perrin's Brownian Motion Research*, 'Studies in History and Philosophy of Science' 2008, vol. 39, No. 3, pp. 312–322 and *A Visual History of Jean Perrin's Brownian Motion Curves*, in: *Histories of Scientific Observation*, eds. L. Daston, E. Lunbeck, Chicago 2011, pp. 156–179.

field, a significant achievement"[886]. The name of the Einstein-Smoluchowski theory is changed in her work to the Einstein-Perrin theory.

Smoluchowski does not feature among the architects of the theory of Brownian motion in a Milton Kerker article on Svedberg which states that Brownian motion was generally considered a key proof of atomism. Kerker was interested in Svedberg's role in justifying Albert Einstein's 1905 theoretical analysis. He believed it paved the way for a key experiment aimed at confirming the existence of atoms and molecules, and Svedberg was the first to claim that he had performed this experiment[887].

Smoluchowski is mentioned in relation to the work on which he later collaborated with Svedberg for seven years. Kerker writes that the joint work led to the creation of a monograph summarising the research, and another publication included an overview of that part of the experimental work that contributed to confirming the molecular structure of matter. Svedberg wanted to confirm his achievements in this field, especially after Perrin's criticism of his work on Brownian motion and in light of the general approval enjoyed by Perrin's work. Kerker believes that Svedberg's research used experimental tests of Marian Smoluchowski's theory of thermal fluctuations[888]. This article makes no mention of shared credit for the discovery.

In the article *A History of Random Processes*[889], American physicist Stephen G. Brush recalls the words of Max Born, according to whom Einstein's theory was not only a breakthrough in the understanding of Brownian motion, but also a turning point in physicists' awareness of the reality of atoms and molecules, comprehension of the kinetic theory of heat, and realisation of the fundamental role of probability in the laws of nature. Unfortunately, the cited text by Born provides another example of Smoluchowski being overlooked. Brush notes that soon after the publication in 1905 of Einstein's work on Brownian motion, Smoluchowski published a paper in which he states that Einstein's results completely agree with his earlier ones, which he arrived at in a different way. He finds his method simpler and more convincing. Of

---

[886] Idem, *A Visual History of Jean Perrin's Brownian Motion Curves*, op. cit., p. 173.
[887] See M. Kerker, *The Svedberg and Molecular Reality*, op. cit., p. 192.
[888] Idem, p. 193.
[889] See S.G. Brush, *History of Random Processes. I. Brownian Movement from Brown to Perrin*, 'Archive for History of Exact Sciences' 1968, vol. 5, No. 1, pp. 14–29.

course, this is a relative issue, but most physicists and chemists of the time agreed with the statement. Smoluchowski's arguments were based on combinatorics and calculations of the approximate mean free path in terms of kinetic theory, while Einstein referred to abstract statistical mechanics and the dispersion equation.

According to Brush, Smoluchowski developed the results of his experiments better theoretically and put much more effort into experimental confirmation of theoretical assumptions. Einstein relied on Planck's research (determination of molecular size from the law of radiation). He argued that the resistance of scientists to the existence of atoms (for example, Ostwald and Mach) resulted from the positivist philosophy they practised. He noted that philosophical prejudices consist in considering theory superior to facts and empiricism[890].

Over time, a legend emerged that reinforced this unfavourable trend for Smoluchowski, according to which between March 18 and December 19, 1905, Albert Einstein wrote five papers that changed the face of physics. They concerned the theory of relativity, the photoelectric effect, and the theory of Brownian motion. The physicist John Stachel, as already mentioned, noted that these five works published by Einstein have caused 1905 to be considered in the general discourse of physicists a "wonder year" analogous to the Newtonian year of 1666, dubbed the *annus mirabilis*. It is widely held that in 1666 Newton wrote a series of papers that laid the foundations of physical and mathematical theories that revolutionised 17th-century science. More detailed research has proven that these works were written between 1665 and 1667 and in 1668. Such minor inaccuracies are of no great moment in the creation of legend. In 1905, Einstein, like Newton, published papers that established the groundwork of a revolution in 20th-century science[891]. Roger Penrose claims the publication of Einstein's first five paper started the fourth revolution in science, manifesting in the way nature is perceived. It is the way with legends that they are not expected to be too closely aligned with the truth; they have a life of their own. In the case of Newton, not everything happened in 1666 (the date itself was certainly not chosen by chance), and in

---

[890] Ibid.
[891] J. Stachel, *Introduction*, in: A. Einstein, *5 works that changed the face of physics*, op. cit., p. 15.

Einstein's case too, a question mark could be placed by the articles analysing Brownian motion.

Einstein's later achievements in physics made him a legend of science, influencing everyone who comes into contact with his legacy. This is an additional obstacle in attempting any discussion, as legends are hard to argue with. Unfortunately, Albert Einstein rarely included citations of other authors in his work, even though he used their results. Perhaps the reason for such behaviour was that at that time he worked in a patent office, outside the academic environment, though he was a brilliant and very ambitious man, as Marian Niemiec writes[892].

Einstein's paper *On the Movement of Small Particles Suspended in Stationary Liquids Required by the Molecular-Kinetic Theory of Heat*, was submitted for publication on May 11[893], 1905, and on June 30, *On the Electrodynamics of Moving Bodies* was published in which he formulated the special theory of relativity and initiated the era of relativistic physics.[894] Einstein changed our perception of reality, our understanding of the fundamental paradigms of time and space, and simultaneously during this same period developed the basic proof for the atomic structure of matter (a subject from a different branch of physics), although he never specifically dealt with it. What enabled him to develop simultaneously and in parallel a topic he had not previously considered?

The very title of Smoluchowski's work—*On irregularities in the distribution of gas molecules and their influence on entropy and the equation of state*—must have intrigued a physicist of Einstein's intellect and knowledge. Even upon casual contact it encouraged closer inspection and Einstein probably read it, the consequence of which was the creation of the paper *On the Movement of Small Particles Suspended in Stationary Liquids Required by the Molecular-Kinetic Theory of Heat*, the contents of which suggest familiarity with the

---

[892] See M. Niemiec, *Marian Smoluchowski – człowiek wszechstronny (Marian Smoluchowski – a multifaceted man)*, 'Pismo Uniwersytetu Opolskiego' ('Magazine of the University of Opole') 2006, No. 3–4 (69–70).

[893] A. Einstein, *On the movement of small particles suspended in a stationary liquid required by the molecular kinetic theory of heat*, op. cit., pp. 549–560.

[894] Idem, *On the Electrodynamics of Moving Bodies*, op. cit., pp. 891–921.

concepts included in Smoluchowski's work. The fact that both articles describe phenomena occurring in different media is of no importance, as Smoluchowski wrote about the analogy of Brownian motion in gas and liquid media:

> Approximate direct calculation of shifts in a liquid medium. The essential concept on which the theory of Brownian motion rests is the assumption that suspension particles (at a certain temperature) behave to a certain degree analogously to [gas – J.G.] molecules, namely that the mean kinetic energy of translational movement made through the medium of the particle's centre of mass equals the mean energy of translational movement of a gas molecule at that temperature. (…) This assumption corresponds to the basic assumption of the kinetic theory of gases and is a direct consequence of Maxwell's principle of energy equipartition. An analogy between gas molecules and emulsion particles has also been established empirically in the other phenomena discussed[895].

After reading Smoluchowski's article, Einstein arrived at several revealing conclusions that were not clearly presented in the Polish author's paper, but contained inspirations for a solution to a problem that had been widely debated in physics in recent years. This testifies to the greatness of his mind, which transposed suggestions about the states and irregularities occurring in gases and remarks about entropy and internal pressure to the processes occurring in gases and liquids and, at the same time, based on minor suggestions from the Polish physicist, expanded his own conception of diffusion by adding an aspect related to Brownian motion. Unfortunately, he did not mention what had led him to this idea or who had inspired these discoveries.

An analysis of Smoluchowski's and Einstein's articles at least substantiates the hypothesis presented earlier that the key proof for the theory of Brownian motion, based on osmotic pressure and described in the May article, had its origins in Smoluchowski's publication.

Einstein, as Stachel writes, published a total of four main papers between 1905 and 1908 on the subject of Brownian motion in liquids. In 1905, *Über die von der molekularkinetischen Theorie der Wärme geforderte Bewegung von in ruhenden Flüssigkeiten suspendierten Teilchen* (*On the Movement of Small Particles Suspended in Stationary Liquids Required by the Molecular-Kinetic*

---

[895] M. Smoluchowski, *On thermodynamic fluctuations and Brownian motion*, op. cit., p. 300.

Theory of Heat) was published, in 1906 and 1907 *Zur Theorie der Brownschen Bewegung* (On the Theory of Brownian Motion) and *Theoretische Bemerkungen über die Brownsche Bewegung* (Theoretical Remarks on Brownian Motion), and in 1908 *Elementare Theorie der Brownschen Bewegung* (Elementary Theory of Brownian Motion). He also prepared a synopsis of a lecture to the Bern Society of Natural Sciences.

During this period, three articles also appeared on related subjects. The first, originally published as a dissertation in 1905, concerned the determination of molecular sizes. The other two (from 1907 and 1908) contain proof of Brownian motion and present its measurement. It is also worth mentioning Einstein's remark on the measurement of Brownian motion following a speech he gave at a meeting in Salzburg in 1909 organised by the Society of German Natural Scientists and Physicians, and other statements made in the course of the discussion[896]. Einstein makes no references in these articles to any records or thoughts from before 1905. In *The Collected Papers of Albert Einstein*, there are no significant records in the years 1902–1905 regarding the problem of Brownian motion. The only paper close to the subject is *A New Determination of Molecular Dimensions*, but in it Einstein does not address the issue of Brownian motion.

Some scientific publications state that in 1905–1906, two theorists—Albert Einstein and Marian Smoluchowski—independently of one another and following different paths, explained the essence of Brownian motion and proved the validity of atomic theory. This appears in various publications from the early 20th century but in the light of the considerations presented here, essential doubts arise as to the correctness of stating the matter in such a way. The thesis that Einstein invented his way of proving the essence of Brownian motion in parallel to and independently of Smoluchowski is problematic. An analysis of the Pole's paper *On irregularities in the distribution of gas molecules and their influence on entropy and the equation of state* and Einstein's *On the Movement of Small Particles Suspended in Stationary Liquids Required by the Molecular-Kinetic Theory of Heat*, supports the presumption that the Polish physicist's contribution to Einstein's work is highly probable.

---

[896] See J. Stachel, *Einstein on Brownian Motion*, op. cit., p. 206.

7.5. Perrin's experimental evidence

Jean Perrin, professor of physics at the Sorbonne, wrote in his 1909 essay *Mouvement brownien et réalité moléculaire* (*Brownian Motion and Molecular Reality*):

> The singular phenomenon discovered by Brown did not attract much attention. It remained, moreover, for a long time ignored by the majority of physicists, and it may be supposed that those who had heard of it thought it analogous to the movement of the dust particles, which can be seen dancing in a ray of sunlight, under the influence of feeble currents of air which set up small differences of pressure or temperature. When we reflect that this apparent explanation was able to satisfy even thoughtful minds, we ought the more to admire the acuteness of those physicists, who have recognised in this, supposed insignificant, phenomenon a fundamental property of matter[897].

It is indisputable that the French physicist belonged to a group of scientists who showed insight. At the 1926 awards ceremony at which Perrin received the Nobel Prize in Physics, Carl Wilhelm Oseen stated that he had "put a definite end to the long struggle regarding the real existence of molecules"[898]. Charlotte Bigg described Perrin's merits in explaining Brownian motion as follows:

> A sheet of squared paper on which three broken lines have been drawn. A connect-the-dots game gone slightly awry, with no pattern obviously recognisable. No scale is inscribed that might provide clues about the size and nature of the object or phenomenon represented here. No indications on the procedure involved in the production of this two-dimensional abstraction. No numbers, letters, or symbols to tell the viewer how to hold the figure, or in what direction the lines run. (...) Yet show this image to a physicist or a mathematician and the response will be immediate: this is Brownian motion. This image, published for the first time in September 1909 by French physical chemist Jean Perrin has acquired iconic status in the physical sciences. It was and is still perceived as an experimental confirmation and a visual equivalent of Albert Einstein's theoretical demonstration, in a paper

---

[897] J. Perrin, *Mouvement brownien et réalité moléculaire* (*Brownian movement and molecular reality*), 'Annales de chimie et de physique' ('Annals of Chemistry and Physics') 1909, vol. 18, pp. 1–114.

[898] See S. Psillos, *Moving Molecules Above the Scientific Horizon*, op. cit., p. 340.

of 1905 (...) Einstein even doubted that the methods he suggested for measuring Brownian motion could be realised experimentally: "I would have thought such a precise study of Brownian motion impossible to realise," he wrote in admiration to Perrin in November 1909[899].

The theoretical formulae devised by Smoluchowski and Einstein were confirmed by the careful measurements of two physicists—Svedberg and Perrin. In this way, one of the most accurate methods known to date of determining the number of molecules was created. As Smoluchowski notes, Perrin's measurements in particular were made very accurately[900]. In general, it can be said that the thermodynamic view is justified and creates an excellent research tool in terms of general (macroscopic) features of phenomena and average events. However, where there are discrepancies with the kinetic theory, in the microscopic details of phenomena, in random deviations from the normal course of events, the superiority of kinetics has been established beyond doubt[901].

On March 9, 1908, Paul Langevin presented to the Academy of Sciences the article *On the Theory of Brownian motion* (*Sur la théorie du mouvement brownien*), in which he gave Einstein's formula in a form very similar to that which appears in Perrin's notebook. Langevin also discussed Marian Smoluchowski's publications and assessed critically the first attempt at experimental verification of Einstein's methods by the Swedish physicist Svedberg. Perrin claimed in 1911 that it was Langevin who first drew his attention to Einstein's research. Einstein had proposed new quantitative methods of measuring Brownian motion and of determining the dimensions of the particles, thus offering novel tools for testing the validity of the kinetic theory. He argued in particular that it was meaningless to measure the instantaneous velocity of individual particles, as previous researchers had done[902].

After Einstein's 1905 publication of a theoretical explanation of Brownian motion, Perrin, together with his assistant Dąbrowski, initiated experiments to verify the announced discovery. The idea of the research was based

---

[899] C. Bigg, *A Visual History of Jean Perrin's Brownian Motion Curves*, op. cit., p. 156.
[900] See M. Smoluchowski, *On thermodynamic fluctuations and Brownian motion*, op. cit., p. 299–300.
[901] See idem, *Themes and issues of today's physics*, op. cit., p. 215.
[902] C. Bigg, *A Visual History of Jean Perrin's Brownian Motion Curves*, op. cit., p. 163.

on using of the concept of the Avogadro constant, which made it possible to calculate the number of particles in a given volume.

Perrin's reasoning was originally based on generalities about Brownian motion. There was a prevailing conviction that molecular collisions would provide explanations enabling principles to be found to explain the nature of these movements. Such characteristic beliefs existed before Perrin, so we do not know why it was Avogadro's number that encouraged Perrin to conduct experimental research. We know, however, that this number tipped the balance of evidence in favour of the atomistic conception of matter. Using the Avogadro constant, Perrin conducted proof to settle a century-old dispute over John Dalton's atomic theory, which argued that chemical compounds always contain the same proportion of elements by mass[903].

Reviews and research inspired by Einstein's work appeared in the following months and years in journals in the fields of physiology, chemistry and physics throughout Europe. However, outside of Germany, the French were probably the quickest to take up ultramicroscopy. Aimé Cotton (1869–1951), a lecturer in physics at the École Normale Supérieure in Paris, and Henri Mouton (1869–1935), a former student of that university working at the Pasteur Institute, not only reviewed the works of Siedentopf and Zsigmondy after their publication, but earlier, as early as June 1903, proposed a modified ultramicroscopic arrangement[904].

In his article *Moving Molecules Above the Scientific Horizon: On Perrin's Case for Realism*, Stathis Psillos considers the issue of the shift that occurred in scientific communities in the years 1908–1912 due to the need to accept the fact that atoms are not merely symbolic concepts, as most researchers had thus far believed, but real and physically extant entities with mass and dimensions. Psillos also focuses on analysis of the theoretical and experimental work carried out by Perrin which allowed accurate determination of Avogadro's number, thus confirming the validity of molecular-kinetic theory. Conscious of the fact that for many of his contemporaries the atomic hypothesis was only a metaphysical concept, Perrin presented the rudiments and successes of the kinetic theory of gases, emphasising that thegasstate

---

[903] See S. Psillos, *Moving Molecules Above the Scientific Horizon*, op. cit., p. 357.
[904] See Ch. Bigg, *Evident Atoms: Visuality in Jean Perrin's Brownian Motion Research*, op. cit., p. 317.

equation, established by Johannes Diderik van der Waals (1837–1923) in 1873, relating gas state parameters: pressure, volume and temperature, confirms atomic theory. Importantly, he argued that Brownian motion, being an example of molecular motion, must obey the laws of that motion. He used this assumption and derived the data to calculate Avogadro's number and to determine the fundamental properties of atoms, such as their mass and dimensions[905].

Perrin stresses that the equation of distribution of emulsion was arrived at independently—and by different means—by Einstein and Smoluchowski, but—as he points out—they did not realise that this equation could be used to perform an experiment confirming the molecular theory of Brownian motion[906].

According to Stathis Psillos, the problem with reconstructing Perrin's experiment is that, firstly, it does not explain Perrin's contribution to the confirmation of the atomic hypothesis, since most of the methods of calculating Avogadro's number were already known before Perrin published his research; secondly, it is not clear in what sense the molecules constitute a direct cause of agreement in the various experimental calculations of Avogadro's number.

Like the other technologies Perrin employed, the ultramicroscope was made to provide evidence in support of the particulate view of matter, thereby showing matter to be amenable to study using the laws of mechanics (kinetic theory), and in turn validating the physical-chemical approach characteristic of his work and that of his closest friends and allies, Pierre and Marie Curie, Paul Langevin, and beyond, the Cavendish Laboratory scientists. In order to obtain a meaningful measurement, Perrin used the formulae and methods developed by Albert Einstein for the determination of the average molecular energy of individual particles. Einstein proposed making the displacement of a particle during a given interval of time the primary observable quantity, rather than instantaneous velocity. He believed the values obtained by Victor Henri (1872–1940), who tried to use ultramicroscopy and cinematography to measure displacements, were untrustworthy due to Henri's erroneous

---

[905] See S. Psillos, *Moving Molecules Above the Scientific Horizon*, op. cit., p. 340.
[906] Idem, p. 353.

estimation of grain diameters[907]. Nancy Cartwright argues that Perrin's reasoning is an example of inference to the most likely cause. However, it does not follow from this that the related phenomena (allowing determination of Avogadro's number) have a cause, and if so, what the cause is, in particular—whether it is the incessant movement of molecules[908].

Psillos cites the explanation of Richard Miller, who argues that Perrin's reasoning was based on truisms specific to Brownian motion, for example that Brownian motion requires a causal explanation and that particle collisions provide that explanation, however, these truisms were known before Perrin. Hence we do not have an explanation of why it was Perrin's determination of Avogadro's number that tilted the balance in favour of the atomic conception of matter.

Such a course of action may suggest that Perrin wanted to show that it is possible to eliminate the molecular hypothesis; that it is scaffolding that could be removed after the actual relationship between empirical phenomena had been established. Note that the 'evident realities' that enter into the functional relations thus established are not merely observable quantities; for instance, the diameter of the Brownian particles or the wavelength of emitted light are not observable but are determinate and measurable, and this is what Perrin insisted on. More importantly, however, Perrin did not take it that the possibility of eliminating the constant N implied that molecules could be dispensed with. Psillos's point is not that philosophical dogma is unreasonable, but rather that adhering to philosophical dogma in the face of mounting scientific evidence against its basic presuppositions is unreasonable and contrary to what he calls "epistemological openness[909]". This is a significant conclusion in the context of Smoluchowski's various statements regarding the truths of science. The Polish scholar's characteristic 'epistemological openness' and non-attachment to the dogmas of science should once again be emphasised.

Meanwhile, Perrin's experiment does not show that the atomic concept is true, but that the basic component of his theory, namely the discontinuous

---

[907] See C. Bigg, *Evident Atoms: Visuality in Jean Perrin's Brownian Motion Research*, op. cit., p. 319.
[908] Idem, p. 358.
[909] Ibid.

structure of matter, is true. It also shows how certain properties of the constituents of matter have been rendered definite and measured. Avogadro's number gave credibility to the atomic concept of matter, becoming an irrefutable argument for the truth of the atomic concept. It remained an open matter to develop proof that molecules are made up of atoms and that the latter have an internal structure. It was known that various assumptions that had helped in modelling a range of phenomena on the basis of the atomic concept (for example, that molecules are elastic spheres) were incorrect and required refinement or replacement. Commitment to the truth of a theory is not about uncompromisingly accepting or rejecting it. Progress and convergence do not require theories to contain the whole truth and nothing but the truth. The atomic conception of matter, in its essentials, has become a stable and permanent part of our evolving scientific image of the world.

It can be assumed, Psillos concludes, that these are strategies that remove doubts, explain the epistemological change of attitude towards hypotheses, and explain and support acceptance of the reality of entities that were initially taken to be ontically suspicious[910].

Advocates of atomism welcomed Perrin's experiments as the first visible evidence of the existence of atoms, or at least of their motion. Walther Nernst wrote in the 1909 edition of his textbook of physical chemistry that in view of the ocular confirmation of the picture the kinetic theory provided of the world of molecules, it must be admitted that this theory starts to lose its hypothetical character. A revealing description of Perrin's work by the Swedish mathematician Magnus Gösty Mittag-Leffler (1846–1927) goes in the same direction. In a letter to Poincaré in 1909, he wrote that it consists of a method of isolating and seeing atoms. Rutherford put it more accurately in the mathematical and physical section of the British Association for the Advancement of Science (BAAS) in 1909. He said the nature of Brownian motion impressed the observer with the idea that the particles are "hurled hither and tither" by the action of forces resident in the solution and that these can only arise from the continuous and incessant movement of the invisible molecules that make up the fluid. William Crookes (1832–1919),

---

[910] Idem, p. 362.

inventor of the spinthariscope[911], a visualisation device, wrote to Perrin in October 1910 inviting him to give a lecture at the Royal Institution as one of the "scientific men of great eminence"[912].

From the above discussions, quotations and considerations, a picture emerges of the scientific reality of physics in the early 20th century. Two issues stand out, the first being the intensive progress of the work, the characteristic dynamic, and the participation in the research of a wide range of scientists trying to prove or disprove the theory Brownian motion. The second is the fact that the strong scientific centre for physics that the University of Vienna was in the 1890s was weakened. Paris, with the Sorbonne at the helm, gained momentum; it was to there that the main weight of discussion in physics and chemistry shifted, with the inherent participation, as in the 19th century, of German universities.

In 1899, Smoluchowski started working at the University of Lviv, which with the shift of discussion to Paris distanced him from the main current of lively scientific discussion. These changes meant that he involuntarily participated less in the ongoing discussions of scientists working on the theory of Brownian motion.

The discovery by Smoluchowski and Einstein marked the beginning of a discourse that culminated in three Nobel Prizes being awarded—to Zsgimondy, Perrin and Svedberg. From 1905, discussion on the subject centred mostly on French and German scientists, with the participation of Swedes, primarily Svedberg, and scholars from the Cavendish Laboratory. Unlike Einstein, Smoluchowski's voice went largely unheard in the discussion many academics were engaged in at the time. Conversely, Einstein constantly took part in the discussion, and in solving the problems that arose in the experimental proof of the atomic composition of matter. Inevitably, over the course of the first five to seven years after the two scientists' publications, Einstein's position as the main, and sometimes only, person responsible for the discovery became firmly established in the minds of scientists. Smoluchowski took part more seldom in these discussions, while Einstein assisted, commented, and inspired the experimental research of scholars investigating

---

[911] An instrument consisting of a fluorescent screen and lenses for the observation and counting of alpha rays. The simplest form of scintillation counter.

[912] C. Bigg, *Evident Atoms: Visuality in Jean Perrin's Brownian Motion Research*, op. cit., p. 320.

Brownian motion (such as in a lecture given on November 2, 1910, at the Physical Society in Zurich, entitled *Über das Boltzmann'sche Prinzip und einige unmittelbar aus demselben fliessende Folgerungen (On Boltzmann's Principle and Some Immediate Consequences Thereof)*[913]. In it, Einstein detailed his point of view on statistical physics and emphasised the role of fluctuation in the context of Boltzmann's entropy[914]. Duplantier notes that this is the most direct derivation of the Brownian diffusion formula.

Consequently, Smoluchowski naturally ceased to be thought of as one of the creators of the theory of Brownian motion. He worked far away, at a small provincial university in Lviv, and from 1912 in Kraków, and his participation in the discovery faded in the memory of those who had known him before, while it did not enter the awareness of scientists who were just embarking on this field of research. Those who do not take part in the discussion cannot be right, and from a certain point Smoluchowski ceased to participate in the ongoing academic discourse. The fact that someone read out one of the Pole's articles on his behalf, as Langevin did, changed nothing. It was Einstein's ongoing scientific activity that cemented the conviction of who should be considered responsible for the discovery.

Smoluchowski's initial lively commentary on his participation in the discovery, now found in various publications, caused no particular stir among physicists at the time. The main interest in the matter of Brownian notion shifted to problems related to proving the theory's truthfulness. These facts—the great distance of Lviv from Paris, and new scientific concepts—probably caused Smoluchowski to become somewhat discouraged over time from actively participating in the discourse, which further alienated him from the mainstream discussion surrounding Brownian motion.

Smoluchowski was a scientist to the bone; he was interested in the substance of science, not in extracurricular discussions. He attached more importance to conducting scientific research and creating new value than to seeking splendour or proving his superiority. He could not take part in the ongoing discussion on Brownian motion, and new ideas emerged that demanded

---

[913] A. Einstein, *Über das Boltzmann'sche Prinzip und einige unmittelbar aus demselben fliessende Folgerungen (On Boltzmann's Principle and Some Immediate Consequences Thereof)*, Lectures for the Physical Society, Zurich 1910.

[914] B. Duplantier, *Brownian Motion*, op. cit., p. 238.

his attention. This does not mean that he completely cut himself off from the discussion in progress; let us not forget that between 1907 and 1914 he established quite lively contacts with Svedberg, the main topic of which was work on the theory of Brownian motion.

During this period, Smoluchowski also carried out extensive and valuable educational work, to which he always attached great importance. This took the form of popular lectures for teachers, doctors and naturalists about the emerging atomic sciences, the content of which we know from three volumes of published *Writings*. On the research side, he was involved in new problems in physics, but that was not all, as he also wrote essays on quite unusual topics, such as *Several notes on the physical foundations of tectonic theories*[915], and also gave a lecture on the issue of womenin the exact sciences[916]. However, he essentially focused on two issues in which he achieved good results, and thanks to which he went down forever in the history of physics. These were problems related to the processes of applying probability theory to physics, and extremely important achievements in the field of coagulation research.

It was no accident that Smoluchowski did not participate in the first Solvay Congress in Brussels in October 1911. It was a meeting of the most outstanding minds of physics and chemistry with the participation of Albert Einstein, Maria Skłodowska-Curie, Max Planck and Ernest Rutherford, Henri Poincaré and Hendrik Lorentz, but was attended mainly by physicists dealing with the dynamically developing atomic theory—hence the topic of the congress was the theory of radiation and the quanta.

Smoluchowski was also involved in this new branch of physics, but focused on a particular aspect of it, which culminated in 1916 with lecture given at the Wolfskehl Congress in Göttingen entitled *Three discourses on diffusion, Brownian movements, and the coagulation of colloid particles*. These lectures have gone down in the history of science and are still considered the best introduction to the issues of coagulation. An article entitled *Attempt for a mathematical theory of kinetic coagulation of colloid solutions*, which appeared

---

[915] M. Smoluchowski, *Several notes on the physical foundations of tectonic theories*, a printout from 'Kosmos' ('Cosmos'), the Journal of the Polish Copernicus Society of Naturalists, 1909, No. XXXIV, pp. 547–579.

[916] *Women in science*; Lecture presented at the Union of Science and Literature in Lviv in 1912, 'Rok Polski' ('Polish Year'), 1917, No. 2, pp. 7–24.

in *Zeitschrift für Physikalische Chemie* (*Journal of Physical Chemistry*), has been cited more than 5,500 times.

The success of his work on the coagulation process was not as spectacular as his discovery the theory of Brownian motion (the Polish physicist largely deprived himself of recognition for this discovery), but to this day adds to Smoluchowski's prestige through the stature he achieved among scientists of his day as the creator of coagulation, and the constantly increasing number of citations in this field.

7.6. Summary

The facts presented above do not provide sufficiently solid arguments to explain the events that occurred in the period between Smoluchowski's earlier elaboration of theses and Einstein's subsequent theorems. They are presumptions, each of which is unconvincing individually but, when looked at holistically, cannot be ignored.

In his 1905 papers, Einstein presented his own model of the behaviour of solution particles and the suspension particles floating in it. However, it is almost certain that he benefited from the results of several years of work by Smoluchowski. At the time of writing the May 1905 article, he need not have been fully aware of it, but—it is worth remembering—Einstein never mentioned that he had read Smoluchowski's previously published article, which is puzzling, especially since he also did not do so following the Polish physicist's publication of the article in 1906, nor did he later address any of his comments.

The fact that Einstein never confirmed that he had read the article *On irregularities in the distribution of gas molecules and their influence on entropy and the equation of state* before publishing his own theories makes one wonder about the reason for this behaviour. If he believed the article was not particularly relevant to his breakthrough on Brownian motion because when reading it he had focused on issues related to diffusion, then his silence on the matter is incomprehensible.

Moreover, Einstein had the opportunity to mention Smoluchowski's article in his December paper, in which he included the names of two physicists who, through direct observation, had confirmed the thesis that Brownian motion is caused by disordered thermal movements of liquid molecules. If he was convinced of the probity of the situation and believed in the originality and primacy of his method, it is difficult to guess why he did not mention Smoluchowski or comment on the suggestions contained in his publication.

During the writing of the May 1905 article, Einstein may not have known that Smoluchowski had worked on the subject of Brownian motion and had had a mathematical solution to the problem for two years, because he did not consider the issue, which was not the main subject of his work. In December 1905, Einstein must have at least guessed at the premises of Smoluchowski's article, because he used them. Even assuming, which is unlikely, that the article he read was not the inspiration for his findings, Smoluchowski's statements and comments must have reached him later. He never addressed them or expressed his opinion. This ambiguity of the situation makes it impossible to be sure what the true circumstances of the discovery were.

Until May 1905, Einstein was not a widely known scientist; he was an patent office employee, he wrote scientific papers, but his scientific achievements comprised only a few not very important publications. He knew that any inspiration by Smoluchowski's work would not be revealed. Indeed, it was extremely unlikely at the time. The only documentary trail were the reviews of three articles from a book of limited circulation which could have reached no more than a small group of scientists.

In May 1905, Einstein could not have predicted that a hundred years later his entire scientific legacy would be meticulously archived, many scientists would scrutinise the work he left with the aid of programmes and computer algorithms, and the fact that he was aware of Smoluchowski's paper would come to light.

Smoluchowski's exceptionally righteous character meant that he never questioned the results of Einstein's work. Nor did he ever raise any objections or make any claims, despite many thought-provoking comments. He was extremely straightforward and not given to hasty judgements, especially when he could not be sure of their veracity. This led to unfavourable situations, such as when he held back in 1903 from publishing his work on Brownian motion due to the conviction that he should first experimentally ensure the validity of the theses it contained. Having various doubts, he neither questioned nor suggested anything publicly. Einstein's respect for Smoluchowski is evidenced by the words often quoted: "Everyone who knew Smoluchowski better valued in him not only a sharp mind, but also a noble, subtle and kind man"[917]. It is not insignificant that Einstein made this statement only after the scientist's death.

On December 14, 1917, a short article by Einstein, *Marian v. Smoluchowski*,[918] appeared in *Die Naturwissenschaften* (The Science of Nature), referring to Smoluchowski's unexpected death and at the same time highlighting several of the deceased's research successes. The article is full of nostalgic recollections as well as respect and recognition for the late scholar. Einstein speaks flatteringly about Smoluchowski—both as a man and as a physicist. However, on a more thorough reading, one gets the impression that something is lacking in the praise. The text was written eleven years after the publication of *Outline of a kinetic theory of Brownian motion and turbid solutions*, the theory of Brownian motion had long since been proven and in the opinion of most scientists, Einstein was solely responsible for the discovery. However, the article's content and the issues raised in it do not touch on this problem and Einstein writes nothing about Smoluchowski's rights to joint credit for the discovery, although that is how it was originally perceived among physicists. He never responded to Smoluchowski's numerous comments on the 1905 articles, even after his death, although there was a meaningful, and perhaps final, opportunity to do so. Remaining on a courteous level, he formulates his opinion, finally closing the issue of research on Brownian motion.

It is puzzling why, following the publication of the 1905 papers, Smoluchowski repeatedly discusses them substantively, comparing the solutions proposed by Einstein with his own, assessing their accuracy and errors, while Einstein never shows any interest in these comments, nor does he make any substantive comments on Smoluchowski's work. Einstein's rare remarks were general in nature, not touching on important issues, which is curious as a few years later he took an active part in discussions regarding the work on proving the theory of Brownian motion, for example with Perrin and Svedberg.

It is worth mentioning that the announcement of equations describing Brownian motion never caused a public dispute between the two scientists over precedence, which was undoubtedly mainly due to Smoluchowski.

---

[917] A. Einstein, *Wspomnienie o Smoluchowskim (Recollections of Smoluchowski)*, 'Problemy' ('Problems') 1972, No. 8 (317), VIII, p. 42.

[918] Idem, *Marian v. Smoluchowski*, 'Die Naturwissenschaften' ('The Science of Nature') 1917, vol. 5, No. 50, pp. 737–738.

Smoluchowski's attitude inspires respect but also provokes a desire to stand up for the truth of the discovery. Einstein spoke extremely flatteringly about Smoluchowski's theory, calling it particularly elegant and illustrative. Although he wrote that Smoluchowski, by showing that a fluid's internal friction constantly slows velocity and that random collisions restore it, managed to explain the phenomenon of Brownian motion quantitatively, as has been stated many times, he never responded substantively to Smoluchowski's comments and he wrote that text after the Pole's death.[919]

Some researchers, such as Stephen G. Brush, also point out various ambiguities that arise through examination of both Einstein's 1905 papers. In the article *A History of Random Processes*, he writes that in *Autobiographical Notes*, published in 1949, Einstein indicates the motives that guided him when he took up research on the theory of Brownian motion at the beginning of the century. He claimed to have developed his theory of statistical mechanics and the molecular kinetic theory based on it, unaware of earlier research by Boltzmann and Josiah Willard Gibbs (1839–1903). He maintained that his aim was to prove that there are facts that would confirm the existence of atoms. Brush notes that—contrary to the above assertion—Einstein was familiar with Boltzmann's treatise, as he had quoted it on the fourth page of his 1902 paper on the kinetic theory of matter. He argues that two aspects of Boltzmann's work, namely the application of statistical mechanics and the statistically-based study of the phenomenon of fluctuation, were of particular interest and were used by Einstein, who claimed in the cited essay that he did not know that Brownian motion was already known. He argued that in the course of the research it became apparent that there was observable movement of suspended microscopic particles and that on this basis he himself developed the theory of statistical mechanics and molecular-kinetic theory[920]. Brush writes that there are some doubts as to whether Einstein's recollections are accurate. Einstein himself notes: "Every reminiscence is colored by today's being what it is, and therefore by a deceptive point of view[921]."

---

[919] See A. Einstein, *Recollections of Smoluchowski*, op. cit., p. 41.
[920] See S.G. Brush, *History of Random Processes*, op. cit., pp. 14–29.
[921] A. Einstein, *Autobiographical Notes*, in: *Albert Einstein, Philosopher-Scientist*, ed. P.A. Schilpp, La Salle IL 1949, p. 47.

In his 1914 essay *On thermodynamic fluctuations and Brownian motion*, Smoluchowski summarised the process of the theory's theoretical construction:

> When in 1905 and 1906, Einstein's theoretical work appeared as well as that of this paper's author, both of which, using entirely different methods of reasoning, arrived at compatible results as to the true essence of Brownian motion, experimentalists took up the subject, applying the only correct research method in this case, involving the compilation of statistics on the displacements achieved by particles over certain times. It was understood that the numbers Exner considered the measure of particle velocities in no way correspond to the velocity of actual movement; they are chance shifts resulting from the geometric composition of a great number of small deflections possessing all possible orientations in space. The contradiction highlighted above has therefore disappeared as the displacement achieved over a unit of time must be of a significantly lower order than the actual velocity of movement, occurring in an immeasurably intricate, zig-zag path[922].

The line of Smoluchowski's reasoning in the research he conducted on Brownian motion is an example of problem-solving in accordance with the principle of the stages of thinking developed by instrumentalists and presented by John Dewey (1859–1952), which assumes a research path from positing a hypothesis to its empirical verification.

In summary, it should be reiterated that over time Smoluchowski came to terms with the imposed narrative. After 1905, there followed a period of experimental research involving many physicists and chemists seeking to prove the theory of Brownian motion, which resulted in a public discourse for several years thereafter over the research methods used, their reliability, and the nuances of the theory discovered. Einstein participated intensively in this discourse, while for various reasons Smoluchowski's input was only sporadic. These circumstances led after just a few years to Einstein figuring as the discovery's architect in literature on the subject. The German scientist never protested when he was given exclusive credit for the theory of Brownian motion, nor did he write anywhere that there was another researcher who had also tackled the problem.

There have been many discoveries in the history of science that are astonishing in their fortuity. In the book *The Quantum Brain*, Amit Goswami

---

[922] See M. Smoluchowski, *On thermodynamic fluctuations and Brownian motion*, op. cit., p. 298–299.

highlights an example to demonstrate how creative thought can take different paths. In a conversation after a seminar on the wave nature of matter led by physicist Louis de Broglie (1892–1987), chemist Peter Debye (1884–1966) suggested to physicist Erwin Schrödinger that if matter is a wave, there must be a mathematical "wave equation" that applies to matter. Debye later forgot about his joke, but his comment inspired Schrödinger to postulate an equation for matter waves (now known as the Schrödinger equation)[923].

Sometimes a small suggestion is all it takes to set a thought in the right direction, especially when geniuses are part of the conversation. Einstein found himself in a similar situation, as an article he happened to read inspired him to formulate a theory that changed the understanding of the fundamentals of the structure of matter. However, the situation in this case was rather different as the theory already existed, it was just not published, and Einstein never publicly commented on the situation.

Einstein must have had doubts as to the transparency of the situation around the discovery because if he had not, he would naturally have cited Smoluchowski's paper and the situation would be known—as in the case of Schrödinger and Debye. The Polish scientist encountered a physicist who, on the one hand, was a genius—perhaps someone else writing a review would not have recognised the importance of Smoluchowski's article—but who on the other was known for being reluctant to share his successes by citing the names of other scientists. In addition, over time he became a legend of physics and even of science as a whole, and legends usually have their own laws.

In 1985, publishers Harper and Row released Mark Kac's autobiography *Enigmas of Chance*. In this autobiographical book, Kac writes about the discovery of Brownian motion:

> One of the two historical papers was by Marian Smoluchowski. The other, which appeared a somewhat earlier and used an entirely different approach to the problem, was by Albert Einstein. It was Smoluchowski's bad luck that he had to share his first great discovery, as well as a number of later ones, including the explanation of the blueness of the sky, with so luminous a figure as Einstein. There is probably no more extreme example of the "Matthew Effect", a wonderfully apt term invented by Robert Merton to describe the all too common phenomenon that the credit for

---

[923] See A. Goswami, *Quantum mind. Scientific Evidence for the Power of Your Thoughts*, trans. A. Rutkowska, Białystok 2014, p. 40.

a discovery made jointly or independently by two investigators of unequal fame is invariably given to the more famous one: For whosoever hath, to him shall be given, and he shall have more abundance: but whosoever hath not, from him shall be taken away even that he hath. *[Matthew 13:12]* During his lifetime, Smoluchowski did not suffer from the Matthew Effect. He was universally recognized as one of the leading theoretical physicists of his day and he received many honors which he richly deserved. But with the passage of time the Matthew Effect took its toll. Few realize today what an important role Marian Smoluchowski played in bringing atoms to life and even fewer that it happened in Lwów [Lviv][924].

With a generous dose of sympathy for Smoluchowski, Mark Kac tries to explain these rather unlucky circumstances, but does not go beyond the discourse that has been created over the last century around the papers from 1904–1906. The appraisals repeated by many scientists aware of the ambiguity of the provenance of the discovery of Brownian motion, after presenting their comments on Einstein, have usually ended with a statement along the lines of: "And for that we admire him, and let us draw a veil of silence over the rest." However, something breaks through even this veil; there are questions that remain unanswered, and there are also answers that do not support the view accepted by the scientific community. The hypothesis has emerged that the chief architect of the discovery of the essence of Brownian motion, as well as evidence for the kinetic-atomic structure of matter, was Marian Smoluchowski. This is only a hypothesis with fairly strong premises, which does not undermine Einstein's achievements in introducing this theory to physics. These achievements cannot be denied. However, it must be concluded that in science both surnames should appear in the name of this theory, as was the case in 1906.

---

[924] M. Kac, *Enigmas of Chance, An autobiography*, London, 1985, pp. 21–21

Conclusion

Time brings surprising changes; a hundred years after his death, Smoluchowski is receiving well-deserved praise, though not in the area in which he could have expected splendour. His work in the field of coagulation, used today in science and industry, has made him the most cited Polish scientist. Meanwhile, in his own country, few people have heard of the physicist Marian Smoluchowski. If he is associated with anything, it is with the name of a street in Gdańsk traversed by thousands of people every day as it is the main road to the Medical University. Even fewer people would be able to outline Smoluchowski's philosophical views as they are virtually unknown, even to scientists studying Smoluchowski's physics. They are difficult to discern because they have not been clearly articulated and may seem incoherent or even inconsistent. These impressions fade at deeper levels of analysis as Smoluchowski drew on a range of concepts rather than clinging to a single philosophical thought. He was interested in theoretical approaches that helped him build his own philosophy of perceiving reality and, above all, broadened the horizons of his research in the field of physics. His philosophy was an organism under constant development, subject to endless change under the influence of conclusions drawn from his own research and from the development of science, as well as from analysis of consecutive philosophical concepts. As such, he did not approach philosophical concepts dogmatically, but rather with an awareness of their constant evolution. He used everything that seemed to him to meet his *utility criterion*.

Even in such far-removed views as Ernst Mach's monism, which rejected reductionism, opposing the use of theoretical considerations in the physical explanation of phenomena, Smoluchowski found elements he could accept. Mach's conviction that relationships should be created between phenomena under study in the form of mathematical functions is one example that confirms this. The case was similar with Wilhelm Ostwald's energeticism, which demanded the removal from science of such metaphysical notions as 'atom', 'force' and 'cause'—concepts that are perfectly scientific from today's point of view. Smoluchowski saw partial rationales in this position, going against the widespread pseudoscientific use of the terms at that time.

The Polish researcher showed no interest in phenomenology, except for the so-called phenomenology of energetics. He did not accept the traditional understanding of intentional cause proposed by neo-Thomism and Aristotelianism. He was a supporter of Darwinism, to which, he argued, science owes its departure from the concept of causality in nature.

He was closer to materialistic monism or attributive monism than to the materialism often attributed to him. Smoluchowski's inclusion within the ambit of materialist philosophy in the 1950s was an unacceptable simplification and over interpretation of his views. He never referred directly to materialistic philosophy in his works; it did not constitute an issue he would have to pore over and analyse in his deliberations.

Dealing mainly with the philosophy of nature and science, Smoluchowski was not original in his philosophical beliefs and views. His philosophy is a combination of the views of positivists, *empiriocritics*, conventionalists, pragmatists, scientismists, instrumentalists, essentialists, and even neo-Kantianists. Despite such a diverse array of philosophical concepts within which Smoluchowski conducted his own philosophical deliberations, a certain coherence of thought emerges from his writings and, more importantly, a certain direction of enquiry, enabling more creative thinking in physical research.

This enquiry did not stem from a need to create his own philosophical concept; on the contrary, it was the experience of scientific work and the re-evaluation of the world of physics through the adoption of successive perspectives—thermodynamics, the kinetic concept, and belief in the atomic structure of matter—that caused a change in his perception of reality and prompted attempts at an unconventional perception of scientific issues and regular change. It forced the rejection of existing philosophies and a search for solutions that would not limit human thought. The philosophical path followed by Smoluchowski is one of searching for a sublime, unlimited vision of nature. Doctrinaire thinking and the adaptation of facts to existing philosophical systems is alien to it.

It is a philosophy in constant motion with no intention of finding 'eternal truths' or seeking a systematic and coherent system. Science is made up mainly of hypotheses that reflect to a greater or lesser degree the actual state of affairs, though its main goal is not the truth about reality but the benefits that come from knowing its laws, which is why we should mostly expect our knowledge to be useful.

According to Smoluchowski, we do not treat the accepted laws of physics as objective truths but merely as useful computational tools. It is Smoluchowski's instrumentalism that precludes interpretation of his philosophical views as being materialistic philosophy. This position involved the belief that one should not become attached to theories and scientific hypotheses as ongoing exploration of the world causes constant and incessant correction of knowledge. Truth is the objective limit of possible research. Man has an instinctive ability to discover those truths that lie within the scope of his research capabilities, and although he may be wrong, in the long run false theories are challenged.

Smoluchowski's category of utility is an interesting and important contribution to philosophical thought on science's relationship to truth and reality. Utility is a universal notion relating to all aspects of human functioning and determining the relative value of every theory, hypothesis or concept. Our actions are accomplished through knowledge of what a community deems more useful, whether it concerns science or other issues related to practical human activity. A theory is accepted and used until its usefulness becomes questionable, until a theory appears to which the community attributes greater usefulness, giving increased possibilities and enabling more effective development. The intrinsic value of Smoluchowski's utility criterion is to bring order, understanding and clarification to a timeless factor that determines and stimulates man's constant pursuit of expansive development.

Smoluchowski treated the study of nature and the co-creation of science as a never-ending story that is written as it is read, hence he never compiled his philosophical views into a single compendium or philosophical system as such a system could never be complete, let alone true. Therefore, in some statements, even programmatic ones, his position often differs from what can be read in his other works. This was connected with a constant quest for formulae that would enable a more complete description of specific laws of nature. Smoluchowski saw philosophy as a tool to better understand the relationships occurring in nature. He was above all attached to the purpose behind a scientist's work, the dominant feature of which should be the search and discovery of scientific truth, unburdened by philosophical limitations, though philosophy should be an important, albeit only heuristically helpful, tool to this end.

Smoluchowski regarded elements of various philosophical concepts as a tool with which he could shape his view of the scientific nature of the composition of matter, nature and the world. For this reason, his philosophical views can be interpreted as significant, but not the most important for the development of philosophy in the 20th century.

The research shows Smoluchowski's contribution to the development of the methodology of science, philosophy of nature and philosophy of science to be sufficiently important that it should be widely recognised.

The Polish physicist gave a new, deeper meaning to the problem of chance in physics, resulting in the introduction of probability calculus into physical research in a much broader scope than it had previously been practised. Smoluchowski's research on fluctuations also influenced the acceptance of the indeterminism of material phenomena and, consequently, the introduction to physics of the concepts of chance and probability. This indirectly influenced philosophical discussions on the issue of determinism and causality. The probability calculus introduced to research in physics contributed to the emergence of new methodologies, such as an equation for the theory of stochastic processes or the continuity equation, which is a mathematical form of the conservation law for continuous media. Before Smoluchowski, probability calculus was used mainly in the study of social and biological phenomena, and in physics as a rule only in the theory of compensating for measurement errors. The Polish physicist introduced the application of this calculus to theoretical physics as a mathematical method widely used in scientific research.

It is also worth highlighting his contributions to methodological analyses on the use of deductive, inductive and analogical methods both in laboratory research and in theoretical inquiries.

A fundamental achievement of Smoluchowski was the scientific explanation of the causes of so-called Brownian motion together with a mathematical justification. This constituted the crowning proof of the atomic-kinetic structure of matter, but at the same time settled a centuries-old philosophical dispute over the nature of the material world. The atomic theory was confirmed by Smoluchowski's research on fluctuation, opalescence and *perpetual motion*. Smoluchowski's hypothesis of critical opalescence confirmed and explained the atomic structure of matter in a clear and definitive way.

Smoluchowski's merits should also be stressed in terms both of his educational work and in spreading knowledge by popularising scientific subjects such as physics, chemistry and mathematics. This is evidenced by his regular talks and popular science lectures for teachers, doctors and naturalists, as well as by his participation in a range of conferences and conventions, and finally, by his publications in specialist journals and writing the *Self-Study Handbook*.

September 5, 2017, saw the hundredth anniversary of Marian Smoluchowski's death. His works are part of modern scientific life. He is a scientist regularly cited in many reputable scientific journals. These citations in global literature are related to his scientific achievements, mainly in the field of physics, however Smoluchowski also earned some stature in the development of the philosophy of science and the philosophy of nature, leaving behind a set of philosophical views leading to a new perception of reality. His philosophical thought was often ahead of his time and was a bold anticipation of modern solutions.

Unfortunately, one sad conclusion comes out of these considerations—coincidence, certain omissions, but also the untimely death of Marian Smoluchowski, meant that his name is not commonly mentioned along with those of the greatest physicists of the 20th century. The writing of this monograph has presented an opportunity for his achievements in the field of philosophy to be discussed, and perhaps also recognised and appreciated.

Supplement
William Sutherland—a quantitative theory of Brownian motion

It is impossible to complete an analysis of Marian Smoluchowski's work on the theory of Brownian motion without mentioning the concept of a physicist from the University of Sydney, William Sutherland (1859–1911). Reflections on the Australian physicist's work can be found in *Brownian Motion, "Diverse and Undulating"*[925] by Bertrand Duplantier of the Université Paris-Saclay, published in the journal *Progress in Mathematical Physics* as a summary of the Séminaire Poincaré entitled 'Einstein 1905–2005', which was held in Paris on April 9, 2005.

Duplantier's deliberations differ slightly from the conclusions presented by this author. The main difference concerns the roles attributable to William Sutherland and Marian Smoluchowski in the explanation of Brownian motion. Also important is the different view of the paths the two physicists took to arrive at knowledge of the essence of Brownian motion.

It should be emphasised that in his paper Duplantier repeatedly mentions the Polish physicist, expressing appreciation for his contribution to the breakthrough and highlighting his achievements. He believes Smoluchowski's name is closely connected with Brownian motion and the theory of diffusion. He quotes Marek Kac, who argued that Smoluchowski's demonstration that the concept of a game of chance underlies our understanding of physical phenomena is a real intellectual feat, adding that scientists should be grateful to the Polish scholar for his original and bold introduction of probability calculus to statistical physics and that he deserves to be mentioned alongside Maxwell, Boltzmann and Gibbs[926].

Duplantier begins his argument about Sutherland by recalling a meeting of the Australasian Association for the Advancement of Science held in January 1904 in Dunedin, New Zealand, at which Sutherland submitted two papers, one of them—*The Measurement of Large Molecular Masses*[927] —concerning Brownian motion.

In it, he proposed the so-called quantitative theory of Brownian motion, based on the principles of statistical mechanics. At the beginning of his remarks, Sutherland points out:

> The determination of the large molecular masses of such substances as the physiologist is occupied with, presents almost insuperable difficulties when attempted by the usual methods of the chemist. The vapour-density is unobtainable, and the

---

[925] B. Duplantier, *Brownian Motion*, op. cit., pp. 201–293.
[926] Idem, p. 239.
[927] W. Sutherland, *The Measurement of Large Molecular Masses*, in: *Report of the 10th Meeting of the Australasian Association for the Advancement of Science, Held Dunedin 1904*, ed. G.M. Thomson, Dunedin 1905, pp. 117–121.

molecular lowering of the freezing point of a solvent produced by massive molecules is so small as to be merged in the experimental error of the freezing-point of the solvent in its purest obtainable state. The method to be proposed and discussed in this paper is founded on the measurement of the coefficient of diffusion of the substance through a solvent. The only difficulty about measuring the velocity of diffusion of a substance of large molecular mass is that the experiment for measuring it has to be prolonged for a time inversely proportional to the velocity in question. (Thomas) Graham (1805–1869), the pioneer in the investigation of diffusion studies, made measurements of the velocity of albumen in water. If, then, we can establish a dynamical relationship between the velocity of diffusion of a substance and the size of its molecule, it will be possible to measure the molecular mass of substances like albumen and its products of more or less complete disintegration[928].

The article was the culmination of several years of work by Sutherland on the problem of Brownian motion.

Duplantier writes that in 1905, Albert Einstein and William Sutherland of Melbourne independently developed a quantitative theory of Brownian motion that enabled Perrin to determine the precise value of Avogadro's number (which he did in his famous experiments of 1908–1909). He claims that Sutherland and Einstein succeeded where many others had failed, and that Marian Smoluchowski simultaneously performed an analysis according to a different, more probabilistic, schema of reasoning (German: *Gedankenweg*), which led him to similar conclusions[929].

Duplantier emphasises that as early as 1902, in his works on ionization, ion velocities and atom sizes, Sutherland considered calculating the size of ions based on Stokes' law[930]. Meanwhile, Einstein and Besso, discussing in 1903 the theory of dissociation, which demands the assumption of molecular aggregates combined with water, opened the way for a straightforward calculation of the size of ions in a solution based on hydrodynamic considerations. Einstein came upon the idea of determining the size of ions with the aid of classical hydrodynamics, as he wrote in a letter to Besso on March 17, 1903, proposing something that, according to Duplantier, seems to be merely a calculation that Sutherland had made earlier. Einstein writes: "Have you

---

[928] Idem, *The Measurement of Large Molecular Masses*, op. cit., p. 117.
[929] See B. Duplantier, *Brownian Motion*, op. cit., p. 218.
[930] W. Sutherland, *Ionization, Ionic Velocities, and Atomic Sizes*, 'Philosophical Magazine' 1902, series 6, vol. 4, pp. 625–645.

already calculated the absolute size of the ions under the assumption that they are spheres and that they are large enough to permit the application of the equations of the hydrodynamics of viscous liquids? Given our knowledge of the absolute size of electrons [charge], this would certainly be a simple matter. I would have done it myself, but I do not have the necessary literature and the time for it; you could also make use of diffusion to learn something about neutral salt molecules in solution"[931].

As the editors of *The Collected Papers of Albert Einstein* note: "This passage is remarkable, because both key elements of Einstein's method for the determination of molecular dimensions, the theories of hydrodynamics and diffusion, are already mentioned, although the reference to hydrodynamics probably covers only Stokes's law"[932]. So do Sutherland's deliberations of 1902 bear only a chance resemblance to Einstein's method?

It is also striking that an earlier letter, from February 11-17, 1903, this time written by Besso to Einstein, clearly shows that the two scientists had discussed Sutherland's work. The letter is made up of two parts—the first concerns experimental data related to the dissociation of bi-ionic molecules, while the second concerns what Besso calls "Sutherland's hypothesis". He states that the "ionic hydrates" theory, as he calls it, temporarily rescues this hypothesis with regard to Ostwald's dilution law. Besso also discusses the role of imperfect semi-permeable membranes as a possible experimental test of Sutherland's hypothesis. In the French edition of the two scientists' correspondence, Pierre Speziali pointed out that in this letter Besso discussed another work by Sutherland, entitled *Causes of Osmotic Pressure and of the Simplicity of the Laws of Dilute Solutions*. After reading the letters from 1903 one cannot help but wonder whether Besso and Einstein were also familiar with Sutherland's theses of 1902 on ionic sizes. In that case, the Australian scientist's suggestion to use Stokes' hydrodynamic law to determine the size of molecules would have been a direct inspiration for Einstein's dissertation and subsequent work on Brownian motion![933]

---

[931] A. Einstein, M. Besso, *Correspondance 1903–1955*, translation, notes and introduction P. Speziali, Paris 1979.
[932] *The Collected Papers of Albert Einstein*, op. cit., pp. 170–182.
[933] B. Duplantier, *Brownian Motion*, op. cit., pp. 219–220.

Sutherland had assumed the existence of atoms, and tackled a practical question; the measurement of large molecular masses. He was interested in these masses because of their role in the chemical analysis of organic substances. Although to this day everyone uses the Sutherland-Einstein equation, it may not have been as widely used then. However, it can be seen from Einstein and Besso's correspondence how extremely important Sutherland's idea was of determining the sizes of ions or molecules by means of classical hydrodynamics[934].

Einstein's chief assumption is the legitimacy of using classical hydrodynamics to calculate the influence of the molecules of dissolved substances, treated as rigid spheres, on the viscosity of the solvent in a diluted solution. His method is well suited to determining the size of solute molecules, which are large in comparison to the molecules of the solvent, so he applied it to the dissolution of sugar molecules. In 1905, Sutherland published a description of a method for determining the masses of large molecules which coincides in part with Einstein's method. Both use the molecular theory of diffusion developed by Nernst[935] based on van't Hoff's analogy between solutions and gases, and Stokes' law of hydrodynamic friction[936].

The derivation of equipartition represents the technically difficult part of Einstein's dissertation. It rests on the assumption that the motion of the fluid can be described by the hydrodynamical equations for stationary flow of an incompressible homogeneous fluid, even in the presence of solute molecules; that the inertia of these molecules can be neglected, that they do not interact, and can be treated as rigid spheres moving in the fluid without slipping, under the sole influence of hydrodynamical stress. The equation results from the conditions of dynamic equilibrium in the fluid. As with Sutherland's work, its derivation requires identification of the force on a single large molecule, which appears in Stokes' law, with the apparent force due to the osmotic pressure[937].

---

[934] Idem, p. 221.
[935] W. Nernst, *Stöchiometrie Verwandtschaftslehre* (*Stoichiometry Relationship Theory*), 'Zeitschrift für Physikalische Chemie' ('Journal of Physical Chemistry') 1888, vol. 2, pp. 613–639.
[936] B. Duplantier, *Brownian Motion*, op. cit., p. 223.
[937] Ibid.

Einstein was in possession of two theories about particles in a fluid. The first was Stokes' hydrodynamic theory, based on the hypothesis that the fluid is a continuous medium which adheres to a large solid surface moving through it, without any turbulence, and where the molecular agitation does not seem to play any role. The second was van 't Hoff's osmotic theory, based on the hypothesis that a particle in solution is similar to any other fluid molecule, and therefore is subject to the same laws of molecular agitation. Duplantier underscores that Einstein's insight was necessary, as well as his profound knowledge of statistical mechanics, to understand and prove that the two points of view were simultaneously valid for particles as large as Brownian particles[938].

In conclusion, it would be most appropriate to recall the opinion of the historian of science Roderick W. Home, who noted that the relation of diffusion and viscosity is generally known as the Einstein relation, not the Sutherland-Einstein relation. In his opinion, this occurred because in the early 20th century, theoretical physics was for the most part a German discipline[939]. Although the theory had already been devised, it was not initially published, however it was taken up by continental researchers who had read Einstein's papers and also noticed that a similar theory had been published in *Philosophical Magazine* in an article entitled *A Dynamical Theory of Diffusion for Non-Electrolytes and the Molecular Mass of Albumin*. In the English-speaking world, where *Philosophical Magazine* was one of the leading journals in this field, few people were involved in German-style theoretical physics. There is ample evidence that experimentally-oriented British physicists failed to appreciate Sutherland's work. This is well illustrated by an obituary in *Nature* magazine which emphasises that his work was known to the scientific world. He had a large readership, but his elaborations of issues also considered by Einstein were combined with bold speculations which always led to extensive comparisons with experimental knowledge[940]. Thus, in Britain, Sutherland did not have an audience that would have been as aware of the importance of his research on the relationship between diffusion and viscosity as some

---

[938] See ibid., p. 226.
[939] See R.W. Home, *Sutherland, William (1859–1911)*, in: *Australian Dictionary of Biography*, vol. 12, http://adb.anu.edu.au/biography/sutherland-william-8719 (access: 30/04/2020).
[940] Ibid.

of Einstein's continental readers would have been. Sutherland was also not helped by the way he presented his arguments, as they were full of intricate calculations concerning the molecular mass of albumin. He would probably have done better to leave these mathematical analyses in his own notes. However, he was firmly focused on the problem of determining the molecular mass of large molecules and clearly saw the dependence of diffusion viscosity as an incidental result on the way to achieving a greater aim, rather than something of value in itself. Perhaps, as Duplantier writes, the time has come for physicists to finally recognise Sutherland's achievements and, as Pais has suggested, rechristen the famous relation with a double name[941].

It should be added that Sutherland was a physicist known in the scientific community, which is confirmed by the fact that he and Gibbs were the only physicists from outside Europe invited to participate in the Boltzmann *Festschrift* in 1904[942]. Not having access to laboratory facilities, his research was limited to theoretical work, chiefly in the field of molecular dynamics. He assumed that the particles that make up matter exert an attractive force on each other in addition to gravity. Although his theory contrasted with that of Ludwig Boltzmann and other physicists of his day who adopted a purely kinematic view, it is now widely accepted. In introducing the idea, contemporary texts usually refer to the 'Sutherland model' and characterise the force in terms of the 'Sutherland potential'. Sutherland's general approach allowed him in 1893 to successfully explain one of the major problems at the time, namely the discrepancies between theory and experiment in relation to the dependence of the viscosity of a gas on its temperature. He showed that the existence of an attractive force between gas molecules would increase the effective diameter of the molecules. This research led him to a discovery[943] that today is called Sutherland's law[944].

The above considerations cast an entirely fresh light on the problem of determining the contribution of individual scientists to the discovery of the essence of Brownian motion. First of all, it should be stressed once

---

[941] B. Duplantier, *Brownian Motion*, op. cit., pp. 221–222.

[942] Idem, p. 221.

[943] In 1893, William Sutherland established the relationship between the dynamic viscosity and the absolute temperature of a perfect gas. This formula is called Sutherland's law.

[944] See R.W. Home, *Sutherland, William (1859-1911)*, op. cit.

again that Smoluchowski developed a theory of Brownian motion in the years 1900–1903, but did not publish it. These facts make clear that both Sutherland's and Einstein and Besso's work essentially started in 1903, at a time when Smoluchowski was already considering making the results of his work public. The second issue requiring consideration is the application of Stokes' law to the research. In Smoluchowski's concept, and in the accounts of many other scientists referred to here, this fact is treated as an incomprehensible error, while in the concepts of Sutherland and of Einstein and Besso presented by Duplantier, alongside Nernst's molecular theory of diffusion and van't Hoff's osmotic theory, Stokes' law was a key factor in finding the cause of Brownian motion. Undoubtedly, a comparative analysis of various research aspects of the discovery of Sutherland, Smoluchowski and Einstein could be interesting, but this requires a separate study.

In regard to Smoluchowski's works, Duplantier is very impressed with the solutions the Pole proposed. He writes that the novelty and originality of Smoluchowski's approach lies in replacing an extremely difficult problem (a Brownian particle colliding in a gas or liquid) with a relatively straightforward stochastic process. Any dynamic event, such as a collision for example, is considered a random event similar to the roll of a die, where the elementary probabilities are (to some extent) determined by underlying mechanical laws. This way of reasoning plays a fundamental role in mechanics and statistical physics today, and—as noted by Kac—it is hard to imagine the degree of Smoluchowski's intellectual boldness in initiating this topic at the beginning of the last century[945].

Duplantier emphasises that Einstein and Smoluchowski defined the action of Brownian motion in the same way. They had previously determined the "average velocity of motion" by following the path of a particle as closely as possible. The values thus obtained were always several microns per second for particles of the order of a micron. But such activity assessments are completely erroneous. The trajectories are confusing and complicated and it is difficult to follow them; the trajectory actually measured is much simpler and shorter than the real one. Similarly, the apparent average velocity of a particle at a given time varies most unpredictably in terms of magnitude and direction,

---

[945] B. Duplantier, *Brownian Motion*, op. cit., p. 241.

and does not tend to reduce as the time needed for observation decreases. Therefore, disregarding the true velocity, which cannot be measured, and the extremely complicated path along which the particle moves at a given time, Einstein and Smoluchowski chose the rectilinear segment connecting the start and end points as characterising the magnitude of the change. This means the line will be longer the more actively the particle moves. The distance will be the displacement of the particle within the time considered.

Duplantier began his review by recalling Exner's early work, even prior to publication of the Sutherland-Einstein-Smoluchowski formula for the average square shift, in which at least one attempt at partial verification of the formula can be seen. Soon after this formula's publication, verification was swiftly taken up by Svedberg, who thought he had accomplished it. Perrin sharply criticised the results and declared Svedberg the "victim of an illusion" with regard to his description[946].

The above considerations show that Duplantier saw Smoluchowski's significant contribution to the development of the theory of Brownian motion, which was and is not very common. However, he was not well acquainted with all the Polish physicist's work and perhaps that is why he did not list Smoluchowski among the creators of the theory of Brownian motion. It is to this researcher's great merit that he made us aware that in the history of Brownian motion, in addition to the European mainstream, an important position was also developed in the antipodes. The combination of Smoluchowski's and Sutherland's research processes, in the context of the final effect in the form of Einstein's 1905 articles, makes it clear that due to the flow of information that existed between scientists, new circumstances are still being revealed regarding this discovery. The full facts will most probably never be known, but research on the issue would be extremely interesting.

***

The analyses presented in this monograph were conducted primarily from the point of view of the philosophy of science and philosophy of nature,

---

[946] Idem, pp. 263–264.

therefore fragments referring directly to physics may seem incomplete to the attentive reader. The book repeatedly discusses issues related to physics, which is obvious due to Marian Smoluchowski's profession, but it should be noted that this is not a scientific work on physics. In a desire to anticipate such comments, the author refers readers to the numerous source materials. It should also be borne in mind that in philosophy the history of philosophy is recognised as a separate discipline, having its own methodology and rich literature, while there is no separate subject in science that can be described as the 'history of physics' or 'history of chemistry'. It is extremely difficult to find a physicist or chemist who would professionally and scientifically verify the historical nuances of the discoveries made. "Scientists involved in physics, chemistry or mathematics basically never read original works from more than 40–50 years ago, even of such fundamental importance as the works of Newton, Maxwell, Boltzmann, Einstein or the creators of quantum mechanics"[947]. Every area of scientific research in modern physics is already quite far removed from its historical origins, and the situation is similarly in chemistry, botany and medicine. This is a natural phenomenon; the exact sciences deal with the future, they are interested in new discoveries, hypotheses and new theories. Their operation is forced by the *criterion of utility*, and historical events are important only insofar as they can be used in the ongoing process of researching new theories.

Acknowledgments

I would like to express my gratitude first and foremost to my friend Prof. Marek Lechniak, to whom I am indebted for a more serious interest in philosophy and forhis beneficial encouragement in my doctoral studies as well as in the writing of my doctoral thesis.

I would like to thank my promoter, Prof. Zenon Roskal, without whose help, valuable comments and constructive criticism neither the doctoral thesis nor this book would have been possible. I could always count on his substantive help.

---

[947] From a publishing review by R. Alicki.

I would also like to extend my thanks to my reviewers, Professors Ludwik Kostro, Jacek Rodzeń and Robert Alicki, for assisting me with valuable advice. I especially thank Ludwik, whose faith in my work was extremely supportive.

I thank Prof. Robert Alicki for his help in analysing Einstein's articles.

Gratitude is also due to Prof. Paweł Horodecki, who always responded in a friendly way to all my requests for advice and help.

I would like to thank Dr Ewa Szumilewicz for the discussions, comments and support.

I would like to thank Krystyna Bembennek and Dagmara Wachna for the effort they put into the editorial work.

I would also like to thank Teresa Ossowska for her help with translations.

I would like to thank Beata Wisniewska for her administrative and clerical work, which was very helpful in creating this book.

Bibliography
Works by Marian Smoluchowski

Smoluchowski M., *Atomistyka współczesna (Contemporary atomic theory)* in: *Pisma Marjana Smoluchowskiego z polecenia Polskiej Akademji Umiejętności (The Writings of Marian Smoluchowski commissioned by the Polish Academy of Arts and Sciences)* vol. 3, Jagiellonian University Press, Kraków 1928, pp. 31–42.

Smoluchowski M., *Atomistyka współczesna. Odczyt wygłoszony na ogól. posiedzeniu XI zjazdu lekarzy i przyrodników w Krakowie (Contemporary atomic theory. A lecture given at the general meeting of the 11th Congress of Physicians and Naturalists in Kraków)*, printout from 'Pamiętnika XI Zjazdu Lekarzy i Przyrodników' ('Commemorative journal of the 11th Congress of Physicians and Naturalists').

Smoluchowski M., *Dwie książki z dziedziny 'filozofii przyrody' (Two books from the field of the 'philosophy of nature)*, 'Ateneum Polskie' 1909, vol. IV, pp. 291–302.

Smoluchowski M., *Dzisiejszy stan teorii atomistycznej (The current state of atomic theory)*, in: *The Writings of Marian Smoluchowski commissioned by the Polish Academy of Arts and Sciences*, vol. 3, Jagiellonian University Press, Kraków 1928.

Smoluchowski M., *Ewolucja teorii atomistycznej* (*The evolution of atomic theory*) in: *The Writings of Marian Smoluchowski commissioned by the Polish Academy of Arts and Sciences*, vol. 3, Jagiellonian University Press, Kraków 1928, pp. 16–30.

Smoluchowski M., *Experimentell nachweisbare, der üblichen Thermodynamik wiedersprechende Molekularphänomene* (*Experimentally verifiable molecular phenomena that contradict conventional thermodynamics*), in: *The Writings of Marian Smoluchowski commissioned by the Polish Academy of Arts and Sciences*, vol. 2, Jagiellonian University Press, Kraków 1927, pp. 226–251.

Smoluchowski M., *Gültigkeitsgrenzen des zweiten Hauptsatzes der Wärmetheorie* (*Limits of applicability of the second law of thermodynamics*), in: *The Writings of Marian Smoluchowski commissioned by the Polish Academy of Arts and Sciences*, vol. 2, Jagiellonian University Press, Kraków 1927, pp. 361–398.

Smoluchowski M., *Kierunki i zagadnienia fizyki dzisiejszej* (*Themes and issues in today's physics*), in: *The Writings of Marian Smoluchowski commissioned by the Polish Academy of Arts and Sciences*, vol. 3, Jagiellonian University Press, Kraków 1928, pp. 205–222.

Smoluchowski M., *Kilka uwag o analogiach fizycznych, zwłaszcza w teoriach prądów elektrycznych, prądów cieplnych i zjawiska dyfuzji* (*Several remarks on physical analogies, especially in theories of electric currents, thermal currents and diffusion phenomena*), 'Wiadomości Matematyczne' ('Mathematical News'), 1918, vol. XXII, pp. 167–176.

Smoluchowski M., *Kilka uwag o fizycznych podstawach teorii górotwórczych* (*Several notes on the physical foundations of tectonic theories*), printout from 'Kosmos' ('Cosmos'), the Journal of the Polish Copernicus Society of Naturalists, 1909, No. XXXIV, pp. 547–579.

Smoluchowski M., *Kobiety w naukach ścisłych* (Women in science; Lecture presented at the Union of Science and Literature in Lviv in 1912, 'Rok Polski' ('Polish Year'), 1917, No. 2, pp. 7–24.

Smoluchowski M., *Liczba i wielkość cząstek i atomów* (*The number and size of molecules and atoms*), in: *The Writings of Marian Smoluchowski commissioned by the Polish Academy of Arts and Sciences*, vol. 3, Jagiellonian University Press, Kraków 1928, pp. 315–329.

Smoluchowski M., *Liczba i wielkość cząstek i atomów (The number and size of molecules and atoms)*, 'Wiadomości Matematyczne' ('Mathematical News') 1913, vol. XVII, pp. 315–329.

Smoluchowski M., *Lord Kelvin*, 'Ateneum Polskie' 1908, vol. I, pp. 212–228.

Smoluchowski M., *Lord Kelvin*, in: *The Writings of Marian Smoluchowski commissioned by the Polish Academy of Arts and Sciences*, vol. 3, Jagiellonian University Press, Kraków 1928, pp. 1–15.

Smoluchowski M., *Maurycy Rudzki jako geofizyk (Maurycy Rudzki as a geophysicist)*, printout from 'Kosmos' ('Cosmos') the Journal of the Polish Copernicus Society of Naturalists, 1916, nr XLI, pp. 105–119.

Smoluchowski M., *O atmosferze Ziemi i planet (On the atmosphere of Earth and the planets)* separate printout from the 'Commemorative book'issued by the University of Lviv to celebrate the 500th anniversary of the University of Kraków, Lviv, circa 1900.

Smoluchowski M., *O drodze średniej cząsteczek gazu i o związku jej z teoryą dyfuzyi (On the mean path travelled by the molecules of a gas and its relation to the theory of diffusion)* 'Rozprawy Wydziału Matematyczno-Przyrodniczego Polskiej Akademji Umiejętności' ('Dissertations of the Faculty of Mathematics and Natural Sciences of the Polish Academy of Arts and Sciences'), series A, vol. XLVI, Kraków 1906.

Smoluchowski M., *O fluktuacjach termodynamicznych i ruchach Browna (On thermodynamic fluctuations and Brownian motion)* Editorial Office of 'Prac Matematyczno-Fizycznych' ('Mathematical-Physical Papers'), Warsaw 1914.

Smoluchowski M., *On thermodynamic fluctuations and Brownian motion*, in: *The Writings of Marian Smoluchowski commissioned by the Polish Academy of Arts and Sciences*, vol. 2, Jagiellonian University Press, Kraków 1927, pp. 268–353.

Smoluchowski M., *O nowszych postępach na polu kinetycznych teoryj materji (On more recent progress in the field of the kinetic theory of matter)*, in: *The Writings of Marian Smoluchowski commissioned by the Polish Academy of Arts and Sciences*, vol. 1, Jagiellonian University Press, Kraków, pp. 279–305.

Smoluchowski M., *O nieregularnościach w rozkładzie cząsteczek gazu i ich wpływie na entropię i na równanie stanu (Über Unregelmässigkeiten in der Verteilung von Gasmolekülen und deren Einfluss auf Entropie und Zustandsgleichung/On irregularities in the distribution of gas molecules and their influence on entropy and the equation of state), trans. B.J. Gawecki, Warsaw 1956.

Smoluchowski M., *O oddziaływaniu wzajemnem kul poruszających się w ośrodku lepkim* (*On the interaction of spheres moving in a viscous liquid*) published by the Academy of Arts and Sceinces, Kraków 1911.

Smoluchowski M., *O pewnem zagadnieniu z teorii sprężystości i o związku jego z wytworzeniem się gór fałdowych* (*On a certain problem in the theory of elasticity and its relationship to the formation of fold mountains*), published by the Academy of Arts and Sciences, Kraków 1909.

Smoluchowski M., *O pewnem zagadnieniu kinetycznej teorii roztworów* (*On a certain problem of the kinetic theory of solutions*) in: *A commemorative book to celebrate the 250th anniversary founding of the University of Lviv by King Jan Kazimierz*, Lviv 1911.

Smoluchowski M., *O pojęciu przypadku i pochodzeniu praw fizyki opartych na prawdopodobieństwie* (*On the concept of chance and the origin of laws of physics based on probability*), 'Wiadomości Matematyczne' ('Mathematical News') 1923, vol. 27, z. 2, pp. 1–26.

Smoluchowski M., *O powstawaniu żył podczas wypływu cieczy*, 'Rozprawy Wydziału Matematyczno-Przyrodniczego Akademii Umiejętności', ('Dissertations of the Faculty of Mathematics and Natural Sciences of the Polish Academy of Arts and Sciences') series III, vol. I, section A, Kraków 1904.

Smoluchowski M., *O przewodnictwie cieplnem ciał sproszkowanych* (*On the thermal conductivity of powders*) published by the Academy of Arts and Sciences, Kraków 1910.

Smoluchowski M., *O przewodnictwie cieplnem gazów według dotychczasowych teoryj i doświadczeń* (*On the thermal conductivity of gases according to previous theories and experiments*) in: *The Writings of Marian Smoluchowski commissioned by the Polish Academy of Arts and Sciences*, vol. 1, Jagiellonian University Press, Kraków 1924, pp. 165–199.

Smoluchowski M., *O wynikach nowszych badań nad promieniowaniem* (*On the results of more recent research on radiation*) printout from 'Cosmos. Journal of the Polish Copernicus Society of Naturalists' 1900, books. II i III, pp. 74–87.

Smoluchowski M., *O zjawiskach aerodynamicznych i połączonych z niemi objawach cieplnych* (*On aerodynamic phenomena and the thermal effects that accompany them*), published by the Academy of Arts and Sciences, vol. XLIII, Kraków 1903.

Smoluchowski M., *On the Practical Applicability of Stokes' Law of Resistance, and the Modifications of it Required in Certain Cases*, University Press Cambridge, Cambridge 1912.

Smoluchowski M., *The Writings of Marian Smoluchowski commissioned by the Polish Academy of Arts and Sciences*, vol. 1, Jagiellonian University Press, Kraków 1924; vol. 2, Kraków 1927; vol. 3, Kraków 1928.

Smoluchowski M., *Poradnik dla samouków. Wskazówki metodyczne dla studiujących poszczególne nauki. Fizyka, Geofizyka, Meteorologia* (*Self-Study Handbook. Methodological guidelines for students of individual sciences. Physics, Geophysics, Meteorology*), vol. I and II, Wydawnictwo A. Heflicha i St. Michalskiego, Warsaw 1917.

Smoluchowski M., *Przedmiot, zadanie, metoda oraz podział fizyki* (*Subject, task, method and the division of physics*), in: *The Writings of Marian Smoluchowski commissioned by the Polish Academy of Arts and Sciences*, vol. 3, Jagiellonian University Press, Kraków 1928, pp. 153–204.

Smoluchowski M., *Przyczynek do kinetycznej teorii transpiracji, dyfuzji i przewodnictwa cieplnego w gazach rozrzedzonych* (*Contributions to the theory of transpiration, diffusion and thermal conduction in rarefied gases*), published by the Academy of Arts and Sciences, Kraków 1910.

Smoluchowski M., *Przyczynek do teorii endosmozy elektrycznej i kilku zjawisk pokrewnych* (*Contribution to the theory of electrical endosmosis and several related phenomena*), 'Rozprawy Wydziału Matematyczno-Przyrodniczego Akademii Umiejętności' ('Dissertations of the Faculty of Mathematics and Natural Sciences of the Polish Academy of Arts and Sciences'), series III, vol. 3, section A, Kraków 1903.

Smoluchowski M., *Przyczynek do teoryi ruchów cieczy lepkich, zwłaszcza zagadnień dwuwymiarowych* (*Contribution to the theory of movement of viscous liquids; in particular of two-dimensional problems*), in: *The Writings of Marian Smoluchowski commissioned by the Polish Academy of Arts and Sciences*, vol. 1, Jagiellonian University Press, Kraków 1924, pp. 539–554.

Smoluchowski M., *Sprawozdania z prac polskich na polu fizyki za lata 1901 i 1902* (*Reports on Polish papers in the field of physics for the years*

1901–1902), printout from 'Cosmos. Journal of the Polish Copernicus Society of Naturalists', books XI–XII, pp. 528–545.

Smoluchowski M., *Teorja kinetyczna opalescencji gazów w stanie krytycznym oraz innych zjawisk pokrewnych* (*Kinetic theory of the opalescence of gases in a critical state and of certain correlative phenomena*), 'Rozprawy Wydziału Matematyczno-Przyrodniczego Akademii Umiejętności' ('Dissertations of the Faculty of Mathematics and Natural Sciences of the Polish Academy of Arts and Sciences'), series A, vol. XLVII, Kraków 1907.

Smoluchowski M., *Teorja kinetyczna opalescencji gazów w stanie krytycznym oraz innych zjawisk pokrewnych* (*Kinetic theory of the opalescence of gases in a critical state and of certain correlative phenomena*), in: *The Writings of Marian Smoluchowski commissioned by the Polish Academy of Arts and Sciences*, vol. 1, Jagiellonian University Press, Kraków 1924, pp. 570–588.

Smoluchowski M., *Teorya kinetyczna gazów, według wykładów prof. dra Smoluchowskiego* (*The kinetic theory of gases, according to a lecture by Prof. Smoluchowski*), stenograph created by P. Jabłoński, published by L. Hołubowicz and W. Bilewski, Lviv 1908.

Smoluchowski M., *Termodynamika (i mechanika ciał sprężystych) podług wykładu prof. dra Smoluchowskiego w półroczu zimowym 1910/11* (*Thermodynamics (and the mechanic of elastic bodies) from a lecture by Prof. Smoluchowski 1910/11*), published by mathematical-physical circle, Lviv, 1911.

Smoluchowski M., *Über Unregelmäßigkeiten in der Verteilung von Gasmolekülen und deren Einfluß auf Entropie und Zustandsgleichung* (*On irregularities in the distribution of gas molecules and their influence on entropy and the equation of state*), in: *Festschrift Ludwig Boltzmann. Gewidmet zum sechzigsten geburtstage* (*Ludwig Boltzmann Festschrift (commemorative book). Dedicated to his sixtieth birthday*), Leipzig 1904, pp. 626–641.

Smoluchowski M., *Uwagi o pojęciu przypadku w zjawiskach fizycznych* (*Notes on the concept of chance in physical phenomena*), in: *The Writings of Marian Smoluchowski commissioned by the Polish Academy of Arts and Sciences*, vol. 3, Jagiellonian University Press, Kraków 1928, pp. 74–86.

Smoluchowski M., *Uwagi o roli przypadku we fizyce* (*Notes on the role of chance in physics*), hand-written manuscript, Jagiellonian Library, signature 9398 IV, k. 3

Smoluchowski M., *Uwagi o roli przypadku we fizyce* (*Notes on the role of chance in physics*), 'Zagadnienia Filozoficzne w Nauce' ('Philosophical Problems in Science') 2017, Kraków Philosophical Society, Kraków 1917, pp. 277–302.

Smoluchowski M., *Wybór pism filozoficznych* (*Selection of Philosophical Writings*), Polish Scientific Publishers PWN, Warsaw 1956.

Smoluchowski M., *Wycieczki górskie w Szkocji*, (*Mountain trips in Scotland*), 'Taternik'. Organ Sekcji Tury- stycznej Polskiego Towarzystwa Tatrzańskiego' ('Taternik, A unit of the Tourism Section of the Polish Tatra Society') 1921, pp. 5–9.

Smoluchowski M., *Zarys dziejów fizyki w Polsce* (*Outline of the history of physics in Poland*), reprint of an article in the *Self-Study Handbook*, 'Problemy' ('Problems') 1952, No 12, pp. 808–811.

Smoluchowski M., *Zarys kinetycznej teorii ruchów Browna i roztworów mętnych* (*Outline of a kinetic theory of Brownian motion and turbid solutions*), in: *The Writings of Marian Smoluchowski commissioned by the Polish Academy of Arts and Sciences*, vol. 1, Jagiellonian University Press, Kraków 1924, pp. 490–514.

Smoluchowski M., *Zarys teorii kinetycznej ruchów Browna i roztworów mętnych* (*Outline of a kinetic theory of Brownian motion and turbid solutions*), 'Rozprawy Wydziału Matematyczno-Przyrodniczego Akademii Umiejętności' ('Dissertations of the Faculty of Mathematics and Natural Sciences of the Polish Academy of Arts and Sciences'), vol. 6, section A, Kraków 1906, pp. 257–281.

Smoluchowski M., *Znaczenie nauk ścisłych w wykształceniu ogólnym* (*The importance of the exact sciences in general education*), speech given during theCongress of the Association of Higher School Teachers, May 27, 1917, 'Muzeum' ('Museum') 1917, vol. XXXII, pp. 286–294.

Smoluchowski M., *Zur kinetischen Theorie der Brownschen Molekularbewegung und der Suspensionen* (*On the Kinetic Theory of the Brownian Motion and of Suspensions*) 'Annalen der Physik' ('Annals of Physics') 1906, vol. 21, pp. 756–780.

Smoluchowski M., *Zur kinetischen Theorie der Brownschen Bewegung und der Suspensionen; von M. von Smoluchowski* (*On the Kinetic Theory of the Brownian Motion and of Suspensions; M. von Smoluchowski*), 'Annalen der Physik' ('Annals of Physics') 1906, vol. 326, No. 14, pp. 756–780.

Smoluchowski M., *Dzisiejszy stan teoryi atomistycznej* (*The current state of atomic theory*), printout from 'Cosmos. Journal of the Polish Copernicus Society of Naturalists' No. XXXVIII, Związkowa Drukarnia we Lwowie (Union Press of Lviv), Lviv 1913, pp. 354–373.

Smoluchowski M., *Dr Władysław Natanson, Odczyty i szkice* (*Dr Władysław Natanson, Lectures and sketches*), Warsaw 1908, 'AteneumPolskie' 1908, vol. II, No. 1, pp. 134–136.

Studies and source texts

A.A. Michelson, [speech at the University of Chicago], after: A.K. Wróblewski, *Historia fizyki* (*The History of Physics*), Polish Scientific Publishers PWN, Warsaw 2006.

Ajdukiewicz K., *Zagadnienia i kierunki filozofii. Teoria poznania, metafizyka* (*Issues and directions of philosophy. Theory of knowledge, metaphysics*), Czytelnik, Warsaw 1983.

Amsterdamski S., Augustynek Z., Mejbaum W., *Prawo, konieczność, prawdopodobieństwo* (*Law, necessity, probability*), Książka i Wiedza, Warsaw 1964.

Aristotle, *Physics*, trans. Robin Waterfield, Oxford University Press, Oxford, 1999.

Aristotle, *Posterior Analytics*, trans. Edward Poste, F. Macpherson, Oxford, 1850.

Aristotle, *Dzieła wszystkie. Analityki wtóre* (*Complete works. Posterior analytics*), trans. K. Leśniak, Polish Scientific Publishers PWN, Warsaw 1990.

Aristotle, *Dzieła wszystkie. Fizyka* (*Complete works. Physics*), trans. K. Leśniak, Polish Scientific Publishers PWN, Warsaw 1990.

Aristotle, *Dzieła wszystkie. Metafizyka* (*Complete works. Metaphysics*), trans. K. Leśniak, Polish Scientific Publishers PWN, Warsaw 1990.

Auerbach F., *Geschichtstafeln der Physik* (*Historical Tables of Physics*), Leipzig 1910.

Avenarius R., *Philosophie als Denken der Welt gemäss dem Prinzip des kleinsten Kraftmasses. Prolegomena zu einer Kritik der reinen Erfahrung* (*Thinking of the World According to the Principle of the Smallest*

Measure of Force. Prolegomena to a Critique of Pure Experience), Leipzig 1870.

Bacon F., *Novum Organum*, trans. J. Wikarjak, Polish Scientific Publishers PWN, Warsaw 1955.

Bacon F., *Novum Organum*, ed. J. Devey, P.F. Collier & Son, New York 1902.

F. Bacon, *Novum Organum 1620*, ed. and trans. Basil Montague, Parry & MacMillan, Philadelphia, 1854

Bafia S., *Fizyka w Quaestiones super octo libros 'physicorum' Aristotelis Jana z Głogowa*, Scriptum, Kraków 2013.

Bernstein H.T., *J. Clerk Maxwell on the History of the Kinetic Theory of Gases, 1871*, 'Isis' 1963, vol. 54, No. 2, pp. 206–216.

Bernstein J., *Einstein and the Existence of Atoms*, 'American Journal of Physics' 2006, vol. 74, No. 10, pp. 863–872.

Białobrzeski C., *Podstawy poznawcze fizyki świata atomowego (Cognitive foundations of the physics of the atomic world)*, Polish Scientific Publishers PWN, Warsaw 1984.

Biela A., *Analogia w nauce (Analogy in science)*, Instytut Wydawniczy PAX, Warsaw 1989.

Biela A., *Analogy in Science: From a Psychological Perspective*, P. Lang, Frankfurt am Main, Berlin, New York, Paris, 1991

Biernacki W., *Nowe dziedziny widma (New spectra domains)*, Drukarnia Granowskiego i Sikorskiego, Warsaw 1898.

Bigg C., *A Visual History of Jean Perrin's Brownian Motion Curves*, w: *Histories of Scientific Observation*, eds. L. Daston, E. Lunbeck, University of Chicago Press, Chicago 2011, pp. 156–179.

Bigg C., *Evident Atoms: Visuality in Jean Perrin's Brownian Motion Research*, 'Studies in History and Philosophy of Science' 2008, vol. 39, No. 3, pp. 312–322.

Birks J.B., *Rutherford at Manchester*, Heywood, London 1962.

Bohr N., *On the Constitution of Atoms and Molecules*, 'Philosophical Magazine and Journal of Science' 1913, vol. 26, No. 1, pp. 1–24.

Born M., *Natural Philosophy of Cause and Chance*, Clarendon Press, Oxford 1949.

Brush S.G., *History of Random Processes. I. Brownian Movement from Brown to Perrin*, 'Archive for History of Exact Sciences' 1968, vol. 5, No. 1, pp. 1–36.

Budzanowski A., *Znaczenie prac Mariana Smoluchowskiego dla fizyki subatomowej*, w: *Marian Smoluchowski. Od teorii atomistycznej do fizyki współczesnej* (*The significance of Marian Smoluchowski's work in subatomic physics. From atomic theory to modern physics*), ed. A. Strzałkowski, PAU, Kraków 2003.

Bunge M., *O przyczynowości. Miejsce zasady przyczynowej we współczesnej nauce* (*Causality: the place of the causal principle in modern science*), trans. S. Amsterdamski, Polish Scientific Publishers PWN, Warsaw 1968.

Calaprice A., *Einstein w cytatach* (*Einstein in quotes*), trans. M. Krośniak, Poltex, Warsaw 2014.

Carroll S., *Stąd do wieczności i z powrotem. Poszukiwanie ostatecznej teorii czasu* (*From eternity to here: The quest for the ultimate theory of time*), trans. T. Krzysztoń, Prószyński Media, Warsaw 2011.

Casimir H., *Haphazard Reality: Half a Century of Science*, Harper Books, New York 1984.

Casti J.L., DePauli W., *Gödel. Życie i logika* (*Gödel. A life of logic*), trans. P. Amsterdamski, Cis, Warsaw 2003.

Chandrasekharta S., *Stochastic Problems in Physics and Astronomy*, 'Reviews of Modern Physics' 1943, vol. 15, No. 1, pp. 1–89.

Christiansen F.V., *Heinrich Hertz's Neo-Kantian Philosophy of Science, and its Development by Herald Høffding*, 'Journal for General Philosophy of Science' 2006, vol. 37, No. 1, pp. 1–20.

Cialdini R., *Wywieranie wpływu na ludzi. Teoria i praktyka* (*Influence: Science and Practice*), trans. B. Wojciszke, Gdańskie Wydawnictwo Psychologiczne, Gdańsk 1994.

Cichocki B., *Nagroda Nobla* (*The Nobel Prize*), 'Delta' 1997, No. 12, p. 10.

Cichocki B., *Albert Einstein – praca o ruchach Browna z 1905 roku* (*Albert Einstein – the 1905 work on Brownian motion*), 'Delta' 2005, No. 6, pp. 4–6.

Cohen B.I., *Introduction to Newton's „Principia"*, Harvard University Press, Cambridge 1971.

Comte A., *Metoda pozytywna w szesnastu wykładach* (*The Positive method positive in sixteen lessons*), trans. W. Wojciechowska, Polish ScientificPublishersPWN, Warsaw 1961.

Copleston F., *Historia filozofii* (*A History of Philosophy*), vol. 8, trans. B. Chwedeńczuk, Instytut Wydawniczy PAX, Warsaw 2006.

Corbalán F., Sanz G., *Poskromienie przypadku. Teoria prawdopodobieństwa* (*The taming of chance. Probability theory*), trans. K. Rejmer, Buka Books Sławomir Chojnacki, Warsaw 2012.

Crombie A.C., *Nauka średniowieczna i początki nauki nowożytnej* (*Medieval and Early Modern Science*), vol. 1, (*Science in the Middle Ages, V–XIII Centuries*), trans. S. Łypacewicz, Instytut Wydawniczy PAX, Warsaw 1960.

Crombie A.C., *Nauka średniowieczna i początki nauki nowożytnej.*(*Medieval and Early Modern Science*), vol. 2, *Nauka w późnym średniowieczu i na początku czasów nowożytnych w okresie XIII–XVII w.* (*Science in the Later Middle Ages & Early Modern Times: XIII–XVII Centuries*), trans. S. Łypacewicz, Instytut Wydawniczy PAX, Warsaw 1960.

Czarnocka M., *Obserwacja bezpośrednia* (*Direct observation*), 'Studia Filozoficzne' 1990, No. 1 (290), pp. 155–167.

Davies P., Brown J., *Superstrings: A Theory of Everything?*, Cambridge 1988.

Dąmbska I., *O poglądach metanaukowych Władysława Natansona i Mariana Smoluchowskiego* (*On the metascientific views of Władysław Natanson and Marian Smoluchowski*), 'Zagadnienia Naukoznawstwa', Wydawnictwo Zakład Narodowy im. Ossolińskich, Wrocław 1979, vol. XV, book. 1, pp. 3–11.

Dawes G.W., *Belief is Not the Issue: A Defence of Inference to the Best Explanation*, 'Ratio. An International Journal of Analytic Philosophy' 2013, vol. 26, No. 1, pp. 62–78.

Domański C., *Statystyka bliżej biznesu*, w: *XIX Ogólnopolska Konferencja Dydaktyczna „Nauczanie przedmiotów ilościowych a potrzeby rynku pracy"* (*Statistics closer to business*, in: *XIX National Didactic Conference 'Teaching quantitative subjects and the needs of the labour market'*), Instytut Ekonometrii Uniwersytetu Łódzkiego, Łódź 2010, pp. 3–11.

G. Dryden, J. Vos, *The New Learning Revolution*, Network Educational Press, Stafford, 2005.

Dryden G., Vos J., *Rewolucja w uczeniu (The learning revolution)*, trans. B. Jóźwiak, Zysk i S-ka, Poznań 2000.

Duplantier B., *Brownian Motion, 'Diverse and Undulating'*, Einstein, 1905–2005: Poincaré Seminar 2005, 'Progress in Mathematical Physics' 2006, vol. 47, pp. 201–293.

Duplantier B., *Le mouvement brownien, „divers et ondoyant"Brownian Motion, 'Diverse and Undulating'*), Séminaire Poincaré 1 (Poincaré Seminar 1), Service de Physique Theorique, (Theoretical Physics Department) France 2005, pp. 155–212. Dyk-Majewska E., *Natchniony geniusz*, 'Pismo Politechniki Gdańskiej' 2008, No. 9, pp. 46–49.

Eftekhari A., *Ludwig Boltzmann (1844–1906)*, https://pdfs.semanticscholar.org/5c96/924ab515da7ebb6cb7601ec916099b03aed0.pdf.

Ehring D., *The Transference Theory of Causation*, 'Synthese'1986, vol. 67, No. 2, pp. 249–258.

Einstein A., *5 prac, które zmieniły oblicze fizyki (5 works that changed the face of physics)*, trans. P. Amsterdamski, University of Warsaw Press, Warsaw 2005.

Einstein A., *Autobiographical Notes*, w: *Albert Einstein, Philosopher-Scientist*, ed. P.A. Schilpp, Open Court, La Salle IL 1949.

Einstein A., *Über die von der molekularkinetischen Theorieder Wärme geforderte Bewegung von in ruhenden Flüssigkeiten suspendierten Teilchen (On the movement of small particles suspended in a stationary liquid required by the molecular kinetic theory of heat)*, 'Annalen der Physik' 1905, vol. 322, No. 8, pp. 549–560.

Einstein A., *Wspomnienie o Smoluchowskim (Recollections of Smoluchowski)*, 'Problemy' 1972, No. 8 (317), p. VIII.

Einstein A., *Ist die Trägheit eines Körpers von seinem Energieinhalt abhängig? (Does the Inertia of a Body Depend Upon its Energy Content?)*, 'Annalen der Physik' 1905, No. 18, pp. 639–641.

Einstein A., *Marian v. Smoluchowski*, 'Die Naturwissenschaften' 1917, vol. 5, No. 50, pp. 737–738.

Einstein A., *Über das Boltzmann'sche Prinzip und einige unmittelbar aus demselben fliessende Folgerungen (On Boltzmann's Principle and Some Immediate Consequences Thereof)*, Lectures for the Physical Society, Zurich 1910.

Einstein A., *Über einen die Erzeugung und Verwandlung des Lichtes betreffendenheuristischen Gesichtspunkt* (*On a Heuristic Point of View about the Creation and Conversion of Light*), 'Annalen der Physik' 1905, No. 17, pp. 132–148.

Einstein A., *Zur Elektrodynamik bewegter Körper* (*On the Electrodynamics of Moving Bodies*), 'Annalen der Physik' 1905, No. 17 pp. 891–921.

Einstein A., *Zur Theorie der Brownschen Bewegung* (*Theory of Brownian motion*), 'Annalen der Physik' 1906, vol. 324, No. 2, pp. 371–381.

Encyklopedia Britannica, vol. 40, Wydawnictwo Kurpisz, Poznań 2006.

Encyklopedia filozofii polskiej (*Encyclopaedia of Polish Philosophy*), Wydawnictwo Polskiego Towarzystwa Tomasza z Akwinu, Lublin 2011.

Ernst Mach's Vienna 1895–1930: Or Phenomenalism as Philosophy of Science, eds. J.T. Blackmore, R. Itagaki, S. Tanaka, Kluwer Academic Publishers, Dordrecht 2001.

Exner F.M., *Notiz zu Brown's Molecularbewegung* (*Notes on Brownian molecular motion*), 'Annalen der Physik' 1900, vol. 307, No. 8, pp. 843–847.

Faraday M., *The Philosopher's Tree: Michael Faraday's Life and Work in His Own Words*, ed. P. Day, CRC Press, London 1999.

Faraday Lecture (1881), In *On the Modern Development of Faraday's Conception of Electricity*, 'Journal of the Chemical Society' 1881

Festschrift Ludwik Boltzmann. Gewidmet zum sechzigsten geburtstage (*Commemorative book dedicated to Ludwig Boltzmann on his sixtieth birthday*), J.A. Barth, Leipzig 1904.

Feynman R.P., Leighton R.B., Sands M., *The Feynman Lectures on Physics. Volume I: Mainly Mechanics, Radiation, and Heat*, Addison-Wesley, Reading MA 1963.

Fraassen B.C. van, *The Scientific Image*, Oxford University Press, Oxford 1980.

Franklin J., *The Science of Conjecture: Evidence and Probability before Pascal*, The Johns Hopkins University Press, Baltimore 2001.

Fuliński A., *Współczesne zastosowania równań Smoluchowskiego* (*Contemporary applications of the Smoluchowski equations*) in: *Marian*

Smoluchowski. *Od teorii atomistycznej do fizyki współczesnej* (*Marian Smoluchowski. From atomic theory to modern physics*), ed. A. Strzałkowski, PAU, Kraków 2003.

Gapaillard J., *How Much Did Le Verrier Err on the Position of Neptune?*, 'Journal for the History of Astronomy' February 2015, No. 46 (1), pp. 48–65.

Gawecki B.J., *Szkice filozoficzne* (*Philosophical Sketches*), Książnica Atlas, Warsaw 1935.

Gawecki B.J., *Zagadnienie przyczynowości w fizyce* (*The issue of causality in physics*), Wydawnictwo PAX, Warsaw 1969.

Gawecki B.J., *Władysław Mieczysław Kozłowski (1858–1935)*, Wrocław 1961.

Gilson E., *Tomizm: wprowadzenie do filozofii św. Tomasza z Akwinu* (*Thomism an introduction to the philosophy of St. Thomas Aquinas*), trans. J. Rybałt, Instytut Wydawniczy PAX, Warsaw 1960.

Glymour C., *On the Possibility of Inference to the Best Explanation*, 'Journal of Philosophical Logic' 2012, vol. 41, No. 2, pp. 461–469.

Godlewski T., *Maryan Smoluchowski. Jego życie i działalność naukowa* (*Maryan Smoluchowski. His life and scientific activity*), Wydawnictwo Redakcyi 'Wiadomości Matematycznych', Warsaw 1919; first edition: 'Wiadomości Matematyczne' 1919, vol. 23, books. 1–3, pp. 1–36.

Gołąb-Meyer Z., *Mariana Smoluchowskiego nauczanie fizyki z perspektywy stulecia* (*Marian Smoluchowski's teaching of physics from a century's perspective*), 'Pismo dla nauczycieli fizyki, przyrody oraz ich uczniów' ('Journal for teachers of physics and nature, and their students') 2003, No. 81, pp. 32–36.

Gosiewski W., *Zasady rachunku prawdopodobieństwa* (*Rules of the calculus of probability*), Skład Główny w Księgarni E. Wendego i Sp., Warsaw 1906.

Gostkowski K., *Kilka wspomnień o Marianie Smoluchowskim* (*Several recollections of Marian Smoluchowski*), 'Postępy Fizyki' 1953, vol. 4, book. 2, pp. 233–236.

Goswami A., *Kwantowy umysł. Naukowe dowody na potęgę twoich myśli* (*Quantum mind. Scientific Evidence for the Power of Your Thoughts*), trans. A. Rutkowska, Studio Astropsychologii, Białystok 2014.

Givindaraju V., Raghavan V.V., Rao D., Big Data Driven Natural Language Processing Research and Applications, w: Handbook of Statistics, Elsevier 2015, vol. 33, pp. 203–238.

Grobler A., *Metodologia nauk (The methodology of the sciences)*, Wydawnictwo Aureus, Znak, Kraków 2006.

Grygiel W.P., *Stephena Hawkinga i Rogera Penrose'a spór o rzeczywistość (Stephen Hawking and Roger Penrose's dispute over reality)*, Copernicus Center Press, Kraków 2014.

Hajduk Z., *Czy sukcesy nauki są cudem? Studium filozoficzno-metodologiczne argumentacji z sukcesu nauki na rzecz realizmu naukowego (Are the Successes of Science a Miracle? A Philosophical-Methodological Study on the Argumentation from the Success of Science on Behalf of Scientific Realism)*, review of 'Roczniki Filozoficzne' 2006, No. 54 (2), pp. 429–432.

Hajduk Z., *Filozofia przyrody. Filozofia przyrodoznawstwa. Metakosmologia (Philosophy of Nature. Philosophy of natural science. Metacosmolgy)*, Towarzystwo Naukowe KUL, Lublin 2007.

Hajduk Z., *Między filozofią przyrody a filozofią nauki (Between the philosophy of nature and philosophy of science)*, 'Roczniki Filozoficzne' 2006, vol. 54, No. 2, pp. 7–15.

Hajduk Z., *O przyczynowości: miejsce zasady przyczynowej we współczesnej nauce. Mario Bunge (Causality: the place of the causal principle in modern science. Mario Bunge)*, trans. S. Amsterdamski, 'Studia Philosophiae Christianae' 1969, No. 5/2, pp. 217–225.

Harman G., *The Inference to the Best Explanation*, 'The Philosophical Review' 1965, No. 74, pp. 88–95.

Hawking S.W. *A Brief History of Time. From the Big Bang to Black Holes*, Bantam Books, New York 1998

Hawking S.W., *Krótka historia czasu. Od wielkiego wybuchu do czarnych dziur (A Brief History of Time. From the Big Bang to Black Holes)*, trans. P. Amsterdamski, Alfa, Warsaw 1990.

Hawking S.W., *Brief Answers to the Big Questions*, Bantam Books, New York 2018

Hawking S.W., *Krótkie odpowiedzi na wielkie pytania (Brief Answers to the Big Questions)*, trans. M. Krośniak, Zysk i S-ka, Warsaw 2019.

Hawking S.W., *Wszechświat w skorupce orzecha* (*The universe in a nutshell*), trans. P. Amsterdamski, Zysk i S-ka, Poznań 2018.

G.W.F. Hegel, Elements of the Philosophy of Right, ed. Allen W. Wood, trans. H.B. Nisbet, Cambridge University Press, Cambridge 2003

Hegel Georg Wilhelm Friedrich, *Zasady filozofii prawa* (*Elements of the philosophy of right*), trans. A. Landman, Polish Scientific Publishers PWN, Warsaw 1969.

Heisenberg W., *Physics and Philosophy: The Revolution in Modern Science*, Penguin, London, 2000

Heisenberg W., *Fizyka a filozofia* (*Physics and Philosophy*), trans. S. Amsterdamski, Książka i Wiedza, Warsaw 1965.

Heisenberg W., *Część i całość. Rozmowy o fizyce atomu* (*Physics and Beyond: Encounters and Conversations*), trans. K. Napiórkowski, Państwowy Instytut Wydawniczy, Warsaw 1987.

Heller M., *Czas i przyczynowość* (*Time and causality*), Towarzystwo Naukowe Katolickiego Uniwersytetu Lubelskiego, Lublin 2002.

Heller M., *Jak być uczonym* (*How to be a scientist*), Znak, Kraków 2009.

Heller M., *Moralność myślenia* (*The morality of thought*), Biblos, Tarnów 1993.

Heller M., *Wszystkim, którzy chcą być filozofami* (*To all who want to be philosophers*), 'Semina Scientiarum' 2008, No. 7, pp. 3–4.

Heller M., Pabjan T., *Elementy filozofii przyrody* (*Elements of the philosophy of nature*), Copernicus Centre Press, Kraków 2014.

Heller M., Życiński J., *Matematyczność przyrody* (*The Mathematicality of nature*), Petrus, Kraków 2010.

Herbut J., *Pragmatyzm* (*Pragmatism*), in: *Universal Encyclopedia of Philosophy*, vol. 8, Polskie Towarzystwo Tomasza z Akwinu, Lublin 2000–2009.

Herschel J.F.W., *A Preliminary Discourse on the Study of Natural Philosophy*, Longman and J. Taylor, London, 1830

Herschel J.F.W., *Wstęp do badań przyrodniczych* (*A Preliminary Discourse on the Study of Natural Philosophy*), trans. T. Pawłowski, Polish Scientific Publishers PWN, Warsaw 1955.

Historia matematyki (*History of mathematics*), vol. 3, *Matematyka osiemnastego stulecia* (*Mathematics of the 18th century*), ed. A.P. Juszkiewicz, Polish Scientific Publishers PWN, Warsaw 1977.

Hoefler A., *Logika propedeutyczna dla szkół średnich* (*Propaedeutic logic for secondary schools*), trans. Z. Zawirski, Nakładem Księgarni Naukowej, Lviv 1927.

Home R.W., *Sutherland, William (1859–1911)*, w: *Australian Dictionary of Biography*, vol. 12, http://adb.anu.edu.au/biography/sutherland-william-8719.

Hossenfelder S., *Zagubione w matematyce. Fizyka w pułapce piękna* (*Lost in Math: How Beauty Leads Physics Astray*), trans. T. Miller, Copernicus Center Press, Kraków 2019.

Hume D., *An Enquiry Concerning Human Understanding*, ed. Eric Steinberg, Second Edition, Hacket, Indianapolis/Cambridge, 1993

Hume D., *Traktat o naturze ludzkiej* (*A Treatise of Human Nature*), trans. C. Znamierowski, Fundacja Aletheia, Warsaw 2005.

Hume D., *Badania dotyczące rozumu ludzkiego* (*An enquiry concerning human understanding*), trans. D. Misztal, T. Sieczkowski, Zielona Sowa, Kraków 2004.

Ingarden R., *Wstęp do fenomenologii Husserla* (*An introduction to Husserl's phenomenology*), trans. A. Półtawski, Polish Scientific Publishers PWN, Warszawa 1974.

Isaacson W., *Einstein. Jego życie, jego wszechświat* (*Einstein. His life, his universe*), W.A.B., Warsaw 2014.

James W., *Pragmatyzm. Nowa nazwa kilu starych metod myślenia. Popularne wykłady z filozofii* (*Pragmatism: A New Name For Some Old Ways Of Thinking. Popular lectures on philosophy*), trans. M. Filipczuk, Zielona Sowa, Warsaw 2004.

James W., *Znaczenie prawdy. Ciąg dalszy pragmatyzmu* (*The meaning of truth. A sequel to pragmatism*), trans. M. Szczubiałka, Warsaw 2000.

Jarocki J., *Historyczne i systematyczne ujęcie monizmu neutralnego* (*Neutral monism—a historical and systematic approach*), 'Przegląd Filozoficzny' 2015, No. 3, pp. 177–193.

Jerusalem W., *Wstęp do filozofii* (*Introduction to philosophy*), published by Księgarnia Polska B. Połonieckiego, Lviv and Warsaw 1926.

Kac M., *Marian Smoluchowski and the Evolution of Statistical Physics*, in: S. Chandrasekhar, M. Kac, R. Smoluchowski, *Polish Men of Science*.

Marian Smoluchowski. His Life and Scientific Work, ed. R.S. Ingarden, Polish Scientific Publishers PWN, Warsaw 1986, pp. 15–21.

Kac M., *Enigmas of Chance, An autobiography*, Harper and Row, New York 1985

Kaczorowska E., Knyziak B.A., *Historia odkrycia promieniotwórczości (The history of the discovery of radioactivity)*, 'Meteorologia. Biuletyn Głównego Urzędu Miar' 2011, No. 2, vol. 6, pp. 20–30.

Kampen N.G. van, *Procesy stochastyczne w fizyce i chemii (Stochastic Processes in Physics and Chemistry)*, ed. Ł.A. Turski, trans. M. Dudyński, M. Ekiel-Jeżewska, D. Śledziewska-Błocka, Polish Scientific Publishers PWN, Warsaw 1990.

Kapczyński M., *Indeks Hirscha zastosowanie oraz metody obliczania (The Hirsch Index – application and calculation method)*, Thomson Reuters (Scientific), 2 lipca 2012, https://puss.pila.pl/uploads/indeks_h_zastosowanie_metody_obliczania.pdf.

Kapuściński W., *Poglądy filozoficzne Mariana Smoluchowskiego (Marian Smoluchowski's philosophical views)*, 'Fizyka i Chemia' 1953, vol. 4 (28), pp. 200–209.

Kelly T., *Hume, Norton, and Induction without Rules*, 'Philosophy of Science' 2010, vol. 77, No. 5, pp. 754–764.

Kerker M., *The Svedberg and Molecular Reality*, 'Isis. A Journal of the Science Society' 1976, vol. 67 (21), pp. 190–216.

Kistler M., *The causal criterion of reality and necessity of the laws of nature*, 'Metaphysica' 2002, No. 3 (1).

Klemensiewicz Z., *Marian Smoluchowski, wspomnienie sprzed lat czterdziestu (Marian Smoluchowski, a memoir from forty years ago)*, 'Kosmos'1958, series B, No. 4, pp. 95–107.

Kołakowski L., *Główne nurty marksizmu (Main Currents of Marxism)*, Polish Scientific Publishers PWN, Warsaw 2009.

Kołakowski L., *Filozofia pozytywistyczna. Od Hume'a do Koła Wiedeńskiego (The Philosophy of positivism. From Hume to the Vienna Circle)*, Polish Scientific Publishers PWN, Warsaw 1966

Korpikiewicz H., *Koncepcja wzrostu entropii a rozwój świata (The concept of entropy growth and the development of the world)*, Wydawnictwo Naukowe UAM, Poznań 1998.

Kosmulski M., *Potencjał ζ i równanie Smoluchowskiego* (*ζ Potential and the Smoluchowski equation*), 'PAUza Akademicka. Tygodnik PAU' 2017, No. 380–381, p. 7.

Kostro L., *Alberta Einsteina koncepcja nowego eteru: jej historia, sens fizyczny i uwarunkowania filozoficzne* (*Albert Einstein's concept of a new ether: its history, physical meaning and philosophical determinants*), Scientia, Gdańsk 1999.

Kostro L., *Transpersonal Cosmic Absolute*, 'Scientific Studies and Research. Studia i Badania Naukowe. Europeistyka' 2012, 6, No. 2, pp. 73–103.

Kotarbiński T., *Elementy teorii poznania logiki formalnej i metodologii nauk* (*Elements of the theory of knowledge, formal logic and methodology of the sciences*), Polish Scientific Publishers PWN, Warsaw 1986.

Kotowa B., *Scjentyzm jako światopogląd nauki* (*Scientism as a worldview of science*), 'Nowa Krytyka' 2004, No. 16, pp. 151–159.

Kozłowski W.M., *Przyczynowość jako podstawowe pojęcie przyrodoznawstwa* (*Causality as a fundamental concept of natural science*), Główny Skład u E. Wende'go i S-ki, Warsaw 1906.

König E., *Die Entwicklung des Kausalproblems in der neueren Philosophie. Studien zur Orientirung über die Aufgaben der Metaphysik und Erkenntnisslehre. Band 1: Die Entwicklung des Kausalproblems von Cartesius bis Kant* (*The development of the causal problem in contemporary philosophyStudies for orientation on the tasks of metaphysics and epistemology. Volume 1: The Development of the Problem of Causality from Descartes to Kant.*), Leipzig 1888.

Kragh H., *Max Planck: the reluctant revolutionary*, 'Physics World' 2000, vol. 13, No. 12, pp. 31–36.

Krajewski W., *Marian Smoluchowski – a forerunner of the chaos theory*, in: *Polish philosophers of science and nature in the 20th century*, 'Poznańskie Studia z Filozofii Nauk i Nauk Humanistycznych' 2001, No. 74, pp. 185–188.

Krajewski W., *Marian Smoluchowski jako filozof i materialista* (*Marian Smoluchowski as a materialist philosopher*), 'Świat i My' 1952, No. 28, p. 3.

Krajewski W., *Marian Smoluchowski jako filozof i materialista* (*Marian Smoluchowski as a materialist philosopher*) 'Widnokrąg' 1952, No. 24, p. 2.

Krajewski W., *Marian Smoluchowski jako filozof i materialista (Marian Smoluchowski as a materialist philosopher)*, 'Życie i Kultura', suplement to the newspaper: 'Głos Szczeciński' 1952, No. 23, pp. 6–7.

Krajewski W., *Marian Smoluchowski jako filozof i materialista (Marian Smoluchowski as a materialist philosopher)*, 'Problemy' 1952, No. 12, pp. 843–845.

Krajewski W., *Marian Smoluchowski jako filozof-materialista (Marian Smoluchowski as a materialist philosopher)*, 'Myśl Filozoficzna' 1952, No. 4, pp. 232–248.

Krajewski W., *Światopogląd Mariana Smoluchowskiego (Marian Smoluchowski's worldview)*, Polish Scientific Publishers PWN, Warsaw 1956.

Krajewski W., *Szkice filozoficzne (Philosophical sketches)*, Książka i Wiedza, Warsaw 1963.

Krajewski W., *Wieki fizyk i filozof materialista (w 80-lecie urodzin Mariana Smoluchowskiego), (The Great Physicist and Materialist Philosopher (On the 80th Birthday of Marian Smoluchowski))* 'Trybuna Ludu' 1952, No. 148, pp. 1 and 4; later reprinted in the daily newspaper 'Słowo Ludu' 1952, No. 129 and in the magazines: 'Świat i My' 1952, No. 28, 'Widnokrąg' 1952, No. 24, 'Życie i Kultura', suplement to the newspaper'Głos Szczeciński' 1952, No. 23, „Nowiny Tygodnia" suplement to the newspaper'Nowiny Rzeszowskie' 1952, No. 23, 'Słowo Tygodnia' 1952, No. 14, 'Problemy' 1952, No. 12.

Krajewski W., *Związek przyczynowy (The causal relationship)*, Polish Scientific Publishing PWN, Warsaw 1967.

Krakowska filozofia przyrody w okresie międzywojennym (Kraków Philosophy of nature in the interwar period), vol. 3, *Smoluchowski – Natanson – inni (Smoluchowski – Natanson – others)*, ed. M. Heller, J. Mączka, Ośrodek Badań Interdyscyplinarnych przy Wydziale Filozoficznym Papieskiej Akademii Teologicznej – Wydawnictwo Diecezji Tarnowskiej (Centre for Interdisciplinary Studies at the Faculty of Philosophy of the Pontifical Academy of Theology Publishing House of the Tarnów Diocese) Biblos, Kraków–Tarnów 2007.

Krąpiec M.A., *Analogia (Analogy)*, in: *Powszechna Encyklopedia Filozofii*, vol. 1, Polskie Towarzystwo Tomasza z Akwinu, Lublin 2000–2009.

Kuhn T., *The Copernican Revolution: Planetary astronomy in the development of Western Thought*, Harvard University Press, Cambridge, Massachusetts and London, England, 1992

Kuhn T., *Struktura rewolucji naukowych* (*The Structure of Scientific Revolutions*), trans. H. Ostromęcka, Aletheia, Warszawa 2009.

Lemańska A., *Determinizm*, http://www.kul.pl/files/57/encyklopedia/lemanska_determinizm.pdf.

Lederman L.M., *The God Particle Et Al*, 'Nature' 2007, No. 448, pp. 310–312.

Lipton P., *Inference to the Best Explanation*, Routledge, London 2007.

Litwinowicz-Droździel M., *Indukcje i przepływy. Michael Faraday – mikrostudium o romantycznej nauce* (*Inductions and Flows. Michael Faraday – a Microstudy of Romantic Science*), 'Wiek XIX. Rocznik Towarzystwa Literackiego im. Adama Mickiewicza' 2015, vol. VIII (L), pp. 89–102.

Lorentz H.A., *Die elektromagnetische Theorie von Maxwell und ihre Anwendung auf bewegte Körper* (*Maxwell's electromagnetic theory and its application to bodies in motion*), 'Archives néerlandaises des sciences exactes et naturelles' 1982, No. 25, pp. 363–551.

Loria S., *Marian Smoluchowski (1872–1917). Wspomnienie i próba charakterystyki* (*Marian Smoluchowski (1872–1917). Recollections and an attempt at characterization*), 'Problemy' 1952, No. 12, pp. 794–806.

Loria S., *Marian Smoluchowski i jego dzieło (1872–1917)* (*Marian Smoluchowski and his work (1872–1917)*), 'Postępy Fizyki' 1953, vol. 4, book. 1, pp. 5–38.

Loria S., *Marian Smoluchowski (1872–1917). Wspomnienie i próba charakterystyki* (*Marian Smoluchowski (1872–1917). Recollections and an attempt at characterization*), 'Rocznik Polskiego Towarzystwa Fizycznego' 1953, vol. 4, book. 1, pp. 794–807.

Lubomirski A., *Henri Poincarégo filozofia geometrii* (*Henri Poincaré's philosophy of geometry*), Zakład Narodowy im. Ossolińskich, Wrocław 1974.

Łakoma E., *O narodzinach pojęcia prawdopodobieństwa* (*On the birth of the concept of probability*), 'Historia Nauki i Techniki' 1989, No. 34 (3), pp. 613–632.

Mach E., *Analiza wrażeń i stosunek sfery fizycznej do psychicznej* (*The analysis of sensations, and the relation of the physical to the psychical*), trans. M. Miłkowski, Polish Scientific Publishers PWN, Warsaw 2009.

Mach E., *Erkenntnis und Irrtum. Skizzen zur Psychologie Forschung* (*Knowledge and Error: Sketches on the Psychology of Enquiry*), Verlag von Johann Ambrosius Barth, Leipzig 1906.

Maiocchi R., *The Case of Brownian Motion*, 'The British Journal for the History of Science' 1990, vol. 23, No. 3, pp. 257–283.

Marian Smoluchowski (1872–1917). Fizyk, taternik, romantyk nauki (*Marian Smoluchowski 1872–1917. Physicist, mountaineer, romantic of science*), catalogue of the temporary exhibition Collegium Maius, 17 May–14 July 2002.

Marshall S.J., *Shaping the University of the Future*, University of Wellington, Wellington 2018.

Maryniarczyk A., *Tomizm* (*Thomism*), in: *Powszechna Encyklopedia Filozofii*, vol. 9, Lublin 2000–2009.

Marx W., Cardona M., *Blasts from the Past*, 'Physics World' 2004, vol. 17, No. 2, pp. 14–15.

Analogia w filozofii (*Analogy in philosophy*), ed. A. Maryniarczyk, K. Stępień, P. Skrzydlewski, Polskie Towarzystwo Tomasza z Akwinu, Lublin 2005.

Maxwell J.C., *A dynamical theory of the electromagnetic field*, 1864, § 20.

Maxwell J.C., *On Faraday's Lines of Force*, 'Transactions of the Cambridge Philosophical Society' 1864, vol. X, part I, No. III, pp. 27–83.

Maxwell J.C., *Theory of heat*, Longman's, Green & Co, London, 1871.

McMullin E., *Wartości w nauce* (*Values in science*), trans. J. Rodzeń, 'Zagadnienia Filozoficzne w Nauce' 1999, vol. XXIV, pp. 7–25.

Metallmann J., *Zagadnienie przypadku* (*The problem of chance*), 'Przegląd Współczesny' 1933, XII, vol. XLIV, pp. 85–95.

Metallmann J., *Zasada ekonomii myślenia, jej historya i krytyka* (*The principle of economy of thought, its history and criticism*), Towarzystwo Naukowe w Warszawie, Warsaw 1914.

Nicholas of Cusa, *O oświeconej niewiedzy* (*On Enlightened Ignorance*), trans. I. Kania, Znak, Kraków 1997.

Millican P., *Hume'a wątpliwości sceptyczne dotyczące indukcji* (*Hume's sceptical doubts concerning induction*), trans. P. Grabarczyk, 'Nowa Krytyka' 2007, vol. 20–21.

Morales J.M.R., *Kościół i nauka. Konflikt czy współpraca?* (*The Church and Science. Conflict or cooperation?*), trans. S. Jędrusiak, WAM, Kraków 2003.

Natanson W., *Przypisek do rozprawy M. Smoluchowskiego p.t. „O fluktuacjach termodynamicznych i ruchach Browna"* (*Annotation to a dissertation by M. Smoluchowski entitled 'On thermodynamic fluctuations and Brownian motion'*), Jagiellonian University Press, Kraków 1927.

Nernst W., *Stöchiometrie Verwandtschaftslehre* (*Stoichiometry Relationship Theory*), 'Zeitschrift für Physikalische Chemie' 1888, vol. 2, pp. 613–639.

I. Newton, *Mathematical principles of natural philosophy*, University of California Press, Berkeley, Los Angeles and London, 2022.

Newton's Dream, ed. M.S. Stayer, McGill-Queen's University Press 1989.

Niemiec M., *Marian Smoluchowski – człowiek wszechstronny* (*Marian Smoluchowski – a multifaceted man*), 'Pismo Uniwersytetu Opolskiego' 2006, No. 3–4 (69–70), pp. 26–30.

Norton J.D., *A Material Theory of Induction*, 'Philosophy of Science' 2003, vol. 70, No. 4, pp. 647–670.

Nyhof J., *Philosophical Objections to the Kinetic Theory*, 'The British Journal for the Philosophy of Science' 1988, vol. 39, No. 1, pp. 81–109.

Orłowski Z., *Zastosowanie analogii w nauczaniu fizyki* (*The use of analogy in teaching physics*), educational service: www.profesor.pl.

Ockham Guilelmus de, *Summa Totius Logicae*, Bavarian state library, 1522, I, 12.

Pascal Blaise, *Myśli* (*Pensées/Thoughts*), trans. T. Żeleński-Boy, Instytut Wydawniczy PAX, Warsaw 1977.

R. Penrose, *The Road to Reality. A Complete Guide to the Laws of the Universe*, Jonathan Cape, London 2004

R. Penrose, *Foreword*, in: A. Einstein, *5 prac, które zmieniły oblicze fizyki* (A. Einstein, *5 works that changed the face of physics*), trans. P. Amsterdamski, University of Warsaw Press, Warsaw 2005.

Perrin J., *Atoms*, D. van Nostrand Company, New York 1916.

Perrin J., *Les Atomes*, Librairie Felix Alcan, Paris 1913.

Perrin J., *Mouvement brownien et réalité moléculaire*, 'Annales de chimie et de physique' 1909, vol. 18, Englsih version: F. Soddy, *Brownian Motion and Molecular Reality*, Tylor and Francis, London 1910, faksymile w: *Classical Scientific Papers: Chemistry*, ed. D.M. Knight, American Elsevier, New York 1968, pp. 1–114.

Piersa H., *Analogia w fizyce (Analogy in physics)*, 'Roczniki Filozoficzne' 2011, vol. 59, No. 2, pp. 239–256.

Piotrowski R., *Demon Maxwella. Dzieje i filozofia pewnego eksperymentu (Maxwell's Demon. The history and philosophy of a certain experiment)*, Wydawnictwo Akademickie Dialog, Warsaw 2011.

Planck M., *Jedność fizycznego obrazu świata. Wybór pism filozoficznych (The unity of the physical world-picture. A selection of philosophical writings)*, trans. R. i S. Kernerowie, Książka i Wiedza, Warsaw 1970.

Planck M., *Theorie der Wärmestrahlung (The Theory of heat Radiation)*, Leipzig 1906.

Planck M., *Thermodynamik (Thermodynamics)*, Leipzig 1897.

Planck M., *Über irreversible Strahlungsvorgänge (On irreversible radiation processes)*, 'Annalen der Physik' 1900, vol. 306, No. I. 1.

Planck M., *Vom Relativen zum Absoluten (From the Relative to the Absolute)*, in: M. Planck, *Jedność fizycznego obrazu świata. Wybór pism filozoficznych (The unity of the physical world-picture. A selection of philosophical writings)*, trans. R. i S. Kernerowie, Książka i Wiedza, Warsaw 1970.

Plato, *Parmenides. Teajtet (Parmenides. Theaetetus)*, trans. W. Witwicki, Antyk, Kęty 2002.

Poincaré H., *Nauka i hipoteza (Science and Hypothesis)*, trans. M.H. Horwitz, published by Jakub Mortkowicz, G. Centnerszwer i S-ka, Warszawa 1908.

Poincaré H., *Nauka i metoda (Science and method)*, tłum. M.H. Horwitz, published by Jakub Mortkowicz, G. Centnerszwer i S-ka, Warsaw 1911.

Poincaré H., *Science and Hypothesis*, The Walter Scott Publishing Co., New York, 1905.

Poincaré H., *Teoria Maxwella i fale Hertza (Maxwell's Theory and Hertzian Oscillations)*, trans. W. Malinowski, Skład Główny Księgarni E. Wende i S-ki, Warsaw 1917.

Poincaré H., *The Value of Science*, Cosimo, New York, 2007

Poincaré H., *The Value of Science*, Authorised translation with introduction by George Bruce Halstead, New York, 1958.

Poincaré H., *Wartość nauki (The value of science)*, trans. L. Silberstein, Studencka Oficyna Wydawnicza ZSP Alma-Press, Warsaw 1988.

Polak P., *„Byłem Pana przeciwnikiem [profesorze Einstein]..." (I was your opponent [Professor Einstein]...)*, Copernicus Center Press, Kraków 2012.

Polak P., *Koncepcja przypadku w pismach Mariana Smoluchowskiego (The concept of chance in the writings of Marian Smoluchowski)*, in: *Krakowska filozofia przyrody w okresie międzywojennym (Kraków philosophy of nature in the interwar period)*, vol. 3, *Smoluchowski – Natanson – inni (Smoluchowski – Natanson – others)*, ed. M. Heller, J. Mączka, Ośrodek Badań Interdyscyplinarnych przy Wydziale Filozoficznym Papieskiej Akademii Teologicznej, Wydawnictwo Diecezji Tarnowskiej (Centre for Interdisciplinary Studies at the Faculty of Philosophy of the Pontifical Academy of Theology Publishing House of the Tarnów Diocese) Biblos, Kraków–Tarnów 2007, pp. 443–460.

Polak P., *Marian Smoluchowski jako filozof w świetle pewnego rękopisu (Marian Smoluchowski as a philosopher in light of a certain manuscript)*, 'Postępy Fizyki' 2009, vol. 60, book. 6, pp. 236–239.

Popper K.R. *Conjectures and Refutations: The Growth of Scientific Knowledge* Routledge, London and New York, 2014

Popper K.R., *Droga do wiedzy. Domysły i refutacje (Conjectures and Refutations: The Growth of Scientific Knowledge)*, trans. S. Amsterdamski, Polish Scientific Publishers PWN, Warsaw 1999.

Popper K.R., *Logika odkrycia naukowego (The Logic of Scientific Discovery)*, trans. U. Niklas, pub. PWN, Warsaw 1997.

Popper K.R., *Objective Knowledge: An Evolutionary Approach*, Edition 2, Clarendon Press, Oxford, 1979

Portel Bueso X., *Supersymmetry Searches at the Tevatron and the LHC*, paper delivered at the 31st International Symposium on Physics in Collision, Vancouver, Canada, August–September 2011.

Psillos S., *Moving molecules above the scientific horizon: on Perrin's case for realism*, 'Journal for General Philosophy of Science' 2011, vol. 42(2), pp. 339–363.

Putnam H., *What is "Realism"?* Proceedings of the Aristotelian Society, vol. 76, 1975, pp. 177–178.

Putnam H., *Philosophical Papers*, vol. 2, *Mind, Language, and Reality*, Cambridge University Press, Cambridge 1975.

Putnam H., *Wiele twarzy realizmu i inne eseje* (*The many faces of realism and other essays*), trans. A. Grobler, Polish Scientific Publishers PWN, Warsaw 2013.

Regt Henk W. de, *Ludwig Boltzmann's Bildtheorie and Scientific Understanding*, 'Synthese' 1999, No. 119(1), pp. 113–134.

Regt Henk W. de, *Philosophy and the kinetic theory of gases*, 'The British Journal for the Philosophy of Science' 1996, No. 47, pp. 31–62.

Reinchenbach H., *Kausalität und Wahrscheinlichkeit* (*Causality and Probability*), 'Erkenntnis' 1930, Nr. 1, pp. 158–188.

Rodzeń J., *Czy sukcesy nauki są cudem? Studium filozoficzno-metodologiczne argumentacji z sukcesu nauki na rzecz realizmu naukowego* (*Are the successes of science a miracle? A philosophical-methodological study on the argumentation from the success of science in favour of scientific realism*), OBI, Tarnów 2005.

Röntgen W.C., *On a New Kind of Rays*, Science 3, no. 59 (1896): 227–31. p. 231

Röntgen W.C., *Über eine neue Art von Strahlen* (*On a new kind of rays*), 'Physikalisch-Medizinische Gesellschaft in Würzburg' 1895, pp. 132–141.

Roskal Z.E., *Mariana Smoluchowskiego ujęcie zasady przyczynowości w badaniach ruchów Browna* (*Marian Smoluchowski's approach to the causality principle in research on Brownian motion*), 'Zagadnienia Filozoficzne w Nauce' 2017, No. 62, pp. 99–126.

Roskal Z.E., *Zasada przyczynowości w kontekście eksperymentalnych badań ruchów Browna* (*The principle of causality in the context of experimental research on Brownian motion*), 'Filo-Sofija' 2017, No. 37, pp. 59–73.

Roskal Z.E., *Marian Smoluchowski's Contributions to the Philosophy of Causation*, 'Organon' 2018, No. 50, pp. 5–18.

Rozpravy Československé akademie věd: Řada společenských věd (*Debates of the Czechoslovak Academy of Sciences: social sciences series*), vol. 80, Prague 1970.

Russell B., *Recent Work on the Principles of Mathematics*, 'International Monthly', Vol. 4, 1901, p. 84.

Sierotowicz T., *Realizm w kontekście nauki (Realism in the context of science)*, 'Filozofia Nauki' 1997, No. 5(1), pp. 27–38.

Sławianowski J.J., *Przyczynowość w mechanice kwantowej (Causality in quantum mechanics)*, Wiedza Powszechna, Warsaw 1969.

Słomski W., *Władysław Krajewski*, in: *Polska filozofia powojenna (Polish post-war philosophy)*, ed. W. Mackiewicz, Agencja Wydawnicza Witmark, Warsaw 2001.

Smith A., *Badania nad naturą i przyczynami bogactwa narodów (An Inquiry into the nature and causes of the wealth of nations)*, vol. II, trans. A. Prejbisz, Polish Scientific PublishersPWN, Warsaw 1954.

Smullyan R., *Na zawsze nierozstrzygnięte. Zagadkowy przewodnik po twierdzeniach Gödla (Forever Unresolved. A mysterious guide to Gödel's theorems)*, trans. J. Pogonowski, Książka i Wiedza, Warsaw 2007.

Skarżyński B., *O Jędrzeju Śniadeckim (On Jędrzej Śniadecki)*, Wiedza Powszechna, Warsaw 1955.

Sommerfeld A., *Zum Andenken an Marian von Smoluchowski (In memory of Marian von Smoluchowski)*, 'Physikalische Zeitschrift' 1917, No. 22, pp. 533–539.

Stachel J., *Einstein on Brownian Motion*, w: *The Collected Papers of Albert Einstein*, vol. 2, *The Swiss Years: Writings, 1900–1909*, ed. J. Stachel, Princeton University Press, Princeton 1989, p. 216.

Stachel J., *Introduction*, in: A. Einstein, *5 prac, które zmieniły oblicze fizyki* (A. Einstein, *5 works that changed the face of physics*), trans. P. Amsterdamski, University of Warsaw Press, Warsaw 2005.

Starzec K., *Marian Smoluchowski – teoria nauki a działalność naukowa (Marian Smoluchoski – scientific theory and scientific practice)*, pp. 387–398 and *Dwie interpretacje myśli Mariana Smoluchowskiego (Two interpretations of Marian Smoluchowski's thought)*, pp. 399–426, in: *Krakowska filozofia przyrody w okresie międzywojennym (Kraków Philosophy of nature in the interwar period)*, vol. 3, *Smoluchowski – Natanson – inni (Smoluchowski – Natanson – others)*, ed. M. Heller, J. Mączka, Ośrodek Badań Interdyscyplinarnych przy Wydziale Filozoficznym Papieskiej Akademii Teologicznej – Wydawnictwo

Diecezji Tarnowskiej (Centre for Interdisciplinary Studies at the Faculty of Philosophy of the Pontifical Academy of Theology Publishing House of the Tarnów Diocese) Biblos, Kraków–Tarnów 2007.

Stawarz M., *Punkt wyjścia filozoficznych rozważań Mariana Smoluchowskiego na temat przypadku i prawdopodobieństwa* (*The starting point of Marian Smoluchowski's philosophical considerations on the subject of chance and probability*), 'Semina Scientiarum' 2008, No. 7, pp. 82–95.

Stawarz M., *Rekonstrukcja i krytyczna analiza poglądów filozoficznych Mariana Smoluchowskiego* (*Reconstruction and critical analysis of the philosophical views of Marian Smoluchowski*), doctoral dissertation, supervisor: dr hab. Paweł Polak, Kraków 2016.

Stöker H., *Nowoczesne kompendium fizyki* (*A modern compendium of physics*), trans. A. Krupski, D. Serwotka, A. Wójtowicz, Polish Scientific Publishers PWN, Warsaw 2010.

Storczak Ł.I., *Дискуссияоприродефизическогознания*, „Вопросыфилософии" 1948, No. 1.

Strzałkowski A., *Marian Smoluchowski. Od teorii atomistycznej do fizyki współczesnej* (*Marian Smoluchowski. From atomic theory to modern physics*), Polska Akademia Umiejętności i Komisja Historii Nauki (Polish Academy of Arts and Sciences and the Commission on the History of Science), Kraków 2003.

Subrahmanyan C., Kac M., Smoluchowski R., *Marian Smoluchowski. His Life and Scientific Work*, Polish Scientific Publishers PWN, Warsaw 1999.

Susskind L., *Bitwa o czarne dziury. Moja walka ze Stephenem Hawkingiem o uczynienie świata przyjaznym mechanice kwantowej* (*The black hole war: My battle with Stephen Hawking to make the world Safe for quantum mechanics*), trans. U. and M. Seweryńscy, Prószyński i S-ka, Warsaw 2011.

Susskind L., *Kosmiczny krajobraz. Dalej niż teoria strun* (*The cosmic landscape: String theory and the illusion of intelligent design*), tłum. U. and M. Seweryńscy, Prószyński i S-ka, Warsaw 2011.

Sutherland W., *A Dynamical Theory of Diffusion for Non-Electrolytes and the Molecular Mass of Albumin*, 'Philosophical Magazine and Journal of Science' 1905, series 6, vol. 9, pp. 781–785.

Sutherland W., *Ionization, Ionic Velocities, and Atomic Sizes*, 'Philosophical Magazine' 1902, series 6, vol. 4, pp. 625–645.

Sutherland W., *The Dielectric Capacity of Atoms*, w: *Report of the 10th Meeting of the Australasian Association for the Advancement of Science*, Held Dunedin 1904, ed. G.M. Thomson, Dunedin 1905, pp. 122–124.

Sutherland W., *The Measurement of Large Molecular Masses*, in: *Report of the 10th Meeting of the Australasian Association for the Advancement of Science*, Held Dunedin 1904, ed. G.M. Thomson, Dunedin 1905, pp. 117–121.

Svedberg T., Inouye K., *Eine neue Methode zur Prüfung der Gültigkeit des Boyle-Gay-Lussac'schen Gesetzes für kolloide Lösungen (A new method for testing the validity of the Boyle-Gay-Lussac law for colloidal solutions)*, 'Zeitschrift für Physikalische Chemie-Stöchiometrie und Verwandtschaftslehre' 1911, H. 77, pp. 145–190.

Średniawa B., *Rola współpracy Mariana Smoluchowskiego i Teodora Svedberga w prowadzonych w pierwszych latach XX wieku badaniach ruchów Browna i fluktuacji (The role of collaboration between Marian Smoluchowski and Theodor Svedberg in studies on Brownian motion and fluctuations conducted in the early years of the twentieth century)*, 'Postępy fizyki' 1991, vol. 42, book. 4.

St Thomas Aquinas, *The Summa Contra Gentiles*, Volume 3, Issue 1, London, 1928.

Szczepanowski S., *Nędza Galicji w cyfrach. Program energicznego rozwoju gospodarstwa krajowego (Poverty of Galicia in Figures and a Programme for the Energetic Development of the Economy of the Country)*, Gubrynowicz i Schmidh, Lviv 1888.

Szumilewicz I., *Koncepcja przyczynowości u Macha w świecie współczesnego determinizmu (The concept of causality in Mach in the world of contemporary determinism)*, 'Studia Filozoficzne' 1959, No 4, pp. 127–145.

Szumilewicz I., *Poincaré*, Wiedza Powszechna, Warsaw 1978.

Tarski A., *Pisma logiczno-filozoficzne. Prawda (Logical-philosophical writings. Truth)*, vol. I, trans. J. Zygmunt, Polish Scientific Publishers PWN, Warsaw 1995.

Tempczyk M., *Ontologia świata przyrody (The ontology of the natural world)*, Universitas, Kraków 2005.

Teske A., *Marian Smoluchowski*, 'Fizyka i Chemia' 1951, year IV, May–June, No. 3 (17), pp. 2–5.

Teske A., *Marian Smoluchowski. Życie i twórczość* (*Marian Smoluchowski. Life and works*), Polish Scientific Publishers PWN, Warsaw 1955.

The Collected Papers of Albert Einstein, vol. 2, *The Swiss Years: Writings, 1900–1909*, ed. J. Stachel, trans. A. Beck, Princeton University Press, Princeton 1989.

Thomson J.J., *Recollections and Reflections*, G. Bell, London 1936.

Thorburn W.M., *The Myth of Occam's Razor*, 'Mind' 1918, vol. XXVII, iss. 3, pp. 345–353.

Turek J., *Materializm* (*Materialism*), in: *Powszechna Encyklopedia Filozofii*, vol. 6, Polskie Towarzystwo Tomasza z Akwinu, Lublin 2000–2009.

Thomson J.J., Nobel Lecture of December 11, 1906, in: *Nobel Lectures: Physics, 1901–1921*, Amsterdam 1967, pp. 145–153.

J.P. Toennies, H. Schmidt-Böcking, B. Friedrich, and J. C. A. Lower, *Otto Stern (1888–1969): The founding father of experimental atomic physics* in: 'Annalen der Physik' ('Annals of Physics') (Berlin) 523, No. 12, 962–964 (2011), p. 1048.

S.M. Ulam, Adventures of a Mathematician, University of California Press, Berkeley and Los Angeles, 1991.

Wielka Encyklopedia Powszechna (*Great Universal Encyclopedia*), vol. 10, Polish Scientific Publishers PWN, Warsaw 1967.

Wigner E.P., *Niepojęta skuteczność matematyki w naukach przyrodniczych* (*The unreasonable effectiveness of mathematics in the natural science*), trans. J. Dembek, 'Zagadnienia Filozoficzne w Nauce' 1991, No. XIII, pp. 5–18.

Wittgenstein L., *Tractatus logico-philosophicus*, trans. B. Wolniewicz, Polish Scientific Publishers PWN, Warsaw 2004.

Wójcik W., *Pewna interpretacja konwencjonalizmu Poincarégo* (*A certain interpretation of Poincaré's conventionalism*), 'Kwartalnik Filozoficzny' 1993, book. 3, pp. 21–43.

Wróblewski A.K., *Długie narodziny elektronu* (*The long birth of the electron*), 'Wiedza i Życie' 1998, No. 5, http://archiwum.wiz.pl/1998/98052300.asp.

Wróblewski A.K., *Historia fizyki* (*The history of physics*), Polish Scientific Publishers PWN, Warsaw 2006.

Wróblewski A.K., *Historia fizyki: od czasów najdawniejszych do współczesności* (*The history of physics: from antiquity to modernity*), Polish Scientific Publishers PWN, Warsaw 2009.

Wszołek S., *Esencjalizm transcendentalny K.R. Poppera* (*K.R. Popper's Transcendental Essentialism*), 'Zagadnienia Filozoficzne w Nauce' 2002, No. 31, pp. 120–132.

Wszołek S., *W obronie realizmu naukowego* (*In defence of scientific realism*), 'Zagadnienia Filozoficzne w Nauce' 2006, No. 38, pp. 152–157.

Zdybicka Z., *Monizm* (*Monism*), in: *Powszechna Encyklopedia Filozofii*, vol. 7, Polskie Towarzystwo Tomasza z Akwinu, Lublin 2000–2009.

Ziemińska R., *Sceptycyzm umiarkowany Davida Hume'a* (*David Hume's moderate scepticism*), 'Przegląd Filozoficzny – Nowa Seria' 2011, No. 4(80), pp. 119–136.

Zsigmondy R.A., *Properties of colloids*, Nobel Lecture, Chemistry 1922–1941, Singapore, New Jersey, London, Hong Kong 1999.

Żegleń U., *Putnam*, in: *Powszechna Encyklopedia Filozofii*, vol. 8, Polskie Towarzystwo Tomasza z Akwinu, Lublin 2000–2009.

Życiński J., *Język i metoda* (*Language and method*), Znak, Kraków 1983.

Большая Советская Енциклопедия, ИзготельствоСоветская Енциклопедя (*The Great Soviet Encyclopedia*, Soviet Encyclopedia Publishing House), Moscow 1976.

ЮХрамов Ю.А (Chramov J.)., *Биографиа физики* (*A Biography of Physics*), Kyiv 1983.

Index of people

A
Abbott Benjamin 176
Adams John C. 105, 178
Ajdukiewicz Kazimierz 277, 281
Alembert Jean Le Rond d' 43
Altarelli Guido 266
Ampère André Marie 337
Amsterdamski Stefan 207, 208, 213, 218, 226, 242, 246, 250, 251
Anaxagoras of Clazomenae 256
Aquinas Thomas 51, 52, 258
Arkani-Hamed Nima 114
Aronson Jerrold 216
Aristarchus of Samos 302
Aristotle 52, 92–95, 97–101, 107, 170, 205, 210, 256, 259, 284
Auerbach Felix 200, 202
Augustynek Zdzisław 207, 218, 226, 242, 246, 251
Avenarius Richard 44, 45, 138, 253, 254, 299

B
Bacon Francis 106, 269
Bacon Roger 96, 97, 99
Bafia Stanisław 98
Becquerel Antoine Henri 190–192, 200
Berkeley George 330
Bernard Claude 352
Bernstein Henry T. 325, 328
Bernstein Jeremy 143, 170, 172, 185, 193, 197, 340, 347
Berthelot Marcellin Pierre 164
Besso Michele 353, 374, 378, 409, 410, 411, 414
Biela Adam 119, 120
Biełopolski Aristarch 106
Biernacki Wiktor 190, 197
Bigg Charlotte 381, 387, 388, 389, 391, 393
Bilczewski Józef 85
Birks John B. 38
Blackmore John T. 41
Bohr Niels 68, 199, 201
Boltzmann Ludwig 20, 40, 72, 81, 143, 146, 147, 149, 167, 169, 182–185, 194, 197, 200, 202, 230, 248, 261–263, 325–333, 335, 336, 340, 342, 343, 348, 349, 351, 360, 361, 368, 369, 375, 378, 379, 381, 394, 399, 408, 413, 416
Bonaparte Napoleon 84
Born Max 18, 165, 271, 308, 382
Boyle Robert 19, 255, 291
Bridgman Percy Williams 381
Brillouin Léon 160
Brody Engelbert 330
Broglie Louis de 401
Brown Julian 181
Brown Robert 172, 340, 341, 354
Bruno Giordano 301
Brush Stephen G. 361, 382, 383, 399
Brzozowski Stanisław 39
Budzanowski Andrzej 33
Burkhardt Heinrich 195
Bunge Mario Augusto 207, 208, 213, 214
Bunsen Robert Wilhelm 177

INDEX OF PEOPLE

C
Calaprice Alice 257
Cardano Girolamo 219
Cardona Manuel 17, 18
Carnap Rudolf 111
Carnot Nicolas Léonard Sadi 147, 172, 182
Cartwright Nancy 391
Casimir Hendrik 164
Cassirer Ernst 50
Carathéodory Constantin 198
Chandrasekhar Subrahmanyan 24, 32, 381
Chmielewski Adam 109
Chramow Jurij 200, 202
Christiansen Frederik V. 261
Chwedeńczuk Bohdan 43
Cialdini Robert 61
Cichocki Bogdan 18, 19, 34, 353, 354, 365, 366, 369
Clauberg Johannes 258
Clausius Rudolf 172, 181, 182, 218, 325, 327, 332
Clifford William K. 145, 153
Cohen Bernard I. 194, 197
Cohen Hermann 50
Comte Auguste 42–45, 165, 167
Copernicus Nicolaus 98, 118, 167, 289, 293, 300–302, 311, 312
Copleston Frederick 43
Corbalán Fernando 217, 219, 221
Cotton Aimé 389
Crombie Alistair Cameron 94, 100, 311, 314
Crookes William 173, 392

D
Dahlgren Erik Wilhelm 192
Dalton John 338, 344, 389
Darwin Karol 43, 56, 83, 84, 293, 299, 311, 314
Daston Lorraine 381
Davies Paul 181
Davy Humphry Bartholomew 173

Dawes Gregory W. 271, 272, 274, 275, 285, 286
Day Peter 173
Dąbrowski Stefan 346, 381, 388
Dąmbska Izydora 31, 43, 107, 108, 303
Debye Peter 18, 401
DembekJacek 114
Dewey John 400
Dirac Paul Adrien Maurice 165, 267
Domański Czesław 110
Doppler Christian 106
Dorn Friedrich Ernst 200
Dryden Gordon 181
Du Bois-Reymond Emil 270
Dudyński Marek 14
Duhem Pierre 49, 138, 144–146, 153, 337
Duplantier Bertrand 24, 394, 408–410, 411–415
Dybowski Benedykt 85

E
Eftekhari Ali 184
Einstein Albert 18, 19, 21–24, 48, 111, 137, 138, 155, 167, 169, 177, 179, 180, 185, 188, 194–198, 200, 201, 202, 251, 257, 272, 280, 289, 293, 294, 300, 302, 311, 325, 344, 348–390, 393–402, 408–417
Ehring Douglas 216
Ekiel-Jeżewska Maria 14
Elkan Yehud 330
Empedocles of Akragas 256
Engels Fryderyk 56, 57, 63, 64, 67, 79
Exner Felix M. 341, 356–359, 362, 367, 370, 379, 400, 415
Exner Franz Serafin 40, 343

F
Faraday Michael 173–180, 203, 338
Fermat Pierre de 219, 220
Feynman Richard Phillips 161
Fick Adolf Eugen 14, 119
Filipczuk Michał 46

Fourier Jean Baptiste Joseph 119
Franklin James 256
Fraassen Bastiaan Cornelis van 299
Fuliński Andrzej 14, 15, 32, 33

G
Galileo Galilei 96, 98–100, 111, 115, 129, 167, 179, 188, 219, 301, 302
Galle Johann Gottfried 106
Gawecki Bolesław J. 215, 230, 231, 232, 234, 235, 237–240, 246, 350
Ghiberti Lorenzo 301
Ghins Michel 297, 298
Gibbs Josiah Willard 399, 408, 413
Gilson Etienne 52
Givindaraju Venu 228
Gleichen-Rußwurm Wilhelm Friedrichvon 341
Glymour Clark 286, 287, 330
Godlewski Tadeusz 28
Goodman Nelson 315
Gosiewski Władysław 227, 228, 234
Gostkowski Kazimierz 30
GoswamiAmit 400, 401
Gombaud Antoine 219
Gouy Louis Georges 344, 355, 356, 362
Gödel Kurt 283, 284, 313
Graham Thomas 409
Grobler Adam 285, 295
Grygiel Wojciech P. 128, 170
Gutenberg Johannes 301

H
Habsburg Maria Theresa 38
Hajduk Zygmunt 89, 115, 124, 213, 297

Hamilton William Rowan 258
Harman Gilbert 286, 287
Hawking Stephen W. 165, 222, 280, 281, 302, 322, 381
Hegel Georg Wilhelm Friedrich 56, 201–203, 207
Heinrich Władysław 234, 240
Heisenberg Werner 68, 142, 246, 267, 271, 313
Heller Michał 33, 52, 113, 114, 232, 233, 277, 285
Helmholtz Hermann von 123, 172–174, 182, 214, 273, 344
Henri Victor 112, 362, 365, 390
Heraclitus of Ephesus 141, 142
Herbut Józef 279
Herschel John Frederick William 89, 92, 93, 95, 96, 102, 103, 104, 115, 116
Hertz Heinrich 106, 178, 186, 187, 261, 262, 263, 337
Hiebert Elfried 330
Hirsch Jorge E. 16
Hoff Jacobus van't 353, 365, 374, 375, 411, 412, 414
Home Roderick W. 412, 413
Horwitz Maksymilian Henryk 130
Hossenfelder Sabine 114, 129, 265, 266, 267
Hoyningen-Huene Paul 276
Höfler Alois 38
Hume David 49, 104, 106, 107, 111, 167, 206, 230–232, 304–312, 315, 320, 321

I
Ingarden Roman 23, 32, 316
Ingenhousz Jan 172, 340

INDEX OF PEOPLE

Inouye Katsuji 19
Isaacson Walter 181
ItagakiRyoichi 41

J
Jarocki Jacek 138, 139
James William 46, 205, 279, 280, 282
John Duns Scotus 258
Jerusalem Wilhelm 141
Jędrusiak Szymon 98
Jolly Philipp von 163
Joule James Prescott 172, 180, 182
Juszkiewicz Adolf Pawłowicz 221

K
Kac Mark 23, 32, 401, 402, 408, 414
Kampen Nicolaas Godfriedvan 13, 14
KaniaIreneusz 98
Kant Immanuel 49, 50, 113, 114, 262
Kapczyński Marcin 17
Kapuściński Władysław 26, 30, 55, 87
Kaufmann Walter 87
Kelly Thomas 320, 321
Kepler Johannes 110, 111, 129, 293, 300, 301, 312
Kerker Milton 159, 350, 362, 382
Khrushchev Nikita Sergeyevich 57
Kirchhoff Gustav 112, 126, 177, 201, 337
Kistler Max 215, 216, 217
Klemensiewicz Zygmunt 31, 148
Kołakowski Leszek 63–65, 76, 166, 167
KonovalovDmitri 112
Korpikiewicz Honorata 336
Kosmulski Marek 15, 16, 23
Kotarbiński Tadeusz 282

Kozłowski Władysław Mieczysław 205, 207, 210, 234–237
Kotowa Barbara 44
König Edmund Wilhelm Hermann 206, 208
Kragh Helge 194
Krajewski Władysław 29–31, 54, 55, 58–82, 85–87, 221, 248, 249
Krąpiec Mieczysław Albert 120
Krośniak Marek 257
Krönig August Karl 325
Kuhn Thomas 93, 94, 111, 170, 272, 288–293
Kumaniecki Kazimierz 240

L
Langevin Paul 365, 388, 390, 394
Langmuir Irving 18
Laplace Pierre Simon de 75, 84, 220–223
Laue Max Theodor Felix von 199
Lavoisier Antoine 171
Lederman Leon M. 267
Lehmann Otto 342
Leibniz Gottfried Wilhelm 114
Leighton Robert B. 161
Lemańska Anna 208, 242, 243, 245, 246
Lenin Vladimir 57, 64, 65, 67, 73, 74
Leo XIII 51
Le Roy Édouard 49, 135
Le Verrier Urbain 105
Lippmann Gabriel 25, 157
Lipton Peter 308–313
Litwinowicz-Droździel Małgorzata 173, 174, 177
Lorentz Hendrik Antoon 24, 155, 187, 188, 232, 233, 395

Loria Stanisław 25–27, 30, 144, 154, 159, 160, 214, 333, 360, 361, 381
Loschmidt Josef 326
Lukács György 64
Lunbeck Elizabeth 381
Lubomirski Andrzej 130, 131
Łypacewicz Stanisław 94

M

Mackiewicz Witold 63
Mach Ernst 41, 45, 69, 112, 126, 138–142, 144, 145–147, 153, 166, 170, 183, 185, 214, 230, 231, 232, 234–241, 246–248, 253, 255, 256, 259–261, 277–279, 299, 328, 330, 340, 342, 347, 348, 350, 351, 379, 383, 403, 404
Maiocchi Roberto 352, 353, 354–356, 357, 359, 364, 365, 367, 381
Marconi Guglielmo 187
Marshall Stephen James 181
Markov Andrey Andreyevich 13, 60, 228
Marx Karl 56, 57, 67, 68, 69, 166
Maryniarczyk Andrzej 51, 120
Maxwell James Clerk 106, 121, 152, 155, 157, 158–161, 167, 175–180, 182, 186–188, 200, 201, 218, 222, 273, 311, 325–329, 331, 332, 338, 346, 372, 385, 408, 416
Mayer Julius Robert von 344, 377
Mączka Janusz 33
McMullin Ernan 303
Meinong Alexius 227, 243, 247
Mejbaum Wacław 207, 218, 226, 242, 246, 251
Merton Robert 401

Metallmann Joachim 211, 234, 238, 251, 253, 254, 255, 260, 264
Michelson Albert A. 164, 181, 202
Miller Tomasz 114
Miłkowski Marcin 140
Mittag-Leffler Magnus Gösta 392
Morales José María Riaza 98
Mouton Henri 389

N

Nägeli Carl Wilhelm von 345
Natanson Władysław 29, 31, 33, 43, 107, 108, 239, 303
Natorp Paul 50
Nicholas of Cusa 98, 300–302
Needham John Turberville 341
Nernst Walther 198, 353, 392, 411, 414
Newton Isaac 100, 110, 111, 167, 169, 177, 178–180, 194, 197, 201, 255, 257, 269, 280, 293, 294, 300, 302, 303, 312, 315, 383, 416
Niemiec Marian 384
Norton John D. 264, 268, 269, 275, 276, 313, 314, 315–322
Nowotniak Justyna 288
Nyhof John 326, 330, 331

O

Ockham William 98, 256–259, 278
Ohm Georg 119
Olszewski Karol 24
Onnes Heike Kamerlingh 199
Oseen Carl Wilhelm 387
Ostromęcka Helena 288

INDEX OF PEOPLE

Ostwald Wilhelm 69, 140, 142, 144–147, 157, 183, 248, 337, 342, 348, 350, 379, 383, 404, 410
Ørsted Hans Christian 102

P

Pabjan Tadeusz 232
Parmenides of Elea 141, 256
Pascal Blaise 219, 220
Pearson Karl 144, 145, 153
Peirce Charles Sanders 46, 279, 282, 285, 286, 352
Penrose Roger 169, 179, 194, 263, 335, 336, 383
Perrin Jean Baptiste 18, 21, 23, 25, 145, 146, 198, 346, 351, 360, 362, 364–367, 380–382, 387–393, 398, 409, 415
Peverone Giovanni Francesco 219
Pieper Josef 98
Pierre Curie 191, 200, 201, 390
Piotrowski Robert 112, 159, 160
Pius X 51
Planck Max 24, 143, 163, 184, 186, 193, 194, 197, 200, 202, 383, 395
Plato 256, 259
Pogonowski Jerzy 283
Poincaré Jules Henri 24, 49, 50, 72, 129–138, 143, 151, 153, 169, 268, 293, 311, 332, 333, 392, 395, 408
Polak Paweł 33, 55, 85–87, 215, 222
Ponce John 258
Popper Karl R. 48, 49, 107, 108, 109, 111, 167, 274, 279, 282, 283, 287, 290–292, 302, 308, 313, 351, 352
Portel Bueso X. 266

Prévost Pierre 171
Prus Bolesław 39
Półtawski Andrzej 316
Prejbisz Antoni 322
Psillos Stathis 145, 146, 387, 389–392
Ptolemy 96, 97, 256, 290, 301–303, 312
Putnam Hilary 293–298, 304
Pytheas of Massalia 306

R

Raghavan Vijay V. 228
Rao Dhana 228
Regnault Henri Victor 112
Regt Henk W. de 121, 261, 262, 325, 328–330, 338
Reichenbach Hans 108, 248, 307
Rejmer Krzysztof 217
Richarz Franz 157
Riemann Georg Friedrich Bernhard 167, 322
Rodzeń Jacek 293, 296, 297, 303, 417
Röntgen Wilhelm Conrad 188, 189, 197
Roskal Zenon E. 44, 214, 215, 216, 230, 416
Rubczyński Witold 240
Rubens Heinrich 193, 200
Russell Bertrand 111, 284
Rutherford Ernest 24, 38, 122, 191, 198, 199, 338, 392, 395
Rutkowska Anna 401
Rybałt Jan 52

S

Sands Matthew 161
Sanz Gerardo 217, 219, 221
Schiller Friedrich von 124

Schilpp Paul Arthur 399
Schrödinger Erwin 40, 271, 401
Seddig Max 362
Seleucus of Seleucia 302
Siedentopf Heinrich 355, 356, 362, 389
Sienkiewicz Henryk 39
Sierotowicz Tadeusz 329, 338
SilbersteinLudwik 441
SimonyiKároly 380
Skarżyński Bolesław 62, 63
Skłodowska-Curie Maria 16, 17, 24, 132, 191, 192, 200, 395
Skrzydlewski Paweł 120
Sławianowski Jan Jerzy 229
Słomski Wojciech 63
Smith Adam 322
Smoluchowski Marian 1, 5, 13–35, 37–93, 95–97, 99–119, 122, 123, 125–138, 140, 142–161, 164, 169, 170–172, 181, 185, 195–198, 205–219, 221–225, 227, 229, 230, 233, 234, 239–251, 253, 255, 256, 265, 269, 270, 273–283, 286, 288, 291, 292, 294, 295–300, 302–304, 307, 308, 310, 315, 317, 320–323, 325, 326, 328–351, 353–364, 367, 368, 370–382, 384–386, 388, 390, 391, 393–409, 414–416
Smoluchowska Zofia 28, 39
Smullyan Raymond 283
Socrates 107
Solvay Ernest 24
Sommerfeld Arnold 143, 184
Soto Domingo de 98
Söderbaum Henrik Gustaf 21, 380
Speziali Pierre 410

Stachel John J. 194, 195, 197, 348, 349, 352, 354, 358, 359, 364, 367, 368, 370, 379, 380, 381, 383, 385, 386
Stalin Joseph 57, 66
Stark Johannes 106
Starzec Krzysztof 33, 37, 54, 57, 70, 87, 88
Stawarz Małgorzata 34, 54, 85, 249
Stayer Marcia Sweet 269
Stefan Josef 40, 182, 184
Stępień Katarzyna 120
Stock Jan Jakub 29
Stokes George Gabriel 353, 362, 364–366, 368, 375, 409, 410–412, 414
Stoney George Johnstone 189
Storczak Ł.I. 31, 59, 60, 251
Stern Otto 199
Strutt John William, Lord Rayleigh 18, 155
Strzałkowski Adam 14, 32
Sutherland William 352, 380, 407–415
Svedberg Theodor 18–21, 24, 31, 32, 153, 157, 159, 344, 346, 360, 362, 380, 382, 388, 393, 395, 398, 415
Szilárd Leó 160
Szczepanowski Stanisław 38, 39
Szumilewicz Irena 50, 130, 134, 135, 136, 259
Szumilewicz Ewa 417
Śledziewska-Błocka Danuta 14
Średniawa Bronisław 20, 31, 32

T

Tanaka Setsuko 41
Tarski Alfred 282, 283, 284
Teske Armin 29, 31, 38, 39, 41, 60, 83, 148

Timiriaziew Klimient Arkadiewicz 80
Thirring Hans 183
Thorburn William M. 257, 258, 259
Thomson George M. 408
Thomson Joseph John 189, 190, 191
Thomson William, Lord Kelvin 25, 148, 179, 180, 182, 327
Turek Józef 56, 70
Twardowski Kazimierz 38, 205
Tyndall John 154, 155, 176

U
Ulam Stanisław M. 251

V
Varley Cromwell Fleetwood 186
Villard Paul Ulrich 200
Vinci Leonardo da 96, 256
Vos Jeanette 181

W
Waals Johannes Diderik van der 369, 390
Warburg Emil 25
Watt James 122
Werner Marx 17
Whewell William 352

Wigner Eugene Paul 114
Wittgenstein Ludwig 85, 282
Witwicki Władysław 256
Wojciechowska Wanda 42
Wojciszke Bogdan 61
Wolniewicz Bogusław 85
Wiechert Emil Johann 186
Wien Wilhelm 186, 193
Wójcik Wiesław 74
Wróblewski Andrzej K. 174, 186, 202, 326
Wszołek Stanisław 48, 296

Z
Zawirski Zygmunt 38
Zdybicka Zofia 141
Zeeman Pieter 163, 187
Zermelo Ernst Friedrich Ferdinand 332, 333, 343
Ziemińska Renata 305, 306
Znamierowski Czesław 305
Zsigmondy Richard A. 18, 19, 24, 362, 389, 393
Zygmunt Jan 283
Żhdanow Andrie Aleksandrovich 60
Żegleń Urszula 294
Życiński Józef 114, 284

Studies in Social Sciences, Philosophy and History of Ideas

Edited by Andrzej Rychard

Vol. 1   Józef Niżnik: Twentieth Century Wars in European Memory. 2013.

Vol. 2   Szymon Wróbel: Deferring the Self. 2013.

Vol. 3   Cain Elliott: Fire Backstage. Philip Rieff and the Monastery of Culture. 2013.

Vol. 4   Seweryn Blandzi: Platon und das Problem der Letztbegründung der Metaphysik. Eine historische Einführung. 2014.

Vol. 5   Maria Gołębiewska / Andrzej Leder/ Paul Zawadzki (éds.): L'homme démocratique. Perspectives de recherche. 2014.

Vol. 6   Zeynep Talay-Turner: Philosophy, Literature, and the Dissolution of the Subject. Nietzsche, Musil, Atay. 2014.

Vol. 7   Saidbek Goziev: Mahalla - Traditional Institution in Tajikistan and Civil Society in the West. 2015.

Vol. 8   Andrzej Rychard / Gabriel Motzkin (eds.): The Legacy of Polish Solidarity. Social Activism, Regime Collapse, and the Building of a New Society. 2015.

Vol. 9   Wojciech Klimczyk / Agata Świerzowska (eds.): Music and Genocide. 2015.

Vol. 10   Paweł B. Sztabiński / Henryk Domański / Franciszek Sztabiński (eds.): Hopes and Anxieties in Europe. Six Waves of the European Social Survey. 2015.

Vol. 11   Gavin Rae: Privatising Capital. The Commodification of Poland's Welfare State. 2015.

Vol. 12   Adriana Mica / Jan Winczorek / Rafał Wiśniewski (eds.): Sociologies of Formality and Informality. 2015.

Vol. 13   Henryk Domański: The Polish Middle Class. Translated by Patrycja Poniatowska. 2015.

Vol. 14   Henryk Domański: Prestige. Translated by Patrycja Poniatowska. 2015.

Vol. 15   Cezary Wodziński: Heidegger and the Problem of Evil. Translated into English by Agata Bielik-Robson and Patrick Trompiz. 2016.

Vol. 16   Maria Gołębiewska (ed.): Cultural Normativity. Between Philosophical Apriority and Social Practices. 2017.

Vol. 17   Anita Williams: Psychology and Formalisation. Phenomenology, Ethnomethodology and Statistics. 2017.

Vol. 18   Mikołaj Pawlak: Tying Micro and Macro. 2018.

Vol. 19   Franciszek Sztabiński / Henryk Domański / Paweł B. Sztabiński (eds.): New Uncertainties and Anxieties in Europe. Seven Waves of the European Social Survey. 2018.

Vol. 20   Adriana Mica / Katarzyna M. Wyrzykowska / Rafał Wiśniewski / Iwona Zielińska (eds.): Sociology of the Invisible Hand. 2018.

Studies in Philosophy, Culture and Contemporary Society

Edited by Bogusław Paź

Vol. 21   Jan Felicjan Terelak: Psychology of the Operator of Technical Devices. 2019.

Vol. 22   Dorota Maria Leszczyna: Del idealismo al realismo crítico. La política como realización en José Ortega y Gasset. 2019.

Vol.	23	Zbigniew Drozdowicz: La république des savants. Sans révérence. Traduit du polonais par Catherine Popczyk. 2019.
Vol.	24	Andrzej Waśkiewicz: The Idea of Political Representation and Its Paradoxes. Translated from Polish by Agnieszka Waśkiewicz and Marilyn Burton. 2019.
Vol.	25	Ilona Błocian / Dmitry Prokudin (eds.): Imagination - Art, Science and Social World. 2019.
Vol.	26	Zbigniew Drozdowicz: Faces of the Enlightenment. Philosophical sketches. 2019.
Vol.	27	Włodzimierz Piątkowski: From Medicine to Sociology. Health and Illness in Magdalena Sokołowska's Research Conceptions. 2020.
Vol.	28	Roman Witold Ingarden: Die Mitschriften von den Vorlesungen Martin Heideggers über die Phänomenologische Interpretation von Kants Kritik der reinen Vernunft (Wintersemester 1927/ 28). Aus dem Manuskript abgeschrieben und das Vorwort verfasst haben: Radosław Kuliniak und Mariusz Pandura. 2020.
Vol.	29	Krzysztof Wielecki / Klaudia Śledzińska (eds.): The Relational Theory of Society. Archerian Studies vol. 2. 2020.
Vol.	30	Zenon Gajdzica / Robin McWilliam / Miloň Potměšil / Guo Ling: Inclusive Education of Learners with Disability - The Theory versus Reality. 2020.
Vol.	31	Jan Felicjan Terelak: Antarctic Winter-Over Syndrome. Narrative Perspective. 2021.
Vol.	32	Nuria Sánchez Madrid / Julia Muñoz Velasco / José Luis Villacañas Berlanga (eds.): El ethos del republicanismo cosmopolita. Perspectivas euroamericanas sobre Kant. 2021.
Vol.	33	Zbigniew Drozdowicz: Academic Culture. Traditions and the Present Days. 2021.
Vol.	34	Jan Felicjan Terelak: Antarctic Isolation as a Mars Habitat Analogue. A Psychological Perspective. 2021.
Vol.	35	Aleksandra Horowska (ed.): The Labyrinths of Leibniz's Philosophy. 2022.
Vol.	36	Bogusław Szuba / Tomasz Drewniak (eds.): Beauty in Architecture. Harmony of Place. 2022.
Vol.	37	Monika Bukowska / Krzysztof Wielecki (eds.): The Transformations of Contemporary Culture and Their Social Consequences. Archerian Studies Vol. 3. 2023.
Vol.	38	Kinga Anna Gajda (ed.): (Non) Commemoration of the Heritage in Eastern Europe. 2024.
Vol.	39	Jan Grzanka: The Forgotten Genius of Physics, a work on Marian Smoluchowski. The Story of Marian Smoluchowski. 2025.
Vol.	40	Bartlomiej Sipiński: Phenomenology of Subjectivity. An Essay on the Dialogue Personalism. 2025.
Vol.	41	Jan Grzanka: Between Physics and Philosophy. The Philosophy of Nature and Philosophy of Physics in the Writings of Marian Smoluchowski. 2026.

www.peterlang.com